Genomics and the Global Bioeconomy

Translational and Applied Genomics Series

Genomics and the Global Bioeconomy

Edited by

Catalina Lopez-Correa
Chief Scientific Officer at Genome Canada, Executive Director of the Canadian COVID19 Genomics Network (CanCOGeN), Canada

Adriana Suarez-Gonzalez
10x Genomics, Vancouver, British Columbia, Canada

Series Editor
George P. Patrinos
Department of Pharmacy, University of Patras School of Health Sciences, Patras, Greece

Department of Pathology, College of Medicine and Health Sciences, United Arab Emirates University, Al-Ain, United Arab Emirates

Zayed Center of Health Sciences, United Arab Emirates University, Al-Ain, United Arab Emirates

Department of Pathology—Bioinformatics Unit, School of Medicine and Health Sciences, Erasmus University Medical Center, Rotterdam, The Netherlands

ELSEVIER

ACADEMIC PRESS
An imprint of Elsevier

Academic Press is an imprint of Elsevier
125 London Wall, London EC2Y 5AS, United Kingdom
525 B Street, Suite 1650, San Diego, CA 92101, United States
50 Hampshire Street, 5th Floor, Cambridge, MA 02139, United States
The Boulevard, Langford Lane, Kidlington, Oxford OX5 1GB, United Kingdom

ISBN 978-0-323-91601-1

For information on all Academic Press publications
visit our website at https://www.elsevier.com/books-and-journals

Publisher: Stacy Masucci
Acquisitions Editor: Peter B. Linsley
Editorial Project Manager: Susan E. Ikeda
Production Project Manager: Sreejith Viswanathan
Cover Designer: Matthew Limbert

Typeset by STRAIVE, India

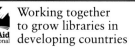

Contents

Part II
Genomic monitoring is revolutionizing our understanding of biodiversity and ecological services

Part IV
Why regulation and policy matter to advance the bioeconomy

Contributors

Numbers in parenthesis indicate the pages on which the authors' contributions begin.

Sachiyo Aburatani (59), Computational Bio-Big Data Open Innovation Laboratory, National Institute of Advanced Industrial Science and Technology (AIST), Tokyo; Bioproduction Research Institute, National Institute of Advanced Industrial Science and Technology (AIST), Sapporo, Japan

Neha Acharya-Patel (103), Department of Biochemistry and Microbiology, University of Victoria, Victoria, BC, Canada

Mawuli Afenyo (125), Texas A&M University, Galveston, TX, United States

Michael J. Allison (103), Department of Biochemistry and Microbiology, University of Victoria, Victoria, BC, Canada

Srijak Bhatnagar (125), Faculty of Science and Technology, Athabasca University, Athabasca, AB, Canada

Evan Bowness (3), University of the Fraser Valley, Abbotsford, BC, Canada

Sen Cai (205), Basic Forestry and Proteomics Center, Haixia Institute of Science and Technology, State Key Laboratory of Ecological Pest Control for Fujian and Taiwan Crops, College of Forestry, Fujian Agriculture and Forestry University, Fuzhou; School of Life Sciences, Capital Normal University, Beijing, China

Xinyi E. Chen (83), Department of Microbiology and Immunology; Department of Computer Science, University of British Columbia, Vancouver, BC, Canada

Burgos-Toro Daniela (25), Department of Biology, National University of Colombia, Bogotá, Colombia

Evan Fraser (3), University of Guelph, Guelph, ON, Canada

Erin J. Gilchrist (177), Molecular Diagnostics, Anandia Laboratories, Vancouver, BC, Canada

Alesandros Glaros (3), University of the Fraser Valley, Abbotsford, BC; University of Guelph, Guelph, ON, Canada

Richard C. Hamelin (139), Department of Forest and Conservation Sciences, University of British Columbia, Vancouver, British Columbia, Canada

Caren C. Helbing (103), Department of Biochemistry and Microbiology, University of Victoria, Victoria, BC, Canada

Casey R.J. Hubert (125), Geomicrobiology Group, University of Calgary, Calgary, AB, Canada

Koji Ishiya (59), Bioproduction Research Institute, National Institute of Advanced Industrial Science and Technology (AIST), Sapporo, Japan

Changmin Jiang (125), Asper Business School, University of Manitoba, Winnipeg, MB, Canada

Samuel King (83), Department of Botany; Department of Zoology, University of British Columbia, Vancouver, BC, Canada

Josiah O. Kuja (153), Department of Botany, School of Biological Sciences, Jomo Kenyatta University of Agriculture and Technology, Juja, Kenya; Department of Biology, Computational and RNA Biology, University of Copenhagen, Copenhagen, Denmark

Deepti Kulkarni (255), Sidley Austin LLP, Washington, DC, United States

Minxuan Li (205), Basic Forestry and Proteomics Center, Haixia Institute of Science and Technology, State Key Laboratory of Ecological Pest Control for Fujian and Taiwan Crops, College of Forestry; College of Life Sciences, Fujian Agriculture and Forestry University, Fuzhou, China

Yuanyuan Liu (205), Basic Forestry and Proteomics Center, Haixia Institute of Science and Technology, State Key Laboratory of Ecological Pest Control for Fujian and Taiwan Crops, College of Forestry, Fujian Agriculture and Forestry University, Fuzhou, China

Emily Marden (255), Sidley Austin LLP, Palo Alto, CA, United States; University of British Columbia, Vancouver, BC, Canada

Eileen M. McMahon (255), Torys LLP, Toronto, ON, Canada

Fernández-Niño Miguel (25), Department of Bioorganic Chemistry, Leibniz-Institute of Plant Biochemistry, Halle (Saale), Germany

Sherly Montaguth-González (287), Agro-Bio Andean Region, Bogotá, Colombia

Anne W.T. Muigai (153), Department of Botany, School of Biological Sciences, Jomo Kenyatta University of Agriculture and Technology, Juja, Kenya

Robert Newell (3), University of the Fraser Valley, Abbotsford; Royal Roads University, Victoria, BC, Canada

Kat Newman (3), University of the Fraser Valley, Abbotsford, BC, Canada

Lenore Newman (3), University of the Fraser Valley, Abbotsford, BC, Canada

Sarah W.S. Ng (83), Department of Microbiology and Immunology; Department of Computer Science, University of British Columbia, Vancouver, BC, Canada

Wataru Ogasawara (59), Department of Bioengineering, Nagaoka University of Technology, Nagaoka, Japan

Jim Philp (305), Organisation for Economic Cooperation and Development (OECD), Paris, France

Teagen D. Quilichini (177), Aquatic and Crop Resource Development Research Centre, National Research Council Canada, Saskatoon, SK, Canada

Melanie Sharman Rowand (255), Torys LLP, Toronto, ON, Canada

Paarsa Salman (83), Department of Botany; Department of Zoology, University of British Columbia, Vancouver, BC, Canada

Janella C. Schwab (83), Faculty of Land and Food Systems, University of British Columbia, Vancouver, BC, Canada

Parneet Sekhon (83), Department of Microbiology and Immunology, University of British Columbia, Vancouver, BC, Canada

Yosuke Shida (59), Department of Bioengineering, Nagaoka University of Technology, Nagaoka, Japan

Tomokazu Shirai (59), Center for Sustainable Resource Science, RIKEN, Yokohama, Japan

Hiroaki Takaku (59), Department of Applied Life Science, Niigata University of Pharmacy and Applied Life Science, Niigata, Japan

Tomohiro Tamura (59), Computational Bio-Big Data Open Innovation Laboratory, National Institute of Advanced Industrial Science and Technology (AIST), Tokyo; Bioproduction Research Institute, National Institute of Advanced Industrial Science and Technology (AIST), Sapporo, Japan

Jun Uetake (153), Field Research Centre for Northern Biosphere, Hokkaido University, Sapporo, Japan

María Andrea Uscátegui-Clavijo (287), Agro-Bio Andean Region, Bogotá, Colombia

Karin Verzijden (255), Axon Lawyers, Amsterdam, Netherlands

Shumin Wang (177), Strain Engineering, Willow Biosciences Inc., Vancouver, BC, Canada

Shijun You (205), Institute of Applied Ecology, Fujian Agriculture and Forestry University; Joint International Research Laboratory of Ecological Pest Control, Ministry of Education, Fuzhou, China

Introduction

Catalina Lopez-Correa

Chief Scientific Officer at Genome Canada, Executive Director of the Canadian COVID19 Genomics Network (CanCOGeN), Canada

Adriana Suarez-Gonzalez

10x Genomics, Vancouver, British Columbia, Canada

Is genomics ready to deliver solutions to solve the world's biggest challenges? The list of challenges related to the bioeconomy is long. The energy transition from fossil fuels to sustainable alternatives needs to happen soon, and some argue it is doable fast. Ensuring that everyone will have something to eat is also a growing challenge as the demand for food quality and diversity will only grow louder. The need for action and innovation to address the growing impact of climate change is urgent. A circular bioeconomy,[a] with genomics as a key driver, offers a way forward to tackle many of these urgent issues we face today.

We are fortunate to live in a time of great progress in genomics and computing. After 20 years of investment on research and innovation, genomics, coupled with advances in computing and artificial intelligence, is now ready to start delivering concrete solutions to help us solve these global challenges. From high-throughput sequencing of whole genomes to measuring gene expression and epigenetic marks at the genome-wide scale and even at the single-cell and spatial level, genomic technologies are advancing and evolving faster than ever making their applications toward a more sustainable and green bioeconomy increasingly tangible. This book highlights a fascinating collection of examples illustrating the concrete applications and impact that these new genomics technologies hold for the next phase of the global bioeconomy.

A key component of the global bioeconomy is the rapid advancement of genomics, which has led to its application in an increasing number of areas, including agriculture, energy, livestock, environmental remediation, and healthcare, among others. Emerging technologies such as CRISPR, stem cells, and single-cell genomics have been instrumental in gaining a deeper

---

a. The OECD (2009) described the bioeconomy as "the set of economic activities in which biotechnology contributes centrally to primary production and industry, especially where the advanced life sciences are applied to the conversion of biomass into materials, chemicals and fuels."

understanding of the complexities of different biological systems. Great advances in fast and cost-effective DNA sequencing technologies, genomic editing techniques, and synthetic biology are allowing us to discover new life-building blocks. This groundbreaking work is creating the foundational knowledge needed to generate important contributions with concrete socioeconomic impact.

The chapters in this book cover a wide range of applications of genomics and its impact in the context of the sustainable development goals (SDGs). The SDGs are a collection of 17 global goals set by the United Nations General Assembly in 2015 for the year 2030. They are the blueprint to achieve a better and more sustainable future for all by addressing some of the global challenges we face, including those related to poverty, inequality, climate, environmental degradation, and prosperity. The topics covered here provide the tools needed to take action on many SDGs from producing more sustainable foods and productive crops to contributing to a zero-emissions future. Framed on the SDG, the chapters cover different applications of genomics, from sequencing, synthetic biology, and gene editing to genomic surveillance with tools like environmental DNA. The topics also span global geographies as well as the regulatory and policy landscapes with examples from Africa, Asia, Europe, North America, and Latin America.

We start this book with one of the most hyped research topics in this century, synthetic biology. Newman et al. provide an overview of protein production using cellular agriculture, a technique that is rapidly nearing commercial scales. This chapter describes genomics and synthetic biology technologies currently being used for protein production, including cell culture-derived proteins and fermentation-derived proteins. Using cell lines with specialized growth media, companies like Finless Foods (California, United States) and Mosa Meats (Maastricht, Netherlands) are producing cultured fish meat and cultured beef meat, respectively. Future Fields (Edmonton, Canada) goes upstream of the process and is using genetically modified fruit flies to produce custom growth factors, which have the potential to address the high cost of humane growth material for cultured meat. The authors also surfaced the potential for these technologies to lead progress toward the SDGs as well as the need for regulation and policy to help ensure widespread environmental and social benefits.

Fernandez-Nino et al. showcase another application of synthetic biology, BioBricks. BioBricks are open-source catalogs of over thousands of standardized DNA components being assembled by engineers and biologists. New and more complex BioBricks are constantly being built and strung together interchangeably to create and modify living cells while expanding and testing our knowledge of cellular function. In their chapter, Fernandez-Nino et al. discussed the most relevant approaches to assembly and design of microbial biofactories relevant for the biotechnology industry. Today, many biotechnology, pharmaceutical, and agriculture companies rely on these synthetic biology tools to develop products. For example, Zymergen (United States) commercializes hyaline, a thin film for electronics made from bio-sourced monomers. These monomers are produced by engineered organisms optimized using artificial intelligence. Similar foundries are emerging globally at an accelerating pace to genetically reprogram cells that are able to make new materials by design like vaccines, medicine, food, and even fuel.

The NEDO Smart Cell Project in Japan demonstrated the use of synthetic biology to enable the shift from fossil fuels to microbial-derived products. Aburatani et al. share the discovery of new microbial genes that control enzyme and oil production using an innovative modeling software able to overcome the limitations of traditional methods that have failed to identify these key gene networks. To improve the productivity of microorganisms, they developed a new network modeling technique called ASENET. ASENET generates a graphical representation of the microbial production systems of interest and enables the artificial design of new microbial hosts. The group has several projects in the pipeline, including the modification of the fungus *Trichoderma reesei,* a well-known industrial cellulase producer and candidate for biomass production. Japanese breeding techniques have successfully developed *T. reesei* strains with high cellulase production capacity. However, these strains produce many cellulases, and a fine-tuned balance between them is a critical factor for biomass production at scale. Although Aburatani et al.'s inferred network revealed how challenging it would be to increase or decrease any single enzyme by itself, they were able to identify good candidate genes in the intricately intertwined gene network. The authors envision a future where Smart Cells are used as sustainable factories for a wide range of materials globally.

The future of the bioeconomy is being propelled by a generation of scientists, engineers, entrepreneurs, and policymakers immersed in a new world of unprecedented biological innovation. Today, there are increasing opportunities for students to apply textbook knowledge to cutting-edge research. A great example is the MIT-founded International Genetically Engineered Machine (iGEM) Foundation, which runs an annual synthetic biology research competition and includes 350+ student research teams around the world. Chen et al., a group of students from the University of British Columbia's iGEM team, showcased in their chapter three projects that highlight the move from ideation to proof-of-concept, early prototyping, all the way to market validation within 10–12 months. The examples include Paralyte, a whole-cell biosensor for shellfish toxin, Probeeotics, an engineered metabolic pathway for bee microbiota, and VPRE, a machine learning model to predict the evolution of SARS-CoV-2.

After an overview of the contributions of synthetic biology, this book moves into a different type of application that is now revolutionizing our understanding of biodiversity: genomic monitoring. Acharya-Patel et al. start this section of this book with an overview of environmental DNA (eDNA) and its potential for monitoring a wide range of organisms and ecosystems. eDNA is the genetic material shed by organisms, micro or macro, into their environment, which can be measured with powerful and cost-effective genomic tools. Today, eDNA sampling is being integrated into large-scale biomonitoring programs and will be useful for informing policy and management decisions in the near future. From tracking invading species like the American bullfrog invading Belgium, to detecting pathogens like SARS-CoV-2 in wastewater for pandemic monitoring, eDNA has been recognized and applied around the world. To fully embrace the potential of eDNA, the authors highlight several opportunities in the field, such as relating eDNA to more conventional methods and standardizing protocols and reporting.

Hubert et al. describe the role of microbial surveys using genomics in de-risking offshore oil exploration and their potential impact on insurance companies. With sea ice melting accelerating at a record pace in recent years, maritime and industrial activity is mounting in the Arctic Ocean. This translates into an ancillary increase in the risk of offshore oil spills, which are likely to result in challenging and expensive remediation efforts in such an unforgiven cold and remote environment. This is where genomic data on baseline microbial communities can play a huge role. Genomic surveys can help predict the potential for biodegradation by identifying the presence of marine bacteria that can eat compounds from oil. This genomic mining work could help inform risk modeling and determinants of insurance premiums for marine shipping but also help understand ecosystem impacts to inform clean-up costs in the event of a marine oil spill.

Genomic surveillance can also be taken to the forests to help predict and prevent pest outbreaks, where invasive species are often not known. Forests are vital for the bioeconomy as they support entire communities by providing building material, food, and energy. Arguably, the most important contribution of our forests is their ability to capture and store carbon, which mitigates climate change. However, forests are constantly challenged by invasions and being pushed beyond thresholds of sustainability. Hamelin shows how a comprehensive genomic biosurveillance approach can help ensure that these thresholds are not crossed. For example, genomics revealed the causal agents of sudden oak death and the ash dieback, two deadly pathogens that were previously unknown to science. Hamelin provides an overview of the different techniques currently being used to identify forest threads from PCR to next-generation sequencing. He also shares an outlook for the future of forest health monitoring as increasing demand will likely result in more access to powerful and cost-effective genomic tools being developed.

Genomic tools are also being used to survey other life-limiting ecosystems highly impacted by climate change, such as the African tropical glaciers, where cold-adapted prokaryotes are potential sources of cryoprotectants for food and pharmaceutical industries. Kuja et al. used a technique called 16S rRNA to survey microorganisms in Lewis Glacier, the largest glacier found in Mount Kenya and one of the best-studied equatorial glaciers. The high abundance of microbial species in the glacier surface and the drastic shift in the foreland soils suggest well-adapted resilient mechanisms to this unique environment. This information can be a game-changer for emerging economies where natural resources could be harnessed for sustainable development. However, climate change is affecting these local microbial reservoirs at an accelerated pace. Kuja and collaborators' work also revealed "seed banks" of inactive organisms, a reservoir of dormant microorganisms with the potential to reactivate when the environmental conditions change, and key indicators of the impact climate change is having on equatorial glaciers. The authors make a call to all African countries to embrace genomic applications to monitor and leverage bioresources from neglected ecosystems by decentralizing genomic platforms across the continent. This chapter illustrates the use and impact of genomics technologies in Africa and its potential application to advance the bioeconomy in emerging economies.

Another area where genomics has played an important role in driving the bioeconomy is agriculture. Whether it is food or fiber, today's crops have been selected or modified using a wide range of ancient methods but also the most modern genomic techniques. Gilchrist et al. focus on a plant grabbing the headlines. Cannabis cultivation spans millennia but despite its long-standing use, scientific research on the crop has been limited. The recent flux of restrictive legislation across the globe is changing this and has opened up new avenues for genomic research in cannabis. Besides a genome sequence published in 2011, a number of more recent publicly available databases have improved cannabis genomic resources. An interesting discovery has been the lack of a genetic basis for the commonly used designations "indica" and "sativa." Although these terms are well accepted in the cannabis industry and even used for medical prescriptions, genomic research has shown that these groups lack conserved genetic profiles as there are significant genetic differences within samples of the same strain. These differences could be exciting for recreational users but might have more serious implications for medical use and patients relying on strain-specific effects. The rapid growth of genomic tools is also supporting the improvement of the cannabis crop to produce and regulate the array of cannabinoids and terpenes in cannabis.

Plant secondary metabolites (PSMs), such as cannabinoids, are key for the bioeconomy as they not only shape plants' odor, color, and taste but also help plants adapt to changes in the environments and resist attacks from pathogens. As if that wasn't enough, PSMs are also very valuable outside the plant world, and are produced by pharmaceutical, cosmetic, and food industries for a wide range of purposes. Li et al. provide a look under the hood of the sophisticated process for PMS production, where traditional genomic tools just fall short. To really understand and leverage these important metabolites, a more holistic view is needed. Multi-omics provides this type of approach where multiple layers of information (i.e., molecules) are quantified, including genomics, transcriptomics (also called gene expression), proteomics, and metabolomics. A case in point is the role of multiomics to uncover the role of flavonoids, which have remained elusive. Rather than relying solely on genomics, researchers integrated metabolic profiles from different types of plants. This metabolomics–genomics-integrated approach was a great first step to enable the discovery of specific genetic markers that improve UV protection, which have been used to generate UV-B resistance of transgenic rice plants.

We close this book with an overview of why regulation and policy matters. To fully deliver on the promise of genomics and the bioeconomy, regulatory authorities across the globe are building policies and strategies to address questions about how to regulate products derived from genome editing and synthetic biology. Marden et al. discuss the flux in regulatory oversight around these products and how fast-paced developments in food science and agricultural production are challenging United States, Canadian, and European Union regulators. Until now, the United States and Canada have taken a product-based approach, largely focused on the end-product, while the EU regulates upstream explicitly on the process of production. However, a key challenge across the board is modernizing legislation at a pace that reflects emerging developments in biotechnology. In the United States for example, products like genome-edited plants can be subject to multiple and sometimes overlapping authorities like the USDA and FDA. On top of that, the lack of a shared set of definitions between agencies referring to biotechnology products can be confusing. United States recently updated USDA regulations relating to permit and food labeling requirements but the FDA's approach to genome editing and synthetic biology in foods remains a work in progress. To address some of these challenges, Health Canada is proposing to update its guidance to clarify the definition of "novel food," which will only include products where the genetic modification derives from the insertion of foreign DNA or when there are other specified changes in the end-product. In the EU, changes are anticipated as the current legislation falls short to regulate products resulting from genome editing but they are unlikely to be rapidly implemented. Marden et al. call on researchers and developers to stay up to date on regulatory developments and to be aware that requirements may differ among jurisdictions.

Uscátegui and Montaguth provide an overview of the regulatory landscape and public perception of biotechnology products in Latin America where the status is far from harmonized and highly debated. While Brazil and Argentina have one of the biggest biotech crop areas in the world, Ecuador declared itself a GMO-free territory in 2008 and Peru released a moratorium against genetically modified organisms in 2011. In countries such as Colombia, Bolivia, and Mexico, the debate is still heated as the efforts to ban transgenic seeds are still on the political agenda. Despite this political turmoil and public opinion polarization, biotechnology holds a big promise for the region. The world's first transgenic wheat tolerant to drought and ammonium glufosinate herbicide was created and approved in Argentina. Recently, Brazil developed and approved a genetically modified (GM) bean variety with resistance to bean golden mosaic virus, a disease with devastating effects for farmers. In Colombia, an alliance between academia and a private union generated an off-patent GM maize variety. To fully realize the impact on the bioeconomy to help farmers and scientists alike, the region needs to tackle misinformation around biotechnology by demonstrating that commercially available GM crops are safe for both humans and the environment.

Policy is fundamental to accelerate the deployment of a sustainable bioeconomy. In 2009, the OECD (Organization for Economic Co-operation and Development) published a policy agenda for developing a bioeconomy, and since then, interest has been growing. In 2012, the United States published its own bioeconomy blueprint, and the EU developed a similar strategy. Today, almost 50 countries incorporate bioeconomy development in their strategies. The global nature of the bioeconomy also stresses the particularly important role to be played by developing countries, not just as a source for "problems" to be solved, but as a source of opportunities to be offered in terms of research and innovation. Philip closes this book with a focus on the downstream economic phenomenon of genomics in the bioeconomy, aimed at generating economic activity. He offers a guide through a myriad of policy challenges from supply-side policy measures, such as building infrastructures, to demand-side measures that help to make a market. By looking beyond genomics R&D, it can be possible to realize the holistic approach of the bioeconomy with policy-making grounded in genomics and biotechnology.

Part I

Synthetic biology as a pillar of the bioeconomy

Chapter 1

Cellular agriculture and the sustainable development goals

Lenore Newman[a], Evan Fraser[b], Robert Newell[a,c], Evan Bowness[a], Kat Newman[a], and Alesandros Glaros[a,b]
[a]*University of the Fraser Valley, Abbotsford, BC, Canada,* [b]*University of Guelph, Guelph, ON, Canada,* [c]*Royal Roads University, Victoria, BC, Canada*

Introduction

One of the most effective ways individual consumers can reduce their direct impact on the environment is through their dietary choices (Machovina et al., 2015; Willett et al., 2019). In particular, "plant-forward" diets (defined as diets that focus on consuming plants and minimizing animal products) and "flexitarianism" (diets that intentionally reduce animal products) have emerged in public and academic debates as important strategies to reduce greenhouse gas (GHG) emissions, conserve habitat and biodiversity, and decrease agricultural water consumption and fertilizer usage. Accordingly, the advent of alternative protein (Tuomisto, 2019) products is positioned as a way of helping reduce humanity's impact on the environment (Mattick, 2018). Such promising protein production techniques include "cellular agriculture," which involves growing animal-analogue proteins in industrial or laboratory settings using cell cultures and tissue-engineering or fermentation-based techniques.

Although there are reasons to be optimistic about the potential contributions of cellular agriculture to developing more sustainable food systems, tremendous complexity and uncertainty remain. Many scholars and practitioners, for instance, point out that animals are an important component of a sustainable agroecosystem, providing valuable ecosystem services on many farms such as nutrient cycling (Parker, 2020). Second, given the technologies around cellular agriculture are extremely nascent and have not yet been taken to scale, it has been impossible to conduct a full life cycle assessment of these technologies (Lynch and Pierrehumbert, 2019; Newman et al., 2021). Third, animal agriculture is a vital source of livelihoods for hundreds of millions of small-scale farmers, many of whom live in the developing world where they have limited

Genomics and the Global Bioeconomy. https://doi.org/10.1016/B978-0-323-91601-1.00010-9

access to markets or high-quality protein (Rota and Urbani, 2021). How might a widespread shift to cellular agriculture affect the livelihoods of the households who currently depend on raising animals?

The optimism around cellular agriculture as a tool to reduce humanity's environmental impact and the complex issues these technologies surface present a need to examine and better understand cellular agriculture's potential role in transitions to sustainable food systems. This chapter will provide a preliminary review of the current evidence on how cellular agriculture may be utilized to promote a broad sustainable development agenda. We employ the Sustainable Development Goals (SDGs) as a lens and use the scientific literature to answer the following research question: how might the widespread adoption of cellular agriculture affect progress toward achieving the SDGs?

This chapter begins with a high-level introduction to the SDGs, discussing their relevance to sustainable diets and alternative proteins. It then discusses cellular agriculture, focusing on two specific set of techniques with strong genomic elements: (1) the use of cell cultures and (2) advanced fermentation, both of which are proposed to grow proteins analogous to what livestock currently produces. The following section explores the potential of both cell culture and advanced fermentation to advancing the SDGs. In the final section, we present a brief research and policy agenda geared at helping enable a successful transition toward more sustainable diets.

Background literature: The sustainable development goals, sustainable diets, and the potential of alternative protein

Humanity has exceeded what is sometimes referred to as a "planetary boundary," particularly in terms of climate change, biodiversity loss, disruptions to nutrient cycles, and land systems change (Rockstrom et al., 2009). The emerging field of planetary health (Myers, 2017) argues that global society has exceeded its safe operating space (Rockstrom et al., 2009; Steffen et al., 2015) and that the social and economic systems that have evolved since the Industrial Revolution are eroding the ecosystems on which we all depend for life. There is widespread consensus that we must remake these systems if they are to become environmentally sustainable. At the same time, addressing poverty and inequality is equally urgent priority. Action is needed at all levels of government, from local to global, to address these concurrent socioeconomic and environmental imperatives (Biermann et al., 2012; Steffen et al., 2015).

In 2015, the United Nations established the "2030 Agenda for Sustainable Development." In 2016, this agenda was linked with 17 "Sustainable Development Goals" (SDGs), designed to create a framework for monitoring progress and exploring unintended trade-offs. While not legally binding, UN member nations agreed to adopt the 2030 Agenda and to develop policies designed to advance the SDGs, as well as to regularly report on their progress toward achieving their targets. Together, the SDGs elaborate a wide range of social, economic, and environmental objectives. Critically, each goal is paired with specific indicators and quantifiable targets.

Many of the SDGs directly or indirectly relate to food systems. Agriculture and food production are among the primary sources of greenhouse gas emissions, most notably methane and nitrous oxide (Willett et al., 2019), and a leading driver of anthropogenic climate change. Agriculture is the primary cause of land systems change, biodiversity loss, and the destruction of natural habitat. The sector is also the largest consumer of freshwater resources, while also being a major source of water contamination. Consequently, anything that may change farming practices is likely to have a direct impact on the SDGs for clean water and sanitation, responsible consumption and production, climate action, life below water, and life on land.

Despite the clear need to reduce the environmental impact of agriculture, it is imperative to also increase food production to feed a growing global population and to achieve the goal of zero hunger, as well as to advance SDGs relating to nutrition and health. Ensuring that high-quality food is both affordable and universally accessible is a necessary step toward addressing inequality and reducing poverty, as well as supporting maternal health and childhood development. However, any policies, technologies, or programs that affect the food system will likely have far-reaching impacts, including possible unintended consequences for people involved in all areas of the food system. It is important to note that many of the world's poorest people are small-scale farmers, and the poorest small-scale farmers typically are female-headed households. Considering the food system holistically is important to ensuring environmental sustainability, achieving zero hunger, reducing poverty and inequality, and advancing gender equality.

At the same time as debates over the SDGs were taking place, another group of scholars were casting a critical eye on the food system (Fraser et al., 2016). Over the past 20 years, a consensus has emerged that many of the world's most pressing social and economic issues intersect with the systems we all depend on for sustenance. Five of the key publications in this regard include a United Nations report entitled "Livestock's Long Shadow" that explores animal agriculture's impact on the environment (Steinfeld et al., 2006), two special issues that explore a wide range of issues and show how food systems and human systems intersect in scholarly journals *Science* (Godfray et al., 2010) and *The Philosophical Transactions of the Royal Society* (Godfray and Garnett, 2014), the Intergovernmental Panel on Climate Change's special report on *Climate Change and Land* (Shukla et al., 2019), and the "Eat-Lancet Report" (Willett et al., 2019) that proposes the components of a nutritionally balanced diet that can be produced sustainably.

While providing a thorough review of this large and ambitious body of literature is beyond the scope of the current chapter, at least two high-level conclusions emerge from this outpouring of scholarship. The first is that one of the most effective ways individual consumers can reduce their direct impact on the planet is through dietary choices, specifically by focusing their buying habits on consuming only products that can be produced with a minimal environmental footprint. A second high-level conclusion is that, in general, reducing livestock consumption and eating a diet high in plant matter should provide a net sustainability benefit. This is because producing livestock requires a quantity of resources disproportionate to either its nutritional or caloric contribution to the global food system.

Despite disproportionate public messaging and media attention surrounding the environmental impacts of consumer choices, simply exhorting people to eat less meat is no substitute for good policy. This focus on individual dietary decision-making shifts the locus of responsibility away from government and onto consumers (Clapp and Cohen, 2009). Similarly, equating livestock consumption with negative environmental outcomes has caused many to point out that there are ways to increase the overall sustainability of livestock-based food systems (Varijakshapanicker et al., 2019). There are extremely unsustainable ways of producing fruits and vegetables. Nevertheless, there remains a scientific consensus that diets high in livestock products have a much larger environmental footprint than more plant-based diets and that there would be substantial benefit to the planet if consumers ate less meat and dairy (Eisler et al., 2014).

It is against a background of this concern over livestock-based diets and general agreement around the need to work toward the SDGs that scholars have begun examining the potential role of alternative protein sources and production techniques in transitioning toward more sustainable food systems (Wu et al., 2014). Such work has been accelerated by the extraordinary explosion of innovative consumer products available today. For example, the new generation of meatless burgers is high in protein but without animal products, and the introduction of these products into consumer markets has intersected with the rising discourse around sustainable diets. Activity in this field has accelerated notably over the past decade. Public interest was piqued in 2013 when researchers in the Netherlands successfully used laboratory techniques to craft a "beef" patty from stem cells (Suthar, 2020). This initial prototype reputedly costs US$300,000, but costs have plummeted since then. Today, the same type of burger can be produced for about $12 (Mosa, 2019). Private and public investment in cellular agriculture has both ballooned, suggesting that a whole new sector of the food economy is currently being developed.

Interest in alternative proteins has been amplified by claims over the potential contributions of these products to sustainable food systems. Companies such as GOOD, Perfect Day, Beyond Meat, and Impossible Foods all position their products as more sustainable alternatives to animal products, offering high-quality analogues that have lower environmental footprints yet do not require consumers to sacrifice taste. Most of these companies' websites explicitly refer to the way their products improve sustainability and some even provide high-level statistics to validate such claims. Impossible Foods' webpage suggests that "every time you eat impossible burger (instead of beef from a cow)," consumers use 96% less land, consume 87% less water, and produce 89% fewer GHGs (Impossible Foods, 2021). However, the livestock industry has started to push back with arguments about the important role animal agriculture plays in pasture management. A nonpeer-reviewed exploration of some of these issues published by the University of Oxford points out that many of these emerging alternative protein systems depend on large industrial supply chains that are extremely capital-intensive (Cusworth, 2021). Accordingly, the authors speculate that alternative protein systems may ultimately hurt the prospects of small-scale farmers, suggesting that promoting alternative protein might lead to trade-offs between the SDGs relating to equity and poverty reduction and those that focus on environmental health and ecosystem services.

Production methods for alternative proteins are rapidly evolving, particularly the novel methods and technologies associated with cellular agriculture. As the field is changing rapidly, there is a dearth of rigorous peer-reviewed studies through which to explore these issues. This chapter contributes to this emerging field and to the academic debate around alternative proteins by describing the methods and technologies associated with a new approach to protein production (i.e., cellular agriculture), evaluating the conditions under which cellular agriculture may lead to progress toward the SDGs, and identifying where it may result in trade-offs.

Key technologies in cellular agriculture

To the layperson, the broad area of alternative proteins must seem bewildering and complex. As an emerging industry, it is difficult to even find a common vernacular through which to describe the different protein production methods and technologies. Test tube burgers, fake meat, clean meat, in vitro meat, animal analogues, plant-based protein, alternative protein, cowless dairy, cruelty-free animal products, and cellular agriculture all compete for space in both public and academic discourses. Each of these terms has strengths and weaknesses, as well as proponents who prefer and lobby for their usage. The vocabulary associated with the emerging technologies of alternative protein is also politically charged, and recently, established industry players have taken to the courts to ensure words like "milk," "dairy," or even "mayonnaise" are only used to refer to goods produced through animal agriculture.

Reviewing the intricacies of these discussions is beyond the scope of this current chapter. Instead, we adopt a relatively simple nomenclature designed to be as politically neutral as possible, using "animal-derived protein" and "plant-derived protein" to refer to the main two avenues through which people obtain protein in their diets. The focus of this chapter is on two related, but distinct, technologies that are together generally considered to be "cellular agriculture." The cellular agriculture system we will investigate occurs when cell cultures are grown on some sort of scaffolding in a bioreactor to create a product analogous to that produced by animal muscle tissue. We will call this approach "cell culture-derived protein." Second, we also explore processes and technologies where proteins analogous to those created by animals are produced through advanced fermentation process that use genetically engineered yeasts or other microorganisms. We will refer to this suite of methods and technologies as "fermentation-derived protein."

Cell culture-derived protein

In essence, cellular agriculture processes begin with harvesting a biopsy of tissue cells from a living animal. Ideally, this would be done under an anesthetic in order to mitigate animal welfare concerns. The next step is to select the stem cells from the tissue sample and then these stem cells are grown in bioreactors where the cells consume an oxygen-rich mixture that includes the basic building blocks of life such as amino acids, glucoses, and growth factors. Sometimes, these cells are also grown on three-dimensional "scaffolds" to give them textures and structures similar to conventional meats. In other systems, products are created through growing cells as thread-like strands that can be braided together with fat and other ingredients. For instance, the first cell-derived burger ever consumed was made up of thousands of individually grown muscle strands that were mixed with a range of other ingredients and seasonings.

Cellular agriculture continues to build on advances made in genomics and synthetic biology. Genomics research in particular can contribute new methods for targeted genetic approaches for cell selection and proliferation for traditional livestock species. Currently, cellular agriculture selects stem cells from tissue biopsy; however, these cells can only replicate themselves a finite number of times. Emerging research could address this issue by developing targeted genetics techniques for selecting cells (such as induced pluripotent stem cells and embryonic stem cells) that can replicate themselves indefinitely in vitro, while still being able to differentiate themselves into the types of meat tissues that are commonly consumed today (Post et al., 2020). Researchers have accomplished this with some success, in growing skeletal muscle from induced pluripotent stem cells from pigs (e.g., Genovese et al., 2017). Such research might ultimately allow for the extended, large-scale replication of cells at high densities, a key requirement for the industrial production of cell culture-derived meat products. These advancements in genomics and livestock cell biology, broadly, can also be applied to other types of cells and tissues required for cellular agriculture production, such as fat (Yuen Jr. et al., 2022). Other roles for genomics, and broadly the omics disciplines (e.g., genomics and proteomics), include ways of optimizing cell selection, growth media creation, and bioreactor design for scaled-up cellular agriculture production. Experts recommend increased use and analysis of large omics datasets to achieve such optimization, alongside the creation of "genome-scale metabolic models." These datasets and models that relate metabolic genes to metabolic pathways for specific livestock would enable greater control and prediction over cell growth pathways (Suthers and Maranas, 2020).

While there have been huge breakthroughs since the original cell-derived hamburger was consumed in 2013, major barriers still exist. First, the growth serum that feeds the multiplying stem cells in the bioreactors is extremely expensive to produce. Traditionally, this growth serum is derived from bovine embryos, a process that is difficult to scale and presents significant ethical issues and animal welfare concerns. As such, numerous companies and research laboratories are attempting to mass-produce plant-derived serum, and the literature seems optimistic that the industry will clear this hurdle in the relatively near future (Selby, 2020). Nevertheless, securing an affordable supply of plant-based serum currently remains a major obstacle.

A second barrier relates to the three-dimensional structure of most animal-derived protein products (Stephens et al., 2018; Campuzano and Pelling, 2019). The cell culture-derived proteins created in bioreactors have none of the form or texture consumers expect in meat. Considerable research is being invested into developing technologies to mold the products of cell culture-derived proteins, including potentially using 3D printing technologies to shape these proteins or developing three-dimensional scaffolds to provide a structure on which cell cultures can grow. Research in the medical sciences, such as innovations in tissue scaffolding (Tijore et al., 2018), offers promising avenues for overcoming these obstacles.

A third major barrier relates to the types of products that can be manufactured through cellular agriculture. Some proteins are easier to manufacture than others, and certain cellular agriculture products are thus easier to produce at scale and commercialize than others. There are currently challenges that need to be overcome to produce cell culture-derived fats (Fish et al., 2020), thereby limiting the potential for developing animal analogues that provide equivalent food experiences as the "original."

These three barriers, combined with the challenge of obtaining regulatory approval for novel consumer products, have led to only a handful of cell culture-derived proteins being available in consumer markets. The products that do exist tend to be hybrid products; for instance, the company JUST received regulatory approval in 2022 to market a synthetic chicken nugget product in Singapore, which consists of a mixture of cellular-derived protein and plant material (Carrington, 2020).

Fermentation-derived protein

Another approach to manufacturing cellular agriculture products involves the use of microorganisms such as yeast to create proteins through fermentation processes (Stephens et al., 2018). Using fermentation to produce protein is a well-established method, being a common technique for synthesizing insulin for diabetics. When insulin was discovered in the early 20th century, most insulin used by diabetics was initially harvested from dogs. Not only did this cause significant animal welfare concerns, but it also created challenges related to maintaining consistent quality and a dependable supply. These problems were overcome in the 1980s when scientists were able to bioengineer microorganisms to produce insulin. Not only did synthetic insulin provide a huge boon in terms of animal welfare, but it also causes fewer allergies and provides much greater volume and consistency of supply.

Advanced fermentation has also been used in food production, prior to the emergence of cellular agriculture. In 1990, the US FDA approved the use of rennet, an enzyme that makes milk separate into curds and whey, produced through advanced fermentation. Prior to 1990, rennet was harvested from the stomach lining of ruminant mammals. Today, most of the world's rennet comes from advanced fermentation processes using bioengineered yeast, providing the world's primary source of rennet for cheesemaking.

Since the breakthroughs involving insulin and rennet in the 1980s and 1990s, the technologies associated with fermentation-derived proteins have evolved rapidly and are now starting to be seen in the marketplace. For instance, Californian start-up company Perfect Day has developed a yeast variety that is able to produce whey protein with long shelf life and is free from hormones, antibiotics, and even lactose (Khan, 2019). Starting in 2020, the company received US FDA approval for these proteins, and there are now commercially available ice creams on the US market that entirely use fermentation-derived whey protein.

Fermentation techniques for cellular agriculture utilize genomics research and knowledge to develop highly efficient and precise production of microorganisms, as well as specified ingredients of interest. Fermentation relies upon microorganisms, which act either as a primary protein ingredient (i.e., biomass fermentation) or as "cell factories" to produce specific functional ingredients (i.e., precision fermentation). Genomic data is invaluable across emerging fermentation techniques, allowing scientists to target specific molecules they want to produce *en masse* and screen for microbial strains for optimal growth of those targeted molecules. For precision fermentation, the introduction of recombinant DNA into microorganisms such as yeast, algae, fungi, or bacteria enables the production of targeted molecules (Stephens et al., 2018). Targeted gene editing and the use of big data to screen for suitable molecules and microbial strains are crucial to advance these precision fermentation tools and techniques (Specht, 2022).

Much like cell culture-derived protein, fermentation-derived protein faces significant technical, consumer, and regulatory challenges. For instance, while companies such as Perfect Day can produce whey protein and fermentation tanks, about 80% of the proteins in cow's milk is casein, a major ingredient in cheese production. To date, a commercially viable casein production has yet to be developed, but the nonprofit research project called *Real Vegan Cheese* claims they are working on this issue and several companies include *New Culture* are researching casein production. Other obstacles to fermentation-derived cellular agriculture include obtaining regulatory approval to market products as "dairy," conflicts with the dairy industry, and consumer concerns over bioengineering. Such obstacles must be overcome if this approach to protein development is to become mainstream. We summarize the key technologies supporting cell culture- and fermentation-derived proteins and outline some of the technological challenges currently being addressed in Table 1.1.

Processes for cell culture- and fermentation-derived protein production are summarized and outlined in Fig. 1.1. We adapted and simplified Kadim et al. (2015) flow process for cell culture. Further, we drew from Perfect Day's (2021) process diagram to outline precision fermentation processes. Note that these are highly simplified process diagrams, and a comprehensive detailing of the technical components of these processes is beyond the scope of this chapter.

Cellular agriculture and the sustainable development goals

To explore how cellular agriculture may be used as a tool for supporting the SDGs, we begin with a framework proposed by Queenan et al. (2017) who divide the SDGs into four broad impact categories: (1) those that address human well-being; (2) those that focus on environment protection; (3) those that focus on infrastructure and the built environment; and (4) a category they term "one health" that refers to food systems assessed in their entirety.

Although all four categories relate to food systems, the first and second are particularly relevant to cellular agriculture and its potential role in food system sustainability. To maintain a reasonable scope, this review focuses on the SDGs that address (1) human well-being and (2) environmental protection. We also add a third impact category—animal welfare—that is not explicitly included in the SDGs but has been proposed by numerous authors (e.g., Garnett et al., 2013) as a key dimension of a sustainable food system. Of course, there are tremendous overlaps in terms of the hoped-for impacts of the SDGs across all three of these dimensions; hence, the framework adopted here simply represents a preliminary heuristic for reflecting on the broad sustainability potential of cellular agriculture.

Human well-being and cellular agriculture

SDGs that relate to human well-being include ending hunger and poverty, promoting good health and well-being, ensuring quality education and gender equality, reducing inequalities, promoting peace justice, and strong institutions (i.e., SDGs 1, 2, 3, 4, 5, 10, and 16). In terms of ending poverty, ensuring good education, promoting strong institutions, or reducing inequality, it is unlikely that advancements in cellular agriculture alone (or any agricultural technology)

TABLE 1.1 Summary of key cellular agriculture technologies and technical challenges.

	Example companies/products	Key technologies	Technological challenges
Cell culture-derived proteins	*Name*: Finless Foods *Product*: cultured fish meat *Website*: https://finlessfoods.com/	Cell lines (e.g., stem cells from tissue biopsy, induced pluripotent stem cells)	Developing safe and efficient cell lines for industrial production
	Name: Future Fields *Product*: cell culture media *Website*: https://www.futurefields.io/	Growth media (including growth factors, amino acids, glucose, minerals, vitamins)	Lowering the cost of growth media for cell culture-derived proteins
		3D scaffolding for complex meat structures	Creating analogous product sensory aspects (look, feel, smell, texture) as animal-derived proteins, through scaffolding technologies and integration with fats among other tissues
	Name: Mosa Meats *Product*: cultured beef meat *Website*: https://mosameat.com/	Bioreactors for cell growth and proliferation	Establishing cheaper and appropriate microorganism and feedstock inputs for production of diverse proteins
Fermentation-derived proteins	*Name*: The EVERY Company *Product*: fermentation-derived egg whites *Website*: https://theeverycompany.com/	Microorganism (e.g., yeast, bacteria, fungi, algae)	
	Name: Perfect Day *Product*: fermentation-derived whey protein *Website*: https://perfectday.com/	Feedstock and growth inputs (e.g., source of glucose)	
		Bioreactors for fermentation processes	
	Name: Formo *Product*: fermentation-derived milk products and cheeses *Website*: https://www.legendairyfoods.de/		

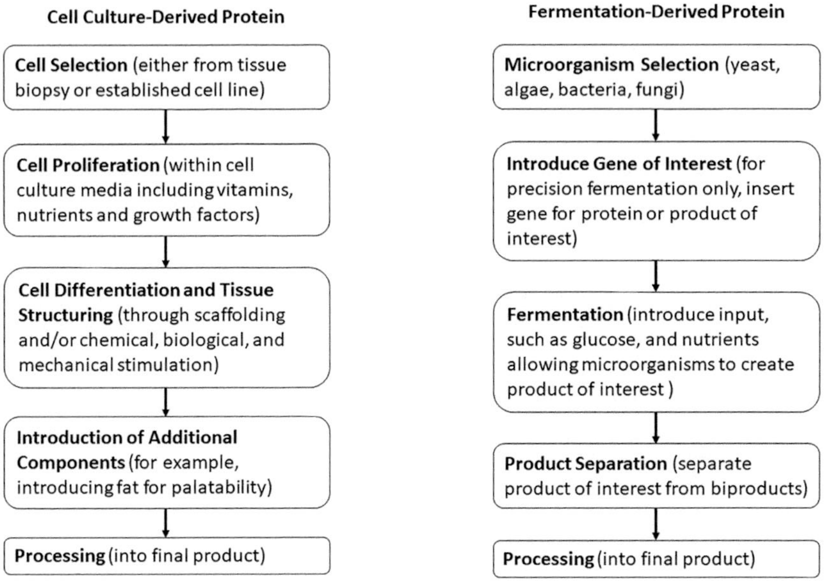

FIG. 1.1 Process diagrams for cell culture- and fermentation-derived proteins. *(Cell culture-derived protein diagram is adapted from Kadim, I., Mahgoub, O., Baqir, S., Faye, B., 2015. Purchas, R. Cultured meat from muscle stem cells: a review of challenges and prospects. J. Integr. Agric. 14(2),222–233, while fermentation-derived protein diagram is adapted from Perfect Day, 2021. Process [webpage]. https://www.perfectdayfoods.com/process/.)*

will play a meaningful role as these are fundamentally social and economic issues rooted in systems of power and politics.

Cellular agriculture may have a role to play in making progress toward the SDG for ending world hunger. Today, the United Nations estimates that there are 82 g of proteins available per capita on the planet. According to Walpole et al. (2012), the average human being weighed 62 kg (using 2005 estimates), and based on general dietary recommendations, people should consume between 0.8 and 1 g of proteins per kilogram of body weight. This suggests that the world already produces ~30%–60% more proteins than is currently needed per capita (although this production is highly unevenly distributed across the globe). However, the world population is expected to grow by 30%–40% by 2100 (Roser et al., 2019), suggesting that there will be a commensurate increase in demand for protein. To avoid further expansion of animal agriculture and associated degradation of habitat, humanity must both find better ways of distributing and utilizing existing proteins and find much more intensive systems for protein production. This supports the argument that the world needs new forms of protein that can be produced without taking up additional land.

In terms of the third SDG related to good health and well-being, cellular agricultural products are sometimes claimed by the literature as offering people healthier dietary choices and safer food products. But for each of these claims, there are vigorous counterclaims, and finding a consensus in the literature is challenging due to two factors. First, there is very little in the way of formal evidence or peer-reviewed research on this topic. Second, most of the claims and counterclaims seem to be made by people with ties to either the cellular agriculture or the livestock industries. There are, however, a small number of scholarly surveys on some of these topics. For instance, Chriki and Hocquette (2020) provide an early critical review of the extent to which cellular agriculture may provide healthier and safer protein. In this review, these authors note that because cellular-derived protein is produced in a fully controlled environment, biosafety and biosecurity concerns should be easier to address than in traditional abattoir settings where food can encounter pathogens such as *Escherichia coli*, *Campylobacter*, and *Salmonella*. Another potential for cellular agriculture to promote healthy or safe food pertains to antibiotic resistance. Today, the majority of the world's antibiotics are used in animal agriculture, which has contributed to the rise of antibiotic-resistant bacteria. While it is possible that cellular agriculture will utilize some antibiotics to stop contamination in bioreactors, at this juncture, it seems unlikely that cellular agriculture would ever use the volume of antibiotics currently deployed in animal agriculture.

Rubio et al. (2020) also provide an academic review of the potential for cellular agriculture to address public health concerns. They point out that the overconsumption of meat drives several very serious public health issues linked with fat and cholesterol that contribute to heart disease. This leads these authors to point out that if cellular agriculture presents consumers with options that are higher in dietary fiber and lower in saturated fat, then there may be public health benefits linked with a transition to cell-based protein systems. However, cellular agriculture products could follow the global trends of aggressively marketed high fat, salt, and sugar foods and beverages (World Health Organization, 2013) by providing products that serve as the basis for low-nutrition, heavily processed foods, marketable due to low cost and gratifying tastes. Currently available alternative protein products are higher in sodium than traditional livestock products.

Discussions over the health impacts of different protein supplies inevitably lead to questions around the nutritional quality of these different proteins (Neacsu et al., 2017). Much of the literature exploring this issue focuses on the digestibility and the amino acid profile in various protein sources. Overall, traditional livestock (and especially beef) has extremely good digestibility and amino acid profiles, whereas alternative sources of protein including plant-based and cellular agriculture need to be blended to provide the same overall nutritional package. In this regard, the literature also shows that different processing techniques can improve the digestibility of alternative proteins so that they match that of traditional livestock (Joshi and Kumar, 2015). Nevertheless, this raises questions of "naturalness" as processing food is problematic in the minds of many consumers (Siegrist, 2008; Frewer et al., 2011).

Overall, the literature on the potential human health impacts of cellular agriculture for consumers is difficult to parse. Partly, the challenge is that these competing health and nutritional claims inevitably privilege incommensurate issues such as food safety versus sodium versus protein digestibility and amino acids. Partly, the challenge is that the technologies associated with cellular agriculture have not yet been deployed at an industrial scale and have not yet been subjected to rigorous testing. Hence, many of the arguments currently expressed in the literature seem conjectural. Overall, this lack of clarity leads Rubio et al. (2020) to conclude that "... comprehensive, baseline nutritional data for cell-based meat is not [yet] publicly available" (p. 5). Regardless of how the industry looks today, it seems likely that over time, the nutritional quality of cell-derived products could rise to address these issues, but that this may require strong regulatory and policy frameworks to ensure that cell-derived proteins do not evolve to become the next generation's "junk food."

Another major dimension to consider in terms of the potential impact that cellular agriculture may have on the SDGs relates to the economic costs and opportunities of the emerging technologies and techniques. Generally speaking, publications on this topic are not peer-reviewed and focus on the opportunities new markets afford. For instance, the Boston Consulting Group (Witte et al., 2021) anticipates that by 2035, between 11% and 22% of the "meat" eaten on the planet will come from alternative sources. BCG expects that plant-derived, advanced fermentation-derived, and cell culture-derived sources will fill this market. Furthermore, BCG estimates that simply developing the capacity to extrude plant-based proteins into a form palatable to consumers will require between $11 and $28 billions of investment. Overall, BCG estimates that alternative proteins could become a $290 billion industry by 2035.

When framed in this way, the development of alternative proteins is presented as an unambiguously wonderful opportunity. But, it is also important to realize that the growth of these new markets may come at a cost to individuals, families, and communities who currently depend on producing animal-derived proteins. Today, the global meat industry is valued at over $1 trillion per annum (Dent, 2020). While much of this is captured by large firms (such as the world's largest meat processor JBS that recently posts ~$50B in annual revenues) (SEC, 2021), producing livestock is also a vital livelihood strategy for poor farmers.

A United Nations' "International Fund for Agricultural Development" 2020 report on the livestock sector argues that development agencies should promote animal agriculture specifically among small-scale farmers as an effective way of promoting the SDGs (Rota and Urbani, 2021). Livestock are an extremely important livelihood assets for rural households and crucial to reducing poverty (Upton, 2004). Livestock agriculture is a source of income for rural households with limited access to markets, as well as a vital source of high-quality dietary protein. This is especially important in areas with undernutrition, protein deficiencies, and gaps in terms of micronutrient availability, especially for young children, where lack of access to high-quality diets can lead to a lifetime of cognitive impairment.

The international development literature also promotes the importance of livestock for environmental reasons, especially in areas with poor-quality soils. In such a context, livestock manure provides an invaluable source of plant nutrients and organic matter. As such, there is a consensus in the development literature that livestock agriculture can be an effective strategy in some parts of the world for improving livelihoods (Swanepoel et al., 2010). As a result, it is important not to assume that promoting alternative proteins and investing in the technologies associated with plant-, cell culture-, or fermentation-derived proteins will necessarily support SDGs in all contexts. There are situations in which animal agriculture will remain a vital component of sustainable food systems and compatible with the SDGs.

Protecting the natural environment and cellular agriculture

The SDGs most directly related to environmental protection include clean water and sanitation, climate action, life below water, and life on land (i.e., SDGs 6, 13, 14, and 15). Cellular agriculture directly relates to all these goals. Indeed, much of the marketing and branding of the current generation of alternative agricultural products such as Impossible Foods, Beyond Meat, and Wildtype focus on the environmental benefits their products imply. For instance, the company Impossible Foods claims that their burgers, which use a mixture of cellular- and plant-derived products, generate 90% fewer greenhouse gas emissions and "significantly less water, land, and energy" than an average US beef patty. Similarly, Wildtype, which produces cellular-derived salmon and opened a tasting facility in San Francisco in 2021, claims that their products will lead to both healthier food and healthier oceans because they "produce enclosed, animal-free systems designed to keep fish in the water and out of farms" (Wildtype, 2021).

The literature on the potential impact of cellular-based technologies to provide environmental benefits is still preliminary but more developed than the literature exploring the potential impact of this technology on human health and well-being. For instance, there are a few life cycle assessments for plant-based beef products that conclude environmental indicators such as nutrient runoff leading to groundwater eutrophication, land use, and greenhouse gas emissions all perform better for alternative protein supplies than traditional livestock (see Mattick et al., 2015; Smetana et al., 2015).

Rubio et al. (2020) point out that mycoprotein (e.g., fungus-derived products such as Quorn) may result in more energy and GHG emissions than cellular- or plant-derived products; however, they also note that these require less land than plant-based sources and the water footprint for plant-derived proteins is extremely dependent on the agricultural system that produces the protein in the first place. Rubio and colleagues conclude that once optimized, cell-derived proteins are expected to require significantly fewer resources than animal-based proteins. However, a 2015 life cycle assessment of cellular-based proteins presents a different conclusion by suggesting that while cellular-derived proteins require less agricultural land, they may need more energy than traditional livestock as biological processes are replaced with industrial ones (Mattick et al., 2015).

Chriki and Hocquette (2020) note that many of the GHG emissions associated with cattle farming involve methane and, although methane has a higher global warming potential than carbon dioxide, it is a much shorter-lived gas that only stays in the atmosphere for a handful of years. By contrast, carbon dioxide stays in the atmosphere for centuries. Hence, if the widespread adoption of cellular agriculture results in increased carbon dioxide emissions but decreased methane emissions, then there may be a short-term benefit in terms of global warming but some unanticipated longer-term consequences. However, even this point is difficult to resolve as understanding the true impact of cell agriculture is also contingent on where the energy to run the cellular agricultural bioreactors comes from. For instance, bioreactors that obtain green energy from renewable sources or are designed to use waste heat from other industrial products will result in reduced carbon footprints.

Overall, the following high-level results seem to stand out from what limited scientific literature exists in terms of the potential for cellular-derived proteins to drive change toward the environmental-facing SDGs. First, finding alternative sources of protein that are not linked with either terrestrial or marine ecosystems should reduce pressure on the Earth's biosphere and protect wildlife. Second, cellular agricultural products can likely be produced with less nutrient runoff and water pollution than conventional livestock. Third, cellular agricultural operations that utilize renewable energy and/or waste heat to operate should result in a net reduction in GHG emissions, although the level of reduction will be contingent on the type of livestock systems they replace and the specific cellular agricultural systems being deployed.

Animal welfare and cellular agriculture

As noted above, the Sustainable Development Goals do not explicitly reference animal welfare and this omission has generated commentary and research by several scholars, such as Olmos Antillón et al. (2021) and Keeling et al. (2019), who point out that ensuring domesticated animals are treated ethically is an important goal and entirely consistent with the ambitions of the SDGs. For instance, although Garnett et al.'s (2013) seminal paper on sustainable intensification predates the SDGs, it explicitly includes animal welfare as a key dimension of a sustainable food system (also see Keeling, 2005).

Discussions around animal welfare are germane to this review because another major claim made by the cellular agriculture industry is that their products can be produced in a "cruelty-free" fashion. In many regards, this is one of the easiest claims made about cellular agriculture to validate as both the direct and indirect effects of cellular agriculture on animal welfare are far smaller than in the animal-based protein industries. With that said, there are still a handful of complexities that warrant being unpacked. First, since alternative protein products may utilize plant-based ingredients, there will be habitat-related problems caused by farming practices (however, given that the livestock industry uses huge amounts of grain for feed, this problem exists in both animal agriculture and alternative protein systems). Second, cell culture-derived proteins currently use donor animals to acquire stem cells and many current cellular-based companies use growth mediums and serums derived from animal products (Chriki and Hocquette, 2020). With that said, today, animal agriculture affects many billion animals each year so while current cellular agricultural practices and technologies still utilize some animal components, the level of impact is minuscule compared to conventional animal agriculture. Furthermore, numerous companies and research laboratories are investigating alternatives to using animal-derived serums for propagating cell cultures (Selby, 2020), and this technological hurdle is expected to be overcome in the relatively near future. In summary, the claims that cellular-based protein systems will result in significantly less animal welfare issues than conventional livestock seem both clear-cut and intuitive.

Conclusion: A brief policy and research agenda

The world faces an imperative sustainability transition, and technology is poised to play an important role. In particular, it seems quite likely that one of the most important aspects of this technological transition will be experienced through the protein sector, where a range of novel technologies, including our ability to harness microorganisms and cell cultures to replace livestock products, will feature prominently. This chapter has surfaced both the potential for this technology to support SDGs and the complexity of achieving these desired outcomes. From this review, we would like to conclude with three high-level points:

1. The emerging technologies and techniques in harnessing microorganisms and cell cultures to generate proteins appear to have great potential for reducing the environmental footprint of our food system, thereby helping promote the Sustainable Development Goals; however, the literature is too nascent and preliminary to be able to conclude definitively how this might unfold.
2. While there are tremendous economic opportunities to establish new enterprises in the broad area of alternative proteins, there are also economic costs associated with this transition that should be managed.
3. There is a significant role for regulation and policymaking to help ensure the technologies associated with alternative proteins generate widespread environmental and social benefits.

Based on these three conclusions, we briefly propose a policy and research agenda for cellular agriculture. A number of technical challenges that need to be researched and addressed before cellular agriculture products will be able to be produced at scale and compete in the marketplace. Such challenges include developing affordable plant-based growth serum, better tools to allow alternative proteins to take on a three-dimensional shape, and an enhanced understanding of the genomics of microorganisms in support of advanced fermentation. The insights from diverse omics fields are increasingly being applied to advance knowledge supporting the commercial-scale potential for cellular agriculture (Ontario Genomics, 2021). Further research in and application of genomics datasets, models, and editing techniques to reprogram or design novel livestock biological systems entirely can contribute to scaling various fermentation and cultivated meat inputs and techniques.

Rigorous research on the environmental impact of different emerging technologies and production systems is also needed. To ensure that the protein revolution indeed reduces the environmental impact of our food systems, thorough and detailed life cycle assessments of the environmental impacts of cellular agriculture at different stages of the production cycle are required. Furthermore, these assessments must be linked directly to policymaking such that governments incentivize specific tools and techniques that allow proteins to be produced with minimal environmental costs.

It will be necessary to both understand and help ameliorate the costs of the transition to alternative proteins in the conventional livestock sector. If alternative proteins reach their potential over the next decade, then they will seriously erode the contribution the livestock sector makes to the economy. While this will ultimately create net benefits in terms of environmental health and food system sustainability, the transition risks undermining a sector of the economy on which hundreds of millions of people currently depend. Helping livestock-dependent communities and families through this transition becomes a policy challenge of the utmost importance, which will involve government support, training and retraining programs, and major investments in rural communities all over the world. Reviewing this suite of policy tools is beyond the scope of the current chapter, but it is likely that valuable lessons could be learned from other sectors of the economy that have undergone massive technological disruption such as mining, forestry, and fisheries where a combination of novel technologies and new consumer demands have disrupted established systems.

Finally, if the global protein production system is facing a major period of disruption as suggested by the literature, then the colleges, universities, and training centers focused on animal agriculture all over the world need to radically rethink how they train the next generation. If the future means we depend less on animal-derived protein and more on plant-, cell culture-, or fermentation-derived proteins, then today's departments of animal bioscience that currently focus most of their energies on topics such as dairy, beef, pork, and poultry management and genomics need to also transition through changing faculty hiring strategies as well as developing new courses and programs of study.

To close, we return to the sustainability challenge of feeding the future in a way that does not further degrade the biosphere. There are no easy answers or single solutions, yet it is very clear that current protein systems significantly undermine planetary health. It seems both inevitable and urgent that we develop and embrace new protein systems to reduce our collective impact on the environment. Much of the imperative for change must come from policymakers and consumers, and we are already seeing signs that both policymakers and consumers worldwide are better understanding the importance of the protein transition. In this, emerging new technologies associated with cell cultures and fermentation greatly increase the scope and scale of what we imagine is possible for transitions to equitable and sustainable food systems.

References

Biermann, F., Abbott, K., Andresen, S., Bäckstrand, K., Bernstein, S., Betsill, M.M., Bulkeley, H., Cashore, B., Clapp, J., Folke, C., Gupta, A., 2012. Navigating the anthropocene: improving earth system governance. Science 335 (6074), 1306–1307.

Campuzano, S., Pelling, A.E., 2019. Scaffolds for 3D cell culture and cellular agriculture applications derived from non-animal sources. Front. Sustain. Food Syst. 3, 38.

Carrington, D., 2020 Dec 2. No-Kill, Lab-Grown Meat to Go on Sale for First Time. The Guardian. Available from: https://www.theguardian.com/environment/2020/dec/02/no-kill-lab-grown-meat-to-go-on-sale-for-first-time.

Chriki, S., Hocquette, J.F., 2020. The myth of cultured meat: a review. Front. Nutr. 7, 7.

Clapp, J., Cohen, M.J. (Eds.), 2009 Sep 30. The Global Food Crisis: Governance Challenges and Opportunities. Wilfrid Laurier Univ. Press.

Cusworth, G., 2021 Aug 3. Conflicts and Consensus: The Role of Legumes in Food Futures. LEAP: Livestock, Environment and People. Available from: https://www.leap.ox.ac.uk/article/conflicts-and-consensus-in-the-role-of-legumes-in-food-futures.

Dent, M., 2020 Mar. 25. The Meat Industry is Unsustainable. IDTechEx. Available from: https://www.idtechex.com/en/research-article/the-meat-industry-is-unsustainable/20231.

Eisler, M.C., Lee, M.R., Tarlton, J.F., Martin, G.B., Beddington, J., Dungait, J.A., Greathead, H., Liu, J., Mathew, S., Miller, H., Misselbrook, T., 2014. Agriculture: steps to sustainable livestock. Nat. News 507 (7490), 32.

Fish, K.D., Rubio, N.R., Stout, A.J., Yuen, J.S., Kaplan, D.L., 2020. Prospects and challenges for cell-cultured fat as a novel food ingredient. Trends Food Sci. Technol. 98, 53–67.

Fraser, E., Legwegoh, A., Krishna, K.C., CoDyre, M., Dias, G., Hazen, S., Johnson, R., Martin, R., Ohberg, L., Sethuratnam, S., Sneyd, L., 2016. Biotechnology or organic? Extensive or intensive? Global or local? A critical review of potential pathways to resolve the global food crisis. Trends Food Sci. Technol. 48, 78–87.

Frewer, L.J., Bergmann, K., Brennan, M., Lion, R., Meertens, R., Rowe, G., Siegrist, M., Vereijken, C.M., 2011. Consumer response to novel agri-food technologies: implications for predicting consumer acceptance of emerging food technologies. Trends Food Sci. Technol. 22 (8), 442–456.

Garnett, T., Appleby, M.C., Balmford, A., Bateman, I.J., Benton, T.G., Bloomer, P., Burlingame, B., Dawkins, M., Dolan, L., Fraser, D., Herrero, M., 2013. Sustainable intensification in agriculture: premises and policies. Science 341 (6141), 33–34.

Genovese, N.J., Domeier, T.L., Telugu, B.P.V.L., Roberts, R.M., 2017. Enhanced development of skeletal myotubes from porcine induced pluripotent stem cells. Sci. Rep. 7 (1), 41833.

Godfray, H.C., Garnett, T., 2014. Food security and sustainable intensification. Philos. Trans. R. Soc., B 369 (1639), 20120273.

Godfray, H.C., Beddington, J.R., Crute, I.R., Haddad, L., Lawrence, D., Muir, J.F., Pretty, J., Robinson, S., Thomas, S.M., Toulmin, C., 2010. Food security: the challenge of feeding 9 billion people. Science 327 (5967), 812–818.

Impossible Foods, 2021. Impact Calculator. Available from: https://impossiblefoods.com/ca/ecological-footprint-calculator.

Joshi, V.K., Kumar, S., 2015. Meat analogues: plant based alternatives to meat products-a review. Int. J. Food Ferment. Technol. 5 (2), 107–119.

Kadim, I., Mahgoub, O., Baqir, S., Faye, B., Purchas, R., 2015. Cultured meat from muscle stem cells: a review of challenges and prospects. J. Integr. Agric. 14 (2), 222–233.

Keeling, L.J., 2005. Healthy and happy: animal welfare as an integral part of sustainable agriculture. Ambio 34 (4), 316–319.

Keeling, L., Tunón, H., Olmos Antillón, G., Berg, C., Jones, M., Stuardo, L., Swanson, J., Wallenbeck, A., Winckler, C., Blokhuis, H., 2019. Animal welfare and the United Nations sustainable development goals. Front. Vet. Sci. 6, 336.

Khan, A., 2019 Jul 17. Perfect Day Releases First Animal-Free Dairy Ice Cream. Medium. Available from: https://medium.com/cellagri/perfect-day-releases-first-animal-free-dairy-ice-cream-3e93006128ba.

Lynch, J., Pierrehumbert, R., 2019. Climate impacts of cultured meat and beef cattle. Front. Sustain. Food Syst. 3, 5.

Machovina, B., Feeley, K.J., Ripple, W.J., 2015. Biodiversity conservation: the key is reducing meat consumption. Sci. Total Environ. 536, 419–431.

Mattick, C.S., 2018. Cellular agriculture: the coming revolution in food production. Bull. At. Sci. 74 (1), 32–35.

Mattick, C.S., Landis, A.E., Allenby, B.R., Genovese, N.J., 2015. Anticipatory life cycle analysis of in vitro biomass cultivation for cultured meat production in the United States. Environ. Sci. Technol. 49 (19), 11941–11949.

Mosa, C.E., 2019 Sept 12. Meat: From €250,000 To €9 Burger Patties. CleanTechnica. Available from: https://cleantechnica.com/2019/09/12/mosa-meat-from-e250000-to-e9-burger-patties/.

Myers, S.S., 2017. Planetary health: protecting human health on a rapidly changing planet. Lancet 390 (10114), 2860–2868.

Neacsu, M., McBey, D., Johnstone, A.M., 2017 Jan 1. Meat reduction and plant-based food: replacement of meat: nutritional, health, and social aspects. In: Sustainable Protein Sources. Elsevier, pp. 359–375.

Newman, L., Newell, R., Mendly-Zambo, Z., Powell, L., 2021. Bioengineering, telecoupling, and alternative dairy: agricultural land use futures in the Anthropocene. Geogr. J. 00, 1–16.

Olmos Antillón, G., Tunón, H., De Oliveira, D., Jones, M., Wallenbeck, A., Swanson, J., Blokhuis, H., Keeling, L., 2021. Animal welfare and the United Nations' sustainable development goals—broadening students' perspectives. Sustainability 13 (6), 3328.

Ontario Genomics, 2021 Nov. Cellular Agriculture: Canada's $12.5 Billion Opportunity in Food Innovation. Retrieved from https://www.ontariogenomics.ca/wp-content/uploads/2021/11/CELL_AG_REPORT_FULL-FINAL.pdf.

Parker, C.F., 2020 Nov 25. Role of animals in sustainable agriculture. In: Sustainable Agricultural Systems. CRC Press, pp. 238–245.

Perfect Day, 2021. Process [webpage]. https://www.perfectdayfoods.com/process/.

Post, M.J., Levenberg, S., Kaplan, D.L., Genovese, N., Fu, J., Bryant, C.J., Negowetti, N., Verzijden, K., Moutsatsou, P., 2020. Scientific, sustainability and regulatory challenges of cultured meat. Nat. Food 1 (7), 403–415.

Queenan, K., Garnier, J., Rosenbaum, N., Buttigieg, S., de Meneghi, D., Holmberg, M., Zinsstag, J., Rüegg, S.R., Häsler, B., Kock, R., 2017. Roadmap to a one health agenda 2030. CAB Rev.: Perspect. Agric. Vet. Sci. Nutr. Nat. Resour. 12 (014), 1–2.

Rockstrom, J., Steffen, W., Noone, K., Persson, A., Chapin III, F.S., Lambin, E., Lenton, T.M., Scheffer, M., Folke, C., Schellnhuber, H.J., Nykvist, B., 2009. Planetary boundaries: exploring the safe operating space for humanity. Ecol. Soc. 14, 1–33.

Roser, M., Ritchie, H., Ortiz-Ospina, E., 2019 May. World Population Growth. Our World in Data. Available from: https://ourworldindata.org/world-population-growth.

Rota, A., Urbani, I., 2021 Jan 1. IFAD Advantage Series: The Small Livestock Advantage: A Sustainable Entry Point for Addressing SDGs in Rural Areas. IFAD.

Rubio, N.R., Xiang, N., Kaplan, D.L., 2020. Plant-based and cell-based approaches to meat production. Nat. Commun. 11 (1), 1.

SEC, 2021. JBS 2020 Management Report and Financial Statements. Available from: https://sec.report/otc/financial-report/275113?__cf_chl_jschl_tk__=pmd_UfIZpw8UaYBHwONBwzmVLukzcLsfSBgimzuxV3E.tFs-1631408686-0-gqNtZGzNAiWjcnBszQd9.

Selby, G., 2020 Dec 1. "Truly animal-free" cellular meat collaboration: CPI and 3DBT examine alternatives to fetal bovine serum. In: Food Ingredients First. Available from: https://www.foodingredientsfirst.com/news/truly-animal-free-cellular-meat-collaboration-cpi-and-3dbt-examine-alternatives-to-fetal-bovine-serum.html.

Shukla, P.R., Skea, J., Calvo Buendia, E., Masson-Delmotte, V., Pörtner, H.O., Roberts, D.C., Zhai, P., Slade, R., Connors, S., Van Diemen, R., Ferrat, M., 2019. IPCC, 2019: Climate Change and Land: An IPCC Special Report on Climate Change, Desertification, Land Degradation, Sustainable Land Management, Food Security, and Greenhouse Gas Fluxes in Terrestrial Ecosystems.

Siegrist, M., 2008. Factors influencing public acceptance of innovative food technologies and products. Trends Food Sci. Technol. 19 (11), 603–608.

Smetana, S., Mathys, A., Knoch, A., Heinz, V., 2015. Meat alternatives: life cycle assessment of most known meat substitutes. Int. J. Life Cycle Assess. 20 (9), 1254–1267.

Specht, L., 2022. The Science of Fermentation. Good Food Institute. Available from https://gfi.org/science/the-science-of-fermentation/#:~:text=Precision%20fermentation%20uses%20microbial%20hosts,for%20producing%20specific%20functional%20ingredients.&text=Precision%20fermentation%20can%20produce%20enzymes,and%20Impossible%20Foods'%20heme%20protein.

Steffen, W., Richardson, K., Rockström, J., Cornell, S.E., Fetzer, I., Bennett, E.M., Biggs, R., Carpenter, S.R., De Vries, W., De Wit, C.A., Folke, C., 2015. Planetary boundaries: guiding human development on a changing planet. Science 347 (6223).

Steinfeld, H., Gerber, P., Wassenaar, T.D., Castel, V., Rosales, M., Rosales, M., de Haan, C., 2006. Livestock's Long Shadow: Environmental Issues and Options. Food & Agriculture Org.

Stephens, N., Di Silvio, L., Dunsford, I., Ellis, M., Glencross, A., Sexton, A., 2018. Bringing cultured meat to market: technical, socio-political, and regulatory challenges in cellular agriculture. Trends Food Sci. Technol. 78, 155–166.

Suthar, T., 2020. Glimpses on lab-grown meat. Food Agric. Spectr. J. 1 (01), 1–3.

Suthers, P.F., Maranas, C.D., 2020. Challenges of cultivated meat production and applications of genome-scale metabolic modeling. AIChE J. 66 (6), e16235.

Swanepoel, F.J., Stroebel, A., Moyo, S., 2010 Nov 9. The Role of Livestock in Developing Communities: Enhancing Multifunctionality. University of the Free State/CTA.

Tijore, A., Irvine, S.A., Sarig, U., Mhaisalkar, P., Baisane, V., Venkatraman, S., 2018. Contact guidance for cardiac tissue engineering using 3D bioprinted gelatin patterned hydrogel. Biofabrication 10 (2), 025003.

Tuomisto, H.L., 2019. The eco-friendly burger: could cultured meat improve the environmental sustainability of meat products? EMBO Rep. 20 (1), e47395.

Upton, M., 2004. The Role of Livestock in Economic Development and Poverty Reduction. FAO.

Varijakshapanicker, P., Mckune, S., Miller, L., Hendrickx, S., Balehegn, M., Dahl, G.E., Adesogan, A.T., 2019. Sustainable livestock systems to improve human health, nutrition, and economic status. Anim. Front. 9 (4), 39–50.

Walpole, S.C., Prieto-Merino, D., Edwards, P., Cleland, J., Stevens, G., Roberts, I., 2012. The weight of nations: an estimation of adult human biomass. BMC Public Health 12 (1), 1–6.

Wildtype, 2021. Available from: https://www.wildtypefoods.com.

Willett, W., Rockström, J., Loken, B., Springmann, M., Lang, T., Vermeulen, S., Garnett, T., Tilman, D., DeClerck, F., Wood, A., Jonell, M., 2019. Food in the Anthropocene: the EAT–lancet commission on healthy diets from sustainable food systems. Lancet 393 (10170), 447–492.

Witte, B., Obloj, P., Koktenturk, S., Morach, B., Brigl, M., Rogg, J., Schulze, U., Walker, D., Koeller, E.V., Dehnert, N., Grosse-Holz, F., 2021. Food for thought: the protein transformation. Ind. Biotechnol. 17 (3), 125–133.

World Health Organization, 2013. Marketing of Foods High in Fat, Salt and Sugar to Children: Update 2012–2013.

Wu, G., Fanzo, J., Miller, D.D., Pingali, P., Post, M., Steiner, J.L., Thalacker-Mercer, A.E., 2014. Production and supply of high-quality food protein for human consumption: sustainability, challenges, and innovations. Ann. N. Y. Acad. Sci. 1321 (1), 1–9.

Yuen Jr., J.S.K., Stout, A.J., Kawecki, N.S., Letcher, S.M., Theodossiou, S.K., Cohen, J.M., Barrick, B.M., Saad, M.K., Rubio, N.R., Pietropinto, J.A., DiCindio, H., Zhang, S.W., Rowat, A.C., Kaplan, D.L., 2022. Perspectives on scaling production of adipose tissue for food applications. Biomaterials 280, 121273.

Chapter 2

Engineering microbial biofactories for a sustainable future

Fernández-Niño Miguel[a] and Burgos-Toro Daniela[b]
[a]*Department of Bioorganic Chemistry, Leibniz-Institute of Plant Biochemistry, Halle (Saale), Germany,* [b]*Department of Biology, National University of Colombia, Bogotá, Colombia*

Identification and design of building blocks for microbial engineering

The omics revolution: An expanded library of biological parts

During the last decade, a substantial improvement in the technologies for the identification and characterization of different biological parts (i.e., genes, transcripts, proteins, and metabolites) has been achieved. For instance, we have observed the rapid evolution of methodologies for sequencing nucleic acids starting from massive sequencing machines used for sequencing by synthesis [e.g., Sanger (Sanger et al., 1977), pyrosequencing (Metzker, 2009), and Illumina (Bronner et al., 2013) sequencing] and moving to the development of real-time portable sequencing devices using nanopore technology for the electrical detection of nucleotides [e.g., MinION (Jain et al., 2016)]. Similarly, many methodologies have also been developed to allow a better detection, quantification, and characterization of peptides (Dallas et al., 2015), lipids (Yang and Han, 2016), proteins (Aslam et al., 2017), and metabolites (Beale et al., 2018) by using next-generation chromatographic columns for liquid and gas chromatography (Napolitano-tabares et al., 2021), tandem mass spectrometry (Wang et al., 2020), and the use of next-generation nuclear magnetic resonance spectroscopy (Li and Gaquerel, 2021). The extensive use of these technologies together with their higher accessibility and cost-effectiveness, has resulted in the production of a large number of data that accumulates in specialized online databases (Table 2.1). Currently, thousands of bits of information produced from metagenomics, transcriptomics, proteomics, and metabolomics experiments can be easily accessed from online databases, thus representing an extensive library for the identification of novel biological parts.

Genomics and the Global Bioeconomy. https://doi.org/10.1016/B978-0-323-91601-1.00003-1

TABLE 2.1 Popular biological databases are frequently used to identify biological parts for the assembly of biosynthetic pathways.

Database	Data type	Main functionality	Type of license	Ref
KEGG	Genomics, metagenomics, metabolomics	Metabolic pathway analysis. Enzyme's identification and genome mining	Licensed/free for academic users	Kanehisa et al. (2021)
BRENDA	Proteomics, metabolomics	Metabolic pathway analysis. Protein structure and gene analysis	Free	Chang et al. (2021)
UniProt/ SwissProt	Proteomics	Protein analysis and annotation	Free	Consortium (2021)
LipidMaps	Lipidomics, metabolomics	Pathways and mass spectrometric analysis of lipids. Lipid's prediction	Free	Fahy et al. (2009)
PRIDE	Proteomics	Protein and peptide analyses and mass spectrometric analysis	Free	Perez-riverol et al. (2019)
PANTHER	Genomics, proteomics	Genome mining. Protein analysis. Ontology and evolutionary gene/protein relationships	Free	Mi et al. (2021)
METLIN	Metabolomics MS/MS data	Metabolite identification and analysis. Metabolite's prediction	Licensed	Smith et al. (2005)
iGEm parts registry	Genomics	BioBricks design/mining	Free	Smolke (2009)
EMBL-EBI DB	Genomics, proteomics, transcriptomics, metabolomics	Genome, transcriptome, proteome, and metabolome analyses. Ontology	Free	Goujon et al. (2010)
GenBank	Genomics	Gene analysis. Sequence similarity search and ontology	Free	Clark et al. (2016)

DDBJ	Genomics	Gene analysis, annotation, and ontology	Free	Fukuda et al. (2021)
Ensembl	Genomics	Gene analysis, annotation, and prediction	Free	Howe et al. (2021)
Rfam	Transcriptomics, genomics	Transcriptome and genome analyses. Annotation.	Free	Kalvari et al. (2021)
InterPro	Proteomics	Protein analysis and sequence prediction	Free	Blum et al. (2021)
DisProt	Proteomics	Protein analysis, ontology, and annotations	Free	Monzon et al. (2020)
CATH	Proteomics	Protein analysis and evolutionary relationships between domains	Free	Knudsen and Wiuf (2010)
BioGRID	Genomics, proteomics	Gene and protein analyses. Ontology	Free	Stark et al. (2006)
REACTOME	Genomics, metabolomics	Ontology, metabolic pathways analysis, and genome mining	Free	Jassal et al. (2020)
SABIO-RK	Metabolomics	Metabolite and metabolic pathways analyses	Licensed/free for academic users	Wittig et al. (2012)

These biological parts are frequently called BioBricks and include protein-coding sequences, promoters, protein domains, ribosome-binding sites, translational units, terminators, and plasmid backbones. They represent biological units (building blocks) that can be used for the assembly of novel biosynthetic pathways and the subsequent engineering of complex biological systems [e.g., biofactories (Montaño López et al., 2021), and biosensors (Su et al., 2011)].

The design of novel artificial (unnatural) biological systems through the systematic integration of BioBricks into plug-and-play modules (pathways) is one of the main goals of synthetic biology (Leggieri et al., 2021). BioBricks are usually compared to electronic components (e.g., capacitors, LED, switch, and transistors) that can be connected to produce a specific activity (phenotype). Unlike electronic components, the amount of BioBricks that have been evolved in biological systems is astonishing. This large amount of data accumulated in the current databases can be used as a library for the identification of BioBricks in different biological systems. The vast amount of information available at all these databases has motivated the creation of specialized databases containing only information related to a certain group of organisms [e.g., the Microbial Genome Database (MBGD) (Uchiyama, 2003), the Comprehensive Yeast Genome Database (CYGD) (Guldener et al., 2005), the Plant Genome Database (PlantGDB) (Dong et al., 2004) and the ARKdb for farmed and other animals (Hu et al., 2001)], a specific type of biomolecule [e.g., nucleotide databases (Arita et al., 2021), protein databases (Chen et al., 2017), and metabolomics Databases (Lai et al., 2017)], and even a particular biological activity/compound property [e.g., FlavorDB—a database of flavor molecules (Garg et al., 2018) and ChEMBL—a database of bioactive molecules with drug-like properties (Gaulton et al., 2015)].

An interesting attempt to create a compressive database of genetic BioBricks (including promoters, repressors, reporters, enhancers, and terminators) has been done by the International Genetically Engineered Machine (iGEM) Foundation. This is an important step for the scientific community as BioBricks databases offer the possibility to identify and use novel building blocks for the design and assembly of biofactories of relevance in Industrial Biotechnology. In 2003, they created the iGEM Parts Registry containing a growing collection of more than 20,000 BioBricks that can be accessed and used to assemble complex biological devices and systems. Very recently, a larger database [BioMaster DataBase (Wang et al., 2021)] has been created by the integration of data from the iGEM Parts Registry and 10 additional databases. More than 40,000 BioBricks can be easily accessed using this database that also provides information related to BioBricks functions, interactions, and associated literature. Although a substantial number of BioBricks can be easily accessed and ordered via iGEM Parts Registry (http://parts.igem.org/) and BioMaster (http://www.biomaster-uestc.cn), its number is very low as compared to the number of BioBricks contained in all the previously mentioned comprehensive databases (Table 2.1). Moreover, comprehensive databases usually include no-analyzed data because of the increasing tendency to deposit raw data from omics studies. This uncharacterized data offers an excellent platform for the identification of novel (unknown) BioBricks whose biological potential has still to be elucidated and boost the design of novel biosynthetic pathways as will be discussed in the next section.

Multi-omics data mining for the identification of novel BioBricks

The discovery of novel BioBricks to assemble artificial biological systems is a major challenge in Synthetic Biology as they offer the possibility to design alternative/unnatural biosynthetic pathways for the heterologous production of valuable products. These BioBricks are fundamental scaffolds for the construction of innovative pathways to synthesize valuable compounds using microbial biofactories, which is a more inexpensive and efficient alternative to the traditional biosynthesis techniques [e.g., organic synthesis (Carreira, 2015), biocatalysts (Bell et al., 2021), and classical extraction techniques (Zhang et al., 2018)]. As mentioned before, the exponential increase and accumulation of raw sequencing data in open-access databases (mainly nucleic acids: genomes and transcriptomes from a vast number of organisms) have recently become a promising alternative for the discovery of novel BioBricks (Ziemert et al., 2016). This has improved the design of biosynthetic pathways by using new sequences from organisms that have not been evaluated for this purpose before, through mining of their genome and transcriptomic data (Baltz, 2018).

The concept of mining genomic data (i.e., genome mining) appeared in 1999 as a strategy to find patterns of similarity between sequences of nucleotides encoding for proteins with recognized function and raw sequencing data by using bioinformatic tools/algorithms for homology search such as the Basic Local Alignment Search Tool (BLAST) (Ziemert et al., 2016). Initially, genome mining was used as an alternative approach to identify sequences of nucleotides related to antibiotic production in different groups of bacteria based on the information of sequences previously associated with antibiotic production during experimental assays (mainly in Streptomyces) (Yang et al., 2019a). Nowadays, its popularity increased (Fig. 2.1) becoming a preferred method for identifying BioBricks involved in the production of different compounds (Albarano et al., 2020).

Two different approaches for genome mining can be defined. The first approach for genome mining, also known as the "classical way" or classical genome mining, aims to identify new BioBricks by using highly conserved sequences associated with the synthesis of a particular metabolite (Ziemert et al., 2016). This approach follows the same principles as reverse genetics: one or several reference enzyme sequences are used to identify, within a genome of interest, homologous sequences that can be related to catalytic domains or highly conserved motifs (Ziemert et al., 2016). Such domains

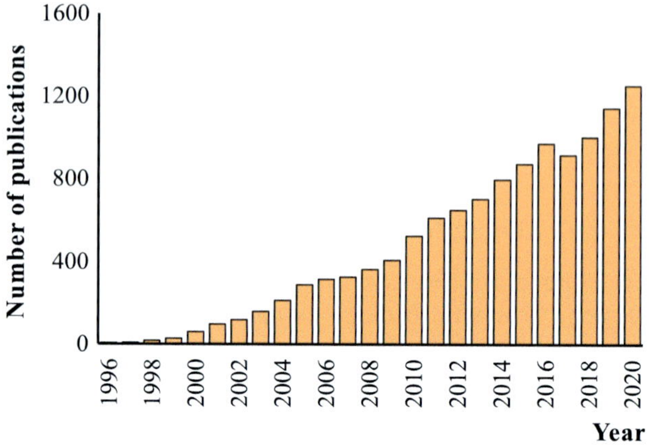

FIG. 2.1 Use of the term "Genome Mining" in publications from January 1980 to December 2020. The total number of publications per year is shown. Information obtained from PubMed (https://pubmed.ncbi.nlm.nih.gov/).

and motifs are then used to find novel candidate enzymes in nonannotated genomes, which can potentially be nominated as new BioBricks for metabolic engineering. This approach is recognized as an alternative to avoid the strenuous processes of experimentation related to the traditional identification of candidate enzymes (i.e., mutagenesis, library analysis, and activity assays). The sequences associated with a certain group of enzymes are clustered in Biosynthetic Gene Clusters (BGCs), which are groups of nonhomologous gene sequences related to the production of specific metabolites (Lin et al., 2020). Accordingly, classical genome mining will not only identify regions associated with catalytic domains for the enzymes of interest, but it will also allow the recognition of new related genes, including regulatory elements, which can potentially be used for the construction of a robust and stable biofactory (Lin et al., 2020). Several BGCs (mainly in bacteria and fungi) have been organized in highly predictable modules (Lin et al., 2020; Starcevic et al., 2008). This has made sequences predictions more reliable as modules are enriched with the addition of new related sequences. The increased popularity of classical genome mining has led to the construction of several databases and bioinformatic tools that aim to search, compare, analyze, and predict novel BioBricks in BGCs. One of the most popular platforms for classical genome mining of bacterial and fungal genomes is AntiSMASH (Blin et al., 2021). This database integrates the genetic information of different microorganisms with the production of many metabolites in a researcher-friendly and curated interface (Blin et al., 2021). Unfortunately, classical genome mining has been mainly applied to bacterial and fungal genomes, and only recently plant genomes are also starting to be explored (Wu, 2019). Thus, some platforms have been recently created to identify BGCs using plant genomes such as PhytoClust (Nadine et al., 2017) and plantiSMASH (Kautsar et al., 2017). Nevertheless, additional research is still required in this field.

The above-mentioned platforms have allowed the discovery of novel BioBricks in unexplored genomes [e.g., the identification of new BioBricks in rare actinobacteria related to the biosynthesis of polyketides, antimicrobial peptides, terpenes, and bacteriocins (Amin et al., 2019)], which improved the selection of enzymes for the biosynthesis of several valuable compounds.

In this way, several enzyme isoforms/variants with the ability to catalyze multiple reactions (i.e., catalytic promiscuity) have been recently identified (Zaparucha et al., 2018; Khersonsky and Tawfik, 2010), thus expanding the current repertory of transformations that can be bio-catalyzed. For example, through genomic mining, Dilokpimol et al. (2018) succeeded in identifying 150 putative sequences associated with the production of glucuronoyl esterases (enzymes that participate in saccharification, a process necessary for biofuel production), in about 250 publicly available genomes in BLAST. To validate this search strategy, 21 sequences (belonging to both basidiomycetes and ascomycetes) were selected and used for heterologous production of this enzyme in *Pichia pastoris*. Eighteen of these sequences achieved enzymatic activity, demonstrating the effectiveness of genomic mining in the search for new sequences associated with enzymes of industrial interest (Dilokpimol et al., 2018). Similarly, Bösch and coworkers used genome mining to find candidate sequences to produce proteusins, which are important for their antiarenaviral activity. After searching Cyanobacterial genomes, the researchers characterized and reconstructed in *Escherichia coli* the genetic block for producing landornamide A from a silent gene in the PCC6506 strain *Kamptonema* sp. The block included enzymes necessary for the bioavailability of that antiviral. This is an example of how using this search method, new compounds of pharmaceutical interest, such as those used for the treatment of choriomeningitis virus, can be found (Bösch et al., 2020).

Genome mining has allowed the exploration of genes in other genomes never before evaluated, finding potentially important genes for the construction of BioBricks. For example, El-Sayed et al. (2020) searched for these genes in endophytic fungi of coniferous plants, finding several candidate genes for the production of taxol, the most effective anticancer agent currently known, that was first described in *Taxus brevifolia*, the Pacific yew tree. This is an example of how candidate genes for the construction of BioBricks can be found in other domains of life, finding possible isoforms of interest for synthetic biology (El-Sayed et al., 2020).

However, a major limitation of classical genome mining is that it does not recognize BGCs that do not fit into the predictable modules. In other words, classical genome mining does not work for unconventional sequences that are out of the clusters defined up to now. To identify those sequences, a different approach for genome mining (known as comparative genome mining) can be used. Unlike classical genome mining, comparative genome mining uses an evolutionary approach for the identification of novel BioBricks (Ziemert et al., 2016). There, lineages of organisms are first constructed, and sequences are compared considering both the phylogenetic relationships among lineages and the biosynthetic pathways. By using comparative genome mining, different BioBricks have been recently identified considering evolutionary events within lineages associated with the production of several secondary metabolites (Ziemert et al., 2016). Like classical mining, several bioinformatic tools have also been developed to simplify comparative genome mining. For example, The Natural Product Domain Seeker (NaPDoS) platform has been created to detect and analyze genes involved in the production of secondary metabolites by phylogenetic clustering of sequences among related species and the prediction of their putative products (Ziemert et al., 2012). Currently, genome mining has been improved by linking information related to other macromolecules such as transcripts, proteins, and metabolites. This has resulted in the identification of novel BioBricks from datasets obtained from organisms exposed to different physiological and environmental conditions. Thus, it has been possible to link the production of a particular compound with a particular environmental/physiological context. The use of a multi-omics mining approach is an important step for the rational design and modification of BioBricks considering the physiological and environmental context to improve their expression, activity, and purification as will be discussed in the next section.

De novo rational design of BioBricks for metabolic engineering

Once identified in databases, BioBricks can be obtained via polymerase chain reaction (PCR) that allows the amplification of the desired BioBrick from isolated DNA (Shetty et al., 2008; Sleight and Sauro, 2013). During PCR, several flanking tails can be added to the BioBricks to simplify subsequent cloning in expression vectors [i.e., by the addition of specific restriction sites (Loenen et al., 2014)], mutagenesis [i.e., by using PCR-based method (Castorena-Torres et al., 2016)], and protein purification [i.e., by the addition of tags (Kimple et al., 2013; Vandemoortele et al., 2019)]. The addition of restriction sites flanking the coding sequence of BioBricks amplified by PCR has been shown to simplify their subsequent assembly into the multiple cloning site of a particular selected/available expression vector. A comprehensive list of restriction endonucleases and their corresponding recognition sequences can be accessed from New England Biolabs (https://international.neb.com/) and Thermo Scientific (https://www.thermofisher.com) product portfolios. In addition to the modification of flanking regions, coding sequences in BioBricks can also be modified from their natural form by using PCR-based mutagenesis. Nowadays, several methods have been developed to perform single and multiple site-directed mutagenesis, including error-prone PCR (Lee and Fried, 2021), the "megaprimer" method (Xu et al., 2003), the one-tube PCR method (Chapnik et al., 2007), and the double polymerase chain reaction method (Edelheit et al.,

2009). These methods have allowed the construction of novel unknown/artificial BioBricks that can be used to further explore protein function/structure relationships, thus increasing BioBricks diversity. The modification of coding sequences in BioBricks also includes the addition of protein tags that can be used to improve secretion competency, affinity, and solubility of recombinant proteins in different hosts, as well as to allow interaction with specificity target macromolecules (Fig. 2.2). Different signals have been extensively used for improving the secretion of heterologous proteins in bacterial (e.g., Staphylococcal protein A, OmpA/T, and hemolysin), yeast (e.g., serum albumin, alpha-mating factor, and *K. lactis* killer toxins), insects (e.g., honeybee melittin and tissue plasminogen activator), mammalian (e.g., human placental alkaline phosphatase) and plants (e.g., extensin secretory signal) and glycomodule signal peptides (Kimple et al., 2013). Similarly, many tags have been engineered and added to recombinant proteins to increase their solubility and affinity, which include His-tag, FLAG, HA-tag, T7-tag, S-tag, NusA, maltose-binding protein, elastin-like peptides, thioredoxin, among others (Kimple et al., 2013). All these tags can be further removed through enzymatic methods that require the addition of cleavage sites by PCR during the BioBrick design phase. Popular cleavage sites for tag removal include recognition sites for intein, sortase A, thrombin, enterokinase, TEV protease, SUMO, aminopeptidases M, carboxypeptidase B, and more (Waugh, 2011).

Besides the popularity of these methods, there are still several limitations when using PCR approaches for the design of BioBricks for metabolic engineering. A common problem is related to the availability of template DNA for PCR. Many protein isoforms are mined from databases and belong to organisms that are difficult to access either due to biological or due to legal reasons (e.g., nonculturable microorganisms, extremophiles, and protect endangered species). In addition, the complexity of posttranscriptional modifications

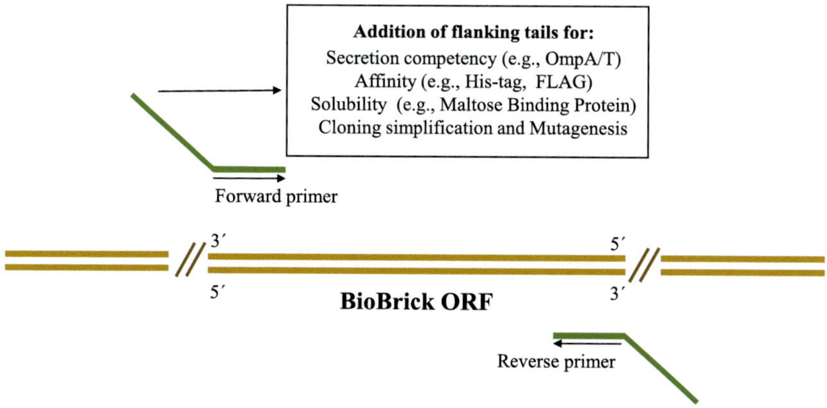

FIG. 2.2 Popular flanking tails added during PCR amplification of BioBricks.

in eukaryotes also represents a challenge for cDNA preparation from isolated RNA as undesired protein isoforms can be obtained. Moreover, isoforms obtained by PCR could also contain recognition sequences for restriction enzymes or proteases that can complicate their subsequent cloning. To overcome these problems, an alternative for BioBricks isolation arose in response to a significant cost reduction of de novo gene synthesis and an increase in its quality.

Very recently, several technologies for de novo gene synthesis have been improved by the modification of traditional phosphoramidite chemistry methods (Vu and Hirschbein, 1991) using column or microarray-based synthesizers (Kosuri and Church, 2014; Hughes and Ellington, 2017). Thus, several gene synthesis companies have emerged (e.g., Biocat, ATUM, GenScript, Genewiz, Idtdna, Biomatik, OriGene, ProteoGenix, Codex DNA, Integrated DNA Technologies, Twist Biosciences, Next-Gen Synthesis, among others), which opened a new market that led to a reduction from approximately \$10 to \$0.10 a base pair during the last decade. The increasing accessibility to customized and de novo synthesized genes has boosted the rational design of BioBricks, whose sequences can be already codon-optimized (Elena et al., 2014), mutated, tagged (Kimple et al., 2013), and even subcloned during the first phase of BioBricks design. Unlike PCR-based methods, the customized synthesis of BioBricks is expected to highly impact and simplify the assembly of metabolic pathways during the construction of artificial biofactories.

Rational engineering of metabolic pathways and modular assembly

A plug-and-play design: Assembly of BioBricks into artificial pathways

The assembly of BioBricks (obtained either via PCR or via de novo synthesis) into artificial pathways represents an additional step through the construction of biofactories, and different approaches have been designed for this purpose (Young et al., 2021). Artificial pathways are frequently assembled into expression vectors that are transformed into a host organism or directly inserted into the host genome depending on their complexity. A comprehensive inventory of expression vectors can be accessed at the Addgene plasmid collection (https://www.addgene.org/), which contains over 103,306 plasmids that can be ordered online. Addgene catalog of plasmids simplifies the selection of expression vectors by providing information regarding the desired expression system (e.g., bacteria, yeast, insect, plant, mammalian), vector type and purpose (e.g., lentiviral, synthetic biology, and CRISPR), and even the depositing institution (more than 1089 institutions around the world). The insertion of biosynthetic pathways into the host genome will be extensively discussed later in the *genome editing* section.

As mentioned before, several methodologies have been designed for the modular assembly of BioBricks into artificial pathways using expression vectors (Young et al., 2021). Traditionally, BioBricks have been assembled into expression vectors by repetitive cycles of enzymatic digestions and ligations using different sets of restriction enzymes and DNA ligases (Bertero et al., 2017). Thus, the addition of restriction sites flanking the open reading frame (ORF) of different BioBricks is a popular technique to enable assembly into the multiple cloning site (MCS) of a particular expression vector. This approach has been used for decades for the construction of different biosynthetic pathways (Bertero et al., 2017). Usually, this process involves the use of multiple restriction enzymes and the selection of expression vectors with an ideal MCS topology for BioBricks assembly. Somehow, this represents an expensive and time-consuming approach that is currently being replaced using alternative cloning methods for pathway assembly. The current alternative methods for pathway assembly can be divided into four different categories: 1. PCR dependent (Tillett and Neilan, 1999), 2. ligation independent (Stevenson et al., 2013), 3. seamless cloning (Zhang, 2021), and 4. recombinational approaches (Park et al., 2015) (Table 2.2).

In PCR-dependent methods, BioBricks are obtained by PCR and ligated into a vector without the use of restriction enzymes (Tillett and Neilan, 1999). Two different polymerases can be used for this purpose: standard Taq DNA polymerases or high-fidelity DNA polymerases. When using standard DNA polymerases for amplifying BioBricks, an adenine residue (A) will be always added to the 3′ end of the amplified BioBrick due to its terminal transferase activity. This A-tail can be then ligated to a vector carrying a single 3′ thymidine (T) complementary overhand by using DNA ligases or specialized DNA topoisomerases. In contrast, when using high-fidelity DNA polymerases to obtain BioBricks, the resulting PCR product does not contain additional residues but results in a blunt-ended PCR product. These blunt-ended BioBricks can be then ligated to a special plasmid which is covalently linked to topoisomerase that enables ligation between the fragments. PCR-dependent methods depend on the ability of specific commercial vectors, which limits the selection possibilities for selectable markers, promoters, RBS, and other regulatory elements to what is commercially available. An alternative approach to avoid the use of DNA ligases is the ligation-independent (LI) method (Stevenson et al., 2013). In LI methods, BioBricks are also obtained by PCR, but with the addition of complementary sequences (overhangs) flanking the BioBricks ORFs (Stevenson et al., 2013). These LI overhangs are complementary to sequences in the target vector with a length higher than 12 bp. Single-stranded overhands in BioBricks and vectors are generated by the 3′-5′exonuclease activity of T4 DNA polymerase in the absence of dNTPs. The reaction starts without dNTPs to promote exonuclease activity but then the activity is shifted back to polymerase by the addition of only one dNTP into the reaction mix. Thus, both polymerase and exonuclease activities of T4 polymerase stall due to the lack of additional

TABLE 2.2 **Popular approaches for the assembly of BioBricks into artificial biosynthetic pathways for the production of valuable products.**

Assembly approach	Definition	Example	Ref
PCR dependent (e.g., TA cloning; blunt-end cloning; TOPO cloning; fast cloning)	BioBricks are obtained using PCR, and then their products are ligated into a vector without using restriction enzymes	Use of TOPO cloning for the construction of BioBrick in *Saccharomyces cerevisiae*, which synthesizes specific synthases for the production of C11 terpenes, important in the pharmacological and food industry	Tillett and Neilan (1999), Ignea et al. (2019)
Ligation independent [e.g., polymerase incomplete primer extension (PIPE); sequence and ligation-independent cloning (SLIC); overlap extension cloning (OEC)]	BioBricks are also obtained with PCR products, but complementary sequences are added to the target vector. Exonuclease activity is promoted in the absence of dNTPs to generate the sticky ends; then, polymerase action is reactivated by adding a single dNTP to the reaction	Use of SLIC to construct a BioBrick in *Escherichia coli*, which synthesizes cyanobacterial hapalosin, an effective agent to improve the success of chemotherapy	Stevenson et al. (2013), D'Angostino and Gulder (2018)
Seamless cloning (e.g., Gibson assembly; HiFi DNA assembly; Golden Gate assembly)	To construct the BioBrick using seamless cloning, three are required: a 5′ exonuclease, a polymerase, and a DNA ligase. It allows no unwanted fragments to be inserted during cloning, maintaining the identity of the block	Use of HiFi DNA assembly for BioBrick engineering in *S. cerevisiae* to synthesize plant phenylpropanoids, important for counteracting bacterial pathogen attacks	Zhang (2021), Ramzi et al. (2018)
Recombinational approaches [e.g., gateway cloning (Invitrogen); creator cloning (BD Clontech)]	Commercial assembly platforms, such as gateway cloning or creator cloning, have made it possible to transfer BioBricks to different vectors without using ligases or restriction enzymes	Use of gateway cloning in a viral vector (narcissus mosaic virus (NMV)) that was then used to infect *Nicotiana benthamiana* plants. This technique was used to promote the production of flavonoids in these plants, which are secondary metabolites with antiinflammatory activity, for example	Park et al. (2015), Zhang et al. (2013)

dNTPs and vectors, and BioBricks are annealed and repaired during transformation. A similar approach that also utilizes flanking complementary overhangs between BioBricks and vectors is based on seamless cloning (Zhang, 2021). In this case, overhangs sequences are created by PCR (with lengths between 15 and 80 bps) and can be designed to assemble multiple BioBricks simultaneously (Zhang, 2021). In seamless cloning, cohesive ends are generated by the addition of 5′exonucleases, followed by the action of a polymerase that fills the gaps between the aligned regions and a DNA ligase that seals the nicks. These methodologies depend on the selection of these three enzymes and errors can be minimized by using high-fidelity enzymes. Finally, recombinational approaches have become a popular alternative for BioBricks assembly with the introduction of three commercial assembly platforms (i.e., Creator, Echo Cloning, and Gateway Cloning) (Park et al., 2015). Recombinational approaches utilize site-specific recombinases that enable the transference of BioBricks between different vectors in the absence of DNA ligases or restriction enzymes (Park et al., 2015). In this method, a recognition site for recombinases is added during the BioBricks design phase to allow their transference between a set of predesigned commercial vectors.

A completely different platform for the assembly of metabolic pathways in *E. coli* (i.e., the ePathBrick platform) has been developed by Xu and co-workers (2012). In this methodology, pairs of restriction enzymes (known as isocaudomers) are used. These enzymes recognize different DNA sequences but generate identical overhangs. Xu and co-workers designed a set of vectors carrying recognition sites for four compatible isocaudomers enzymes (*Avr*II, *Xba*I, *Spe*I, and *Nhe*I) (Xu et al., 2012). These vectors can be utilized to allow BioBricks assembly at defined locations between regulatory elements. Moreover, BioBricks and vectors can be reused and easily manipulated to obtain different expression topologies/gene conformations (operon, pseudooperon, and monocistronic), thus allowing control of their genetic expression (Xu et al., 2012). ePathBrick vectors with different topologies (e.g., different selectable markers) can be accessed and ordered from the Addgene catalog.

Currently, a combination of assembly approaches is often utilized to assemble BioBricks into artificial biosynthetic pathways. An important aspect to consider, which is concomitant with BioBricks assembly, is the creation of topologies (i.e., regulatory networks) in the biosynthetic pathway that allows controlling the genetic expression of BioBricks as will be discussed in the next section.

Controlling genetic expression through regulatory networks

Controlling the genetic expression of BioBricks is a crucial step during pathway design. Many regulatory networks have been elucidated during the last decades to understand how different organisms regulate their genetic expression (Agapakis and Silver, 2009; Singh, 2014). Several regulatory networks have been modified and adapted from their natural configuration and used as a strategy to control the genetic expression in artificial biosynthetic pathways (Agapakis and Silver, 2009; Singh, 2014). Thus, the use of regulatory networks during pathway design represents a frequent approach to control and fine-tune desired phenotypes. Currently, six different and popular regulatory mechanisms can be selected to control the genetic expression in metabolic engineering as shown in Fig. 2.3.

The most basic regulatory mechanism used to control the genetic expression of BioBricks is the simple regulation (Fig. 2.3A). Simple regulation can be performed at three different levels of gene expression: (1) transcription, (2) translational, and (3) posttranslational levels (Agapakis and Silver, 2009; Singh, 2014). Transcriptional simple regulation can be achieved by using a set of different gene promoters that can be selected to regulate the gene expression of genes under their control. Promoters are usually divided into two different categories: regulated (also known as inducible) and unregulated (also known as constitutive) (Gilman and Love, 2016; Haugen et al., 2008). Inducible promoters have been identified as regulatory elements able to activate or repress the genetic expression in the presence/absence of inducer compounds or environmental stimulus that affects their associated transcription factors (Gilman and Love, 2016; Haugen et al., 2008). In contrast, constitutive promoters are active only depending on the availability of RNA polymerase and not affected by transcription factors or any external stimulus (Gilman and Love, 2016; Haugen et al., 2008). Libraries of constitutive promoters have been engineered via mutagenesis to expand their diversity, which allows the selection of promoters with a different degree of expression in bacteria (Chen et al., 2018; De Mey et al., 2007), yeast (Redden et al., 2015; Redden and Alper, 2015; Nevoigt et al., 2006), and plants (Cai et al., 2020). A compressive catalog of inducible and constitutive promoter sequences in prokaryotic and eukaryotic cells can be accessed at the Registry of Standard Biological Parts of the iGEM Foundation (http://parts.igem.org/Promoters/Catalog). This catalog constitutes a valuable source for promoter selection and an essential tool for biosynthetic pathway design. For instance, promoters can be selected to activate or repress the genetic expression in the presence of metal ions (e.g., Cu/Fe/Co), sugar analogs (e.g., IPTG), microbial communication signals (e.g., quorum sensing molecules), and even in response to thermal changes. Recently, the use of multi-regulated promoters with the ability to respond to both activation and repression stimuli has become a popular practice to reduce pathway complexity retaining the ability to control the genetic expression by using a single multi-regulated regulatory element (Chen et al., 2018).

As mentioned before, genetic expression can also be controlled by translational regulation during pathway engineering (Agapakis and Silver, 2009; Singh, 2014). Three popular approaches are frequently used to regulate the gene expression based on the mRNA-specific control: (1) control via antisense RNA (Zhang and Zhang, 2018), (2) control via ribonucleases (Carrier and Keasling,

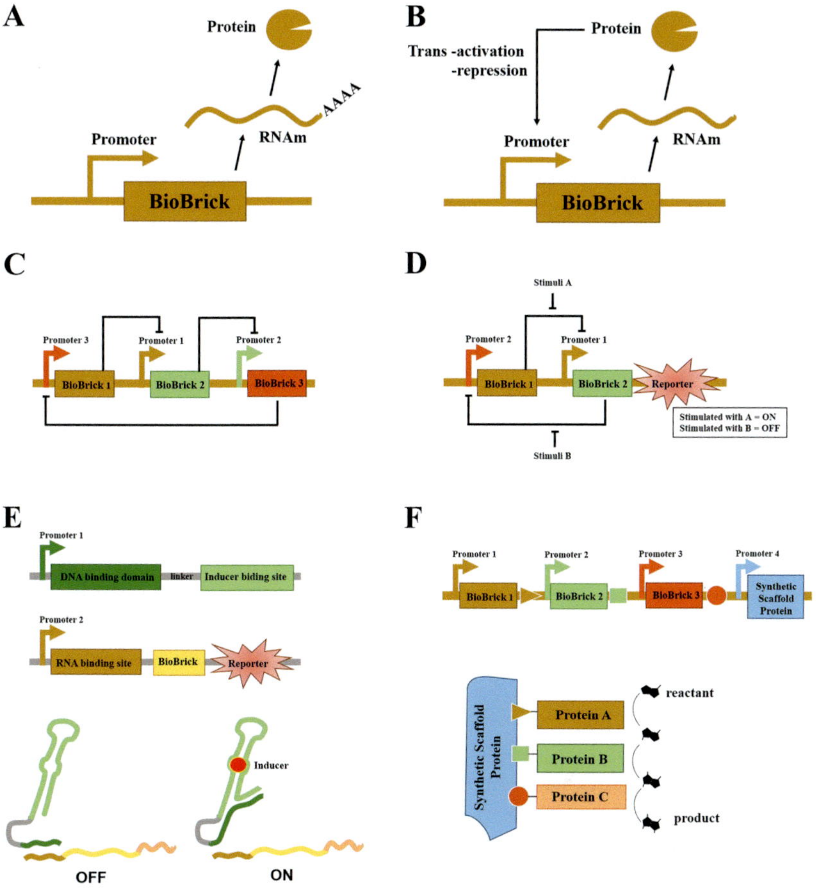

FIG. 2.3 Popular regulatory networks are used to control BioBricks expression. Different regulatory topologies frequently used to control the genetic expression of BioBricks during pathway design: (A) simple regulation, (B) autoregulation, (C) oscillator system, (D) genetic toggle switch, (E) riboswitch, and (F) synthetic scaffolding. A detailed description of each regulatory element is provided in the main text.

1997), and (3) control via RNA aptamers (Hartig, 2018). Antisense RNAs have been extensively used as mechanisms to inhibit RNA translation by binding to their commentary RNAs, thus avoiding their binding with ribosomes and their subsequent translation (Zhang and Zhang, 2018). This control via antisense RNAs can be also used to target complementary RNA sequences and stimulate their degradation by ribonucleases, thus also affecting translation. Finally, translation can be also controlled by the addition of RNA aptamers that are RNA sequences with high-affinity binding domains to specific molecules (Hartig, 2018). Such molecules can be then used to alter RNA structure, thus activating or repressing the genetic expression.

The last level of simple regulation can be performed at the posttranslational level. As already mentioned in the section "De novo rational design of BioBricks," different tags can be added to the BioBricks ORF to modify the resulting protein, thus affecting its secretion, affinity, and solubility (Fig. 2.2). Moreover, posttranslational regulation also includes the design of hybrid proteins for an improved expression, visualization with reporters, cellular localization, scaffolding, and even fine-tune their intracellular degradation (i.e., protein half-life) (Yu et al., 2014).

In addition to simple regulation, alternative approaches to control the genetic expression during pathway design have been also proposed (Fig. 2.3). One of the most popular approaches involves the use of autoregulation networks (Agapakis and Silver, 2009; Singh, 2014). Autoregulation networks result from the positive or negative feedback between the expressed proteins and their promoters (Fig. 2.3B). Thus, autoregulated proteins can either induce or inhibit their expression via trans-activation or trans-repression, respectively. In other words, once autoregulated proteins reach a threshold concentration, they can inhibit or promote their expression at the transcriptional level. Several trans-activation/repression domains have been identified and used to fine-tune the expression level of different proteins in response to their concentration (Agapakis and Silver, 2009). Autoregulation modules have been also used to avoid protein accumulation in complex pathways and even to induce oscillatory behaviors (Agapakis and Silver, 2009).

Recently, oscillatory modules (Fig. 2.3C) have also been used to induce biological artificial rhythms, which have been used to coordinate the genetic expression of proteins over time [e.g., the "repressilator" system (Elowitz and Leibler, 2000)] and even to synchronize the genetic expression between bacterial populations (Fernandez-Niño et al., 2017). Using similar principles of trans-activation/repression, different kinds of regulatory networks with higher complexity have been engineered and known as Genetic Toggle Switches (Fig. 2.3D). Genetic Toggle Switches are regulatory circuits with two mutually repressing domains, where two different expression states can be achieved/tuned depending on the addition of different external stimuli (Agapakis and Silver, 2009; Singh, 2014). Several Genetic Toggle Switches have been modeled to study their behavior in silico together with experimental validation using reporter genes (Gardner et al., 2000). They have been proposed as a programmable "self-contained" regulatory element that can be used to regulate the expression of different metabolites produced via complex biosynthetic pathways (Mannan and Bates, 2021; Bothfeld et al., 2017). The expression of two different expression states (ON/OFF) can also be achieved by using different kinds of regulatory networks known as riboswitches (Fig. 2.3E).

Riboswitches are regulatory elements that utilize the aptamer domains in RNA to control the genetic expression of BioBricks at the RNA level (Mansy, 2010). Riboswitches are composed of two separated expression units, one of them containing a regulatory RNA element (Fig. 2.3E). The regulatory

RNA element contains a DNA binding together with an inducer binding domain. In the absence of inducer molecules, the DNA binding domains of the regulatory RNA element bind to the RNA binding site upstream of the transcribed BioBricks (Fig. 2.3E), which results in the inhibition of BioBricks translation (switch OFF). Once a particular inducer molecule is added to the cell environment, the regulatory RNA element is released from the DNA binding domain (Fig. 2.3E), thus allowing BioBricks translation (switch ON). Very recently, riboswitches have been proposed as a valuable tool to control the production of high-value complex molecules in green alga models by using chimeric plant/algae aptamers (Mehrshahi et al., 2020). Nevertheless, the use of riboswitches to control the genetic expression of these complex molecules in other model organisms still has to be explored. Such molecules usually require the action of multiple enzymes in complex enzymatic cascades. Improving the current production of these compounds is, therefore, a major challenge in microbial metabolic engineering.

A popular approach to allow the colocalization of several enzymes belonging to a particular enzymatic cascade is the use of synthetic protein scaffolds (Fig. 2.3F). The current advances in the design of synthetic protein scaffolds have been recently reviewed by Vanderstraeten and Briers (2020). BioBricks can be fused/linked to protein domains with the ability to bind to orthogonal domains in the synthetic protein scaffolds (Fig. 2.3F). This allows colocalization of enzymes inside the cell, which has been shown to enhance product formation, control spatial organization, and stoichiometry, and avoid toxic intermediates (Vanderstraeten and Briers, 2020). Although the use of synthetic protein scaffolds is expected to boost the design of microbial biofactories in a near future, currently this is achieved by fine-tuning the host metabolome via genome editing as will be discussed in the next section.

Fine-tuning metabolism in microorganisms for an enhanced product expression

Chassis selection for metabolic engineering

During the last decade, several microorganisms (mainly bacteria and yeast) have been established as hosts (biological chassis) to produce different valuable compounds of relevance in industrial biotechnology (Table 2.3). Their remarkable ability to proliferate under growing conditions that are easy to scale up to the industrial level has made them excellent platforms for the efficient production of different compounds, including pharmaceuticals (Dhakal et al., 2019; Keasling et al., 2021), food ingredients (McNeil et al., 2013), biofuels (Kung et al., 2012), biomaterials (Moradali and Rehm, 2020; Vázquez and Villaverde, 2013), and agricultural products (Sutherland, 1997), in a cost-effective way. The use of microbial biofactories thus represents an important step toward environmental protection as they promote sustainability and

TABLE 2.3 Examples of popular microbial hosts (biological chassis) for metabolic engineering and their most representative products.

Biofactory	Popular use/products	Ref
Escherichia coli	Secondary metabolites (e.g., terpenoids, polyketides, phenylpropanoids, and alkaloids)	Yang et al. (2019b)
Streptomyces spp.	Antibiotics, anthelminthic agents, antitumor agents, antifungal agents, and herbicides	Bekker et al. (2014)
Corynebacterium glutaminicum	Industrial production of amino acids	Jung et al. (2010)
Mycobacterium sp.	Dehydrogenases, hydroxylases, and hormones (e.g., testosterone and androstenone)	Fernández-Cabezón et al. (2017), Yao et al. (2014), Zhang et al. (2021)
Vibrio natriegens	1,3-propanediol and violacein	Thoma and Blombach (2021)
Bacillus subtilis	Ethanol, terpenoids, carotenoids, D-lactic acid, and hyaluronic acid	Guan et al. (2015), Romero et al. (2007), Awasthi et al. (2018), Nishizaki et al. (2007), Westbrook et al. (2018)
Kluyveromyces lactis	L-ascorbic acid; beta-galactosidase, chymosin, interleukin 1-beta, interferon-alpha, insulin precursors	César et al. (2013), Spohner et al. (2016)
Rhodococcus ruber	Acrylamide, hormones (e.g., testosterone)	Sun et al. (2016), Guevara et al. (2019)
Rhodococcus opacus	Biofuel production from lignocellulose	Kurosawa et al. (2013, 2015)
Thermoanaerobacterium thermosaccharolyticum	n-Butanol and 1,2-propanoediol	Bhandiwad et al. (2013, 2014), Altaras et al. (2001)
Pseudomonas fluorescens	Vanillin and indole-3-acetic	Di et al. (2011), Kochar et al. (2011)
Clostridium acetobutylicum	Butanol and acetone	Calero and Nikel (2019)

TABLE 2.3 Examples of popular microbial hosts (biological chassis) for metabolic engineering and their most representative products—cont'd

Biofactory	Popular use/products	Ref
Pseudomonas putida	Modified polyhydroxyalkanoates and polymers	Walker and Keasling (2002), Poblete-castro et al. (2012, 2013)
Saccharomyces cerevisiae	Bioethanol, n-butanol, isoprenoids, fatty acids, lactic acid, xylitol, succinic acid	Ostergaard et al. (2000)
Pichia pastoris	Complex secondary metabolites, glycosylated therapeutic proteins, cytochrome P450 enzymes, and salicylic acid	Peña et al. (2018)
Yarrowia lipolytica	Lipid and nanolipid derivatives, lipid-based biofuels, and oleochemicals	Abdel-mawgoud et al. (2018)
Trichoderma reesei (=*Hypocrea jecorina*)	Cellulose, hemicellulose, and bioethanol	Seiboth et al. (2012), Derntl et al. (2015)
Aspergillus oryzae	L-Malic acid and L-malate	Brown et al. (2013), Liu et al. (2017)
Pichia kudriavzevii	Itaconic acid, D-xylonate, D-lactic acid, and ethanol	Yuan and Guo (2017), Sun et al. (2020), Toivari et al. (2013), Sohn et al. (2018)
Kluyveromyces marxianus	Antitumoral agents, xylitol, and the discovery of new promoters and terminators for metabolic engineering	Tseng et al. (2019), Stalidzans and Kokina (2017), Lang et al. (2020), Kumar et al. (2021)

reduce the deleterious impacts of the current industries on the environment (Adarme-vega et al., 2012; Abate et al., 2015). Microbial biofactories have been successfully used to overcome the production of compounds with limited resources [e.g., fuels (Keasling et al., 2021; Kung et al., 2012)] and to replace traditional products with eco-friendly and biodegradable alternatives. Currently, a large variety of microorganisms can be selected for metabolic engineering; thus, defining the appropriate selection criteria is an important step during the design of biofactories.

An ideal microbial chassis should complete several relevant characteristics to become an ideal host for metabolic engineering (Calero and Nikel, 2019; Liu et al., 2020). Thus, the genome of an ideal chassis should be fully sequenced and annotated, and linked to a deeper physiological characterization, including transcriptomics, proteomics, metabolomics, and fluxomics data. In addition, it would be beneficial to have access to its reconstructed metabolic model and access to mutants previously characterized during both in silico and in vitro analyses. Finally, an ideal microbial chassis should be a well-known and characterized microorganism with a preestablished genetic toolbox for metabolic engineering, including transformation protocols, genome editing tools, and mutagenesis-associated techniques. Even when an ideal chassis has been selected, some challenges still need to be overcome to completely exploit its potential to produce valuable products (Calero and Nikel, 2019; Liu et al., 2020). For instance, the insertion of the artificial pathway into the selected chassis can interfere with the host metabolism affecting microbial growth (Calero and Nikel, 2019; Liu et al., 2020). The insertion of artificial pathways can affect the substrate/co-factors' balance and even some products or intermediates can be toxic to the host (Calero and Nikel, 2019; Liu et al., 2020). Several approaches can be used to overcome these problems by using in silico metabolic modeling and genome editing as will be discussed in the next section.

Genome editing for enhanced product formation

As mentioned before, the insertion of artificial pathways into the selected host organism can strongly affect its metabolism and phenotypic response. The in silico reconstruction of metabolic models based on multi-omics data is generally accepted as an efficient approach to estimate the effect of artificial pathway insertions (Saha et al., 2014; Baart and Martens, 2012), together with the prediction of phenotypic changes in response to gene knockouts, overexpression, and the alteration in the metabolic flux balance (Gu et al., 2019; Gillaspy and Senger, 2015). A summary of the most popular tools for in silico modeling of metabolic pathways is presented in Table 2.4. These tools have been used to improve the production capabilities of host organisms considering genomic instability, metabolic flux balance, and phenotypic variability. For example, in silico genome-scale metabolic reconstructions are currently recognized as popular approaches to model cell growth and predict the behavior of gene deletions/insertions into the overall host metabolism in the presence of an artificial biosynthetic pathway (Saha et al., 2014; Baart and Martens, 2012; Gu et al., 2019).

When combined with experimental validation, in silico predictions are useful tools to reduce/suppress interaction between the artificial pathway and the host metabolism in a process called minimization (Gillaspy and Senger, 2015; Jeschek et al., 2017). This has been used to identify target genes in the host organism that can be mutated, deleted, or replaced to optimize bioproduction

TABLE 2.4 Popular tools and software for in silico modeling of metabolic pathways.

Tool	Description	Ref
Method of Minimization of Metabolic Adjustment (MOMA)	Identifies metabolic pathways that must be eliminated to optimize the production of a specific compound. Recognizes genes to be deleted and predicts lethality in those decisions	Segré et al. (2002)
Regulatory On/Off Minimization (ROOM)	Predicts the stability of the metabolic models after gene deletion, reducing the number of flux changes within the system relative to wild type	Shlomi et al. (2005)
OptKnock	Identifies and suggests the deletion of certain regulatory metabolic pathways to overproduce a specific chemical compound	Burgard et al. (2003)
OptStrain	Identifies metabolic pathways that must be suppressed or added to microbial metabolic networks for overproduction of a specific chemical compound	Pharkya et al. (2004)
OptReg	Determines which reactions must be activated, inhibited, or eliminated for the production of a given chemical compound	Sandberg et al. (2019)
Optimal Metabolic Network Identification (OMNI)	Identifies the most efficient set of reactions for the production of a specific metabolite. Finds the changes that must be made to make the in silico predictions match the experimental data	Fong et al. (2006)
COBRA Toolbox	Simulates, analyzes, and predicts metabolic phenotypes using genome-scale models. Includes tools such as network gap filling, metabolic engineering, omics-guided analysis, and visualization	Schellenberger et al. (2011)
Complex Pathway Simulator (COPASI)	Simulates and analyzes biochemical networks and their dynamics, including stoichiometric evaluation of reactions, optimization of model components, defines whether reactions are fast or slow, and has tools to optimize and evaluate chaos within the simulation	Hoops et al. (2006)
GROWMATCH	Reconstructs genome-scale metabolic models, resolving inconsistencies between in silico predictions and in vivo data	Kumar and Maranas (2009)

by genome editing. Consequently, genome editing is usually required to fine-tune and optimize the production of valuable compounds in microbial biofactories (Mohammad and Hassan, 2018).

Currently, the use of CRISPR-guided nucleases is becoming the method of preference for genome edition as it has been extensively reviewed in the literature (Gupta and Musunuru, 2014; Bhardwaj and Nain, 2021; Nidhi et al., 2021; Ahmad et al., 2018; Cui et al., 2018; Ebrahimi and Hashemi, 2020). CRISPR RNA-guided nucleases utilize both a synthetic single-guide RNA and a Cas9-type endonuclease to recognize and cleave specific regions along the genome in a highly efficient way after a short time of design ranging from 2 to 5 days. Nowadays, genome editing with CRISPR has been successfully used to edit the genome of bacteria (Vento et al., 2019; Vigouroux and Bikard, 2020) and yeast (Stovicek et al., 2017). Different CRISPR-Cas systems have been recently developed to expand the editing and regulation possibilities in other organisms (Nidhi et al., 2021; Ebrahimi and Hashemi, 2020), which still requires a higher effort for genome sequencing and annotation to avoid potential off-targets and boost the potential of fine-tuning metabolism in host microorganism for an enhanced product expression.

In addition to CRISPR, two different approaches have been used during the past two decades for genome editing: (1) zinc-finger nucleases (ZFNs) (Urnov et al., 2010) and (2) TAL effector nucleases (TALENs) (Bogdanove and Voytas, 2012; Kim and Kim, 2014). ZFNs are engineered DNA-binding proteins containing two functional domains that are fused to create genomic scissors site-directed DNA cleavage (Urnov et al., 2010). The first domain (a DNA binding domain) is comprised of two-finger modules forming a zinc finger protein that is fused to the second domain (a DNA cleaving domain) with the ability to cleave at the recognition site of the restriction enzyme Fok1 (Urnov et al., 2010). Like ZFNs, TALENs are engineered proteins that also contain a Fok1 DNA-cleaving domain but fused to a TAL DNA-binding effector domain (Bogdanove and Voytas, 2012; Kim and Kim, 2014). ZFNs and TALENs have been extensively used to edit the genome of several organisms including bacteria, yeast, insect, plants, and mammals (Gupta and Musunuru, 2014; Bhardwaj and Nain, 2021). Site-directed DNA cleavages obtained by ZFNs and TALENs are usually followed by targeted mutagenesis, gene replacement, gene correction, transcriptional regulation, multiplex gene targeting, gene knock-out/in, and reporter genome insertion. The design of modules and motifs for ZFNs and TALENs is a laborious process that can range from weeks to months depending on the nature of the desired genome edition (Gupta and Musunuru, 2014; Bhardwaj and Nain, 2021).

The use of biofactories for a sustainable future

The use of microbial biofactories for the production of valuable products represents a huge step toward sustainability as many of these biofactories offer the possibility to transform industrial residues into valuable compounds by using artificial biosynthetic pathways. Thus, several microbial biofactories have been developed during the last years as an alternative for the production of many valuable compounds including pigments, drugs, biosurfactants, food additives, enzyme inhibitors, hormones, herbicides, secondary metabolites, biofuels, and oleochemicals, as summarized in Table 2.3. The production of many of these compounds was achieved by integrating regulatory elements with enzymes engineered to avoid trans-repression, and by fine-tuning the bacterial metabolism to increase co-factors availability and metabolic balance by following approaches previously discussed in this chapter. Interestingly, by using the popular bacterial platform *E. coli* as chassis (Yang et al., 2021; Chen et al., 2013), several biofactories have been constructed to produce different compounds, including secondary metabolites (e.g., terpenoids, polyketides, phenylpropanoids, and alkaloids), and to produce engineered enzymes for textile, food, paper, and detergent industries in a sustainable way (Yang et al., 2021; Chen et al., 2013) (using industry by-products as substrates), thus reducing the biological impact of these industries. Like bacteria, several fungi have also been engineered and are currently used as biofactories to produce different valuable products (Table 2.3). One of the most popular fungal platforms for microbial engineering is the yeast *Saccharomyces cerevisiae* (Kavšček et al., 2015). This popular and well-characterized platform has been engineered to produce many valuable products including bioethanol (Hajar et al., 2017) (the major bioproduct in industrial biotechnology), food additives [e.g., polyphenols and weak organic acids (Chrzanowski, 2020)], and drugs [e.g., artemisinic acid (Ro et al., 2006)]. *S. cerevisiae* biofactories have been designed by using the metabolic engineering approaches previously discussed in this chapter including in silico modeling of metabolic pathways, protein scaffolding, and genome editing.

In this chapter, we have discussed the most relevant approaches to assemble and design microbial biofactories of relevance for industrial biotechnology. As discussed, the number of tools for BioBricks design, pathway assembly, and metabolic fine-tuning has substantially increased during the last few years. Its number is expected to further increase in response to the development of novel methodologies for metabolic engineering such as the design of novel cloning strategies, new bioinformatics platforms for genome mining and metabolic modeling, and the development of new genome editing tools. Consequently, the extensive use of biofactories for the bioproduction of high-valuable bioproducts will represent a big step toward sustainability in industrial biotechnology.

As mentioned before, bioproduction using biofactories is recognized as an inexpensive, eco-friendly, and efficient alternative to the traditional approaches to obtain high-valuable bioproducts. Thus, engineering and reprogramming microbial biofactories will be important for the sustainable production of different biological compounds that are difficult to obtain by traditional approaches such as fuels, food ingredients, and pharmaceuticals. In addition, the use of toxic catalysts or organic solvents is expected to decrease; meanwhile, this kind of technology improves. Nevertheless, additional research is required to evaluate the economic impact of the implementation of these technologies and to design new industrial processes that allow valorization of industrial residues by engineered biofactories. In addition, more policies are yet to be designed to promote its utilization as a sustainable alternative to the current industry. This chapter is expected to represent an incentive for researchers in industry and academia interested in the construction of novel biofactories for a sustainable future.

References

Abate, S., Lanzafame, P., Perathoner, S., Centi, G., 2015. New sustainable model of biorefineries: biofactories and challenges of integrating bio- and solar refineries. ChemSusChem, 2854–2866. https://doi.org/10.1002/cssc.201500277.

Abdel-mawgoud, A.M., Markham, K.A., Palmer, C.M., Liu, N., Alper, H.S., Liu, N., et al., 2018. Metabolic engineering in the host Yarrowia lipolytica Ahmad. Metab. Eng. https://doi.org/10.1016/j.ymben.2018.07.016.

Adarme-vega, T.C., Lim, D.K.Y., Timmins, M., Vernen, F., Li, Y., Schenk, P.M., 2012. Microalgal biofactories: a promising approach towards sustainable omega-3 fatty acid production. Microb. Cell Factories 11, 1–10.

Agapakis, C., Silver, P., 2009. Synthetic biology: exploring and exploiting genetic modularity through the design of novel biological networks. Mol. BioSyst. 5. https://doi.org/10.1039/b901484e.

Ahmad, H.I., Ahmad, M.J., Asif, A.R., Adnan, M., Iqbal, M.K., Mehmood, K., 2018. A review of CRISPR-based genome editing: survival, evolution and challenges. Curr. Issues Mol. Biol. 28, 47–68.

Albarano, L., Esposito, R., Ruocco, N., Costantini, M., 2020. Genome mining as new challenge in natural products discovery. Mar. Drugs 18, 1–17.

Altaras, N.E., Etzel, M.R., Cameron, D.C., 2001. Conversion of sugars to 1, 2-Propanediol by *Thermoanaerobacterium thermosaccharolyticum* HG-8. Biotechnol. Prog. 17, 52–56.

Amin, D.H., Abolmaaty, A., Borsetto, C., Tolba, S., Abdallah, N.A., Wellington, E.M.H., 2019. In silico genomic mining reveals unexplored bioactive potential of rare actinobacteria isolated from Egyptian soil. Bull. Natl. Res. Cent. 43, 1–9. https://doi.org/10.1186/S42269-019-0121-Y.

Arita, M., Karsch-mizrachi, I., Cochrane, G., 2021. The international nucleotide sequence database collaboration. Nucleic Acids Res. 49, 121–124. https://doi.org/10.1093/nar/gkaa967.

Aslam, B., Basit, M., Nisar, M.A., Khurshid, M., Rasool, M.H., 2017. Proteomics: technologies and their applications. J. Chromatogr. Sci. 55, 182–196. https://doi.org/10.1093/chromsci/bmw167.

Awasthi, D., Wang, L., Rhee, M.S., Wang, Q., Chauliac, D., Ingram, L.O., et al., 2018. Metabolic engineering of *Bacillus subtilis* for production of D-lactic acid. Biotechnol. Bioeng., 453–463. https://doi.org/10.1002/bit.26472.

Baart, G., Martens, D., 2012. Genome-scale metabolic models: reconstruction and analysis. Methods Mol. Biol. 799, 107–126.

Baltz, R.H., 2018. Synthetic biology, genome mining, and combinatorial biosynthesis of NRPS-derived antibiotics: a perspective. J. Ind. Microbiol. Biotechnol. 45, 635–649. https://doi.org/10.1007/s10295-017-1999-8.

Beale, D.J., Pinu, F.R., Kouremenos, K.A., Poojary, M.M., Narayana, V.K., Boughton, B.A., et al., 2018. Review of Recent Developments in GC – MS Approaches to Metabolomics-Based Research. vol. 14 Springer US, https://doi.org/10.1007/s11306-018-1449-2.

Bekker, V., Dodd, A., Brady, D., Rumbold, K., Bekker, V., Dodd, A., et al., 2014. Tools for metabolic engineering in Streptomyces. Bioengineered 5979. https://doi.org/10.4161/bioe.29935.

Bell, E.L., Finnigan, W., France, S.P., Hepworth, L.J., Lovelock, S.L., Hayes, M.A., et al., 2021. Biocatalysis. Nat. Rev. Methods Prim. 46, 1–21. https://doi.org/10.1038/s43586-021-00044-z.

Bertero, A., Brown, S., Vallier, L., 2017. Methods of cloning. In: Basic Science Methods for Clinical Researchers. Elsevier Inc, pp. 19–40, https://doi.org/10.1016/B978-0-12-803077-6.00002-3.

Bhandiwad, A., Guseva, A., Lynd, L., 2013. Metabolic engineering of *Thermoanaerobacterium thermosaccharolyticum* for increased n-butanol production. Adv. Microbiol. 2013, 46–51.

Bhandiwad, A., Shaw, A.J., Guss, A., Guseva, A., Bahl, H., Lynd, L.R., 2014. Metabolic engineering of *Thermoanaerobacterium saccharolyticum* for n -butanol production. Metab. Eng. 21, 17–25. https://doi.org/10.1016/j.ymben.2013.10.012.

Bhardwaj, A., Nain, V., 2021. TALENs—an indispensable tool in the era of CRISPR: a mini review. J. Genet. Eng. Biotechnol. 19.

Blin, K., Shaw, S., Kloosterman, A.M., Charlop-powers, Z., Van Wezel, P., Medema, M.H., et al., 2021. antiSMASH 6.0 : improving cluster detection and comparison capabilities. Nucleic Acids Res., 0–7. https://doi.org/10.1093/nar/gkab335.

Blum, M., Chang, H., Chuguransky, S., Grego, T., Kandasaamy, S., Mitchell, A., et al., 2021. The InterPro protein families and domains database: 20 years on. Nucleic Acids Res. 49, 344–354. https://doi.org/10.1093/nar/gkaa977.

Bogdanove, A.J., Voytas, D.F., 2012. TAL effectors: customizable proteins. Science (80-) 1843. https://doi.org/10.1126/science.1204094.

Bösch, A.N., Mariana, B., Greczmiel, U., Gugger, M., Oxenius, A., Lisa, A., et al., 2020. Landornamides, antiviral ornithine-containing ribosomal peptides discovered by proteusin mining. Angew. Chem. https://doi.org/10.1002/ange.201916321.

Bothfeld, W.H., Kapov, G., Tyo, K., 2017. A glucose-sensing toggle switch for autonomous, high productivity genetic control. ACS Synth. Biol. https://doi.org/10.1021/acssynbio.6b00257.

Bronner, I.F., Quail, M.A., Turner, D.J., Swerdlow, H., 2013. Improved protocols for illumina sequencing. Curr. Protoc. Hum. Genet., 1–42. https://doi.org/10.1002/0471142905.hg1802s79.

Brown, S.H., Bashkirova, L., Berka, R., Chandler, T., Doty, T., Mccall, K., et al., 2013. Metabolic engineering of Aspergillus oryzae NRRL 3488 for increased production of L-malic acid. Appl. Microbiol. Biotechnol., 8903–8912. https://doi.org/10.1007/s00253-013-5132-2.

Burgard, A.P., Pharkya, P., Maranas, C.D., 2003. OptKnock: a Bilevel programming framework for identifying gene knockout strategies for microbial strain optimization. Biotechnol. Bioeng. https://doi.org/10.1002/bit.10803.

Cai, Y., Kallam, K., Tidd, H., Gendarini, G., Salzman, A., Patron, N.J., 2020. Rational design of minimal synthetic promoters for plants. Nucleic Acids Res. 48, 11845–11856. https://doi.org/10.1093/nar/gkaa682.

Calero, P., Nikel, P.I., 2019. Minireview chasing bacterial chassis for metabolic engineering: a perspective review from classical to non-traditional microorganisms. Microb. Biotechnol. 12, 98–124. https://doi.org/10.1111/1751-7915.13292.

Carreira, E., 2015. Introduction: frontiers in organic synthesis. Chem. Rev. 8945. https://doi.org/10.1021/acs.chemrev.5b00464.

Carrier, T.A., Keasling, J.D., 1997. Controlling messenger RNA stability in bacteria : strategies for engineering gene expression. Biotechnol. Prog. 13.

Castorena-Torres, F., Peñuelas-Urquides, K., de Leon, M., 2016. Site-directed mutagenesis by polymerase chain reaction. In: Polymerase Chain Reaction for Biomedical Applications. IntechOpen.

César, J., Rosa, C., Colombo, L.T., Caroline, M., Alvim, T., Avonce, N., et al., 2013. Metabolic engineering of Kluyveromyces lactis for L-ascorbic acid (vitamin C) biosynthesis. Microb. Cell Factories 12, 1–13.

Chang, A., Jeske, L., Ulbrich, S., Hofmann, J., Koblitz, J., Schomburg, I., et al., 2021. BRENDA, the ELIXIR core data resource in 2021: new developments and updates. Nucleic Acids Res. 49, 498–508. https://doi.org/10.1093/nar/gkaa1025.

Chapnik, N., Sherman, H., Froy, O., 2007. A one-tube site-directed mutagenesis method using PCR and primer extension. Anal. Biochem. 372.

Chen, X., Zhou, L., Tian, K., Kumar, A., Singh, S., Prior, B.A., et al., 2013. Metabolic engineering of *Escherichia coli*: a sustainable industrial platform for bio-based chemical production. Biotechnol. Adv. 31, 1200–1223. https://doi.org/10.1016/j.biotechadv.2013.02.009.

Chen, C., Huang, H., Wu, C.H., 2017. Protein bioinformatics databases and resources. Methods Mol. Biol. 1558. https://doi.org/10.1007/978-1-4939-6783-4.

Chen, Y., Ho, J.M.L., Shis, D.L., Gupta, C., Long, J., Wagner, D.S., et al., 2018. Tuning the dynamic range of bacterial promoters regulated by ligand-inducible transcription factors. Nat. Commun., 1–8. https://doi.org/10.1038/s41467-017-02473-5.

Chrzanowski, G., 2020. Saccharomyces cerevisiae—an interesting producer of bioactive plant polyphenolic metabolites. Int. J. Mol. Sci. 21 (19), 1–18.

Clark, K., Karsch-mizrachi, I., Lipman, D.J., Ostell, J., Sayers, E.W., 2016. GenBank. Nucleic Acids Res. 44, 67–72. https://doi.org/10.1093/nar/gkv1276.

Consortium, T.U., 2021. UniProt: the universal protein knowledgebase in 2021. Nucleic Acids Res. 49, 480–489. https://doi.org/10.1093/nar/gkaa1100.

Cui, Y., Xu, J., Cheng, M., Liao, X., Peng, S., 2018. Review of CRISPR/Cas9 sgRNA design tools. Interdiscip. Sci. Comput. Life Sci. 10, 455–465. https://doi.org/10.1007/s12539-018-0298-z.

D'Angostino, P., Gulder, T., 2018. Direct pathway cloning (DiPaC) combined with sequence- and ligation-independent cloning (SLIC) for fast biosynthetic gene cluster refactoring and heterologous expression. ACS Synth. Biol. https://doi.org/10.1021/acssynbio.8b00151.

Dallas, D., Guerrero, A., Parker, E.A., Robinson, R., Gan, J., German, B., et al., 2015. Current peptidomics: applications, purification, identification, quantification and functional analysis. Proteomics 15, 1–42. https://doi.org/10.1002/pmic.201400310.This.

De Mey, M., Maertens, J., Lequeux, G.J., Soetaert, W.K., Vandamme, E.J., 2007. Construction and model-based analysis of a promoter library for E. coli: an indispensable tool for metabolic engineering. BMC Biotechnol. 14, 1–14. https://doi.org/10.1186/1472-6750-7-34.

Derntl, C., Kiesenhofer, D.P., Mach, R.L., Mach-aigner, A.R., 2015. Novel strategies for genomic manipulation of Trichoderma reesei with the purpose of strain engineering. Appl. Environ. Microbiol. 81. https://doi.org/10.1128/AEM.01545-15.

Dhakal, D., Kim, E., Koffas, M., Kim, E., 2019. Engineering the microbial platform for the production of biologics and small-molecule medicines. Front. Microbiol. 10, 9–11. https://doi.org/10.3389/fmicb.2019.02307.

Di, D., Luziatelli, F., Negroni, A., Grazia, A., Fava, F., Ruzzi, M., 2011. Metabolic engineering of *Pseudomonas fluorescens* for the production of vanillin from ferulic acid. J. Biotechnol. 156, 309–316. https://doi.org/10.1016/j.jbiotec.2011.08.014.

Dilokpimol, A., Mäkelä, M.R., Cerullo, G., Zhou, M., Varriale, S., Gidijala, L., et al., 2018. Fungal glucuronoyl esterases: genome mining based enzyme discovery and biochemical characterization. New Biotechnol. 40, 282–287. https://doi.org/10.1016/j.nbt.2017.10.003.

Dong, Q., Schlueter, S.D., Brendel, V., 2004. PlantGDB, plant genome database and analysis tools. Nucleic Acids Res. 32, 354–359. https://doi.org/10.1093/nar/gkh046.

Ebrahimi, V., Hashemi, A., 2020. Challenges of in vitro genome editing with CRISPR/Cas9 and possible solutions: a review. J. Pre-Proofs 144813. https://doi.org/10.1016/j.gene.2020.144813.

Edelheit, O., Hanukoglu, A., Hanukoglu, I., 2009. Simple and efficient site-directed mutagenesis using two single-primer reactions in parallel to generate mutants for protein structure-function studies. BMC Biotechnol. 8, 1–8. https://doi.org/10.1186/1472-6750-9-61.

Elena, C., Ravasi, P., Castelli, M.E., Peirú, S., Menzella, H.G., 2014. Expression of codon optimized genes in microbial systems: current industrial applications and perspectives. Front. Microbiol. 5, 1–8. https://doi.org/10.3389/fmicb.2014.00021.

Elowitz, M.B., Leibler, S., 2000. A synthetic oscillatory network of transcriptional regulators. Nature 403, 335–338.

El-Sayed, A., El-Sayed, M., Rady, A., Zein, N., Enan, G., Shindia, A., et al., 2020. Exploiting the biosynthetic potency of taxol from fungal endophytes of conifers plants; genome mining and metabolic manipulation. Molecules 25, 1–21.

Fahy, E., Subramaniam, S., Murphy, R.C., Nishijima, M., Raetz, C.R.H., Shimizu, T., et al., 2009. Update of the LIPID MAPS comprehensive classification system for lipids. J. Lipid Res. 50, S9–14. https://doi.org/10.1194/jlr.R800095-JLR200.

Fernández-Cabezón, L., Galán, B., García, J., 2017. Engineering *Mycobacterium smegmatis* for testosterone production. Microb. Biotechnol. 10, 151–161. https://doi.org/10.1111/1751-7915.12433.

Fernandez-Niño, M., Giraldo, D., Gomez-Porras, J.L., Dreyer, I., Barrios, A.F.G., Arevalo-Ferro, C., 2017. A synthetic multi-cellular network of coupled self-sustained oscillators. PLoS One 12, 1–11. https://doi.org/10.1371/journal.pone.0180155.

Fong, S.S., Palsson, B.Ø., Herrga, M.J., 2006. Identification of genome-scale metabolic network models using experimentally measured flux profiles. PLoS Comput. Biol. 2. https://doi.org/10.1371/journal.pcbi.0020072.

Fukuda, A., Kodama, Y., Mashima, J., Fujisawa, T., 2021. DDBJ update: streamlining submission and access of human data. Nucleic Acids Res. 49, 71–75. https://doi.org/10.1093/nar/gkaa982.

Gardner, T.S., Cantor, C.R., Collins, J.J., 2000. Construction of a genetic toggle switch in Escherichia coli. Nature 403, 339–342.

Garg, N., Sethupathy, A., Tuwani, R., Nk, R., Dokania, S., Iyer, A., et al., 2018. FlavorDB: a database of flavor molecules. Nucleic Acids Res. 46, 1210–1216. https://doi.org/10.1093/nar/gkx957.

Gaulton, A., Kale, N., Van Westen, G.J.P., Bellis, L.J., Bento, A.P., Davies, M., et al., 2015. A large-scale crop protection bioassay data set. Sci. Data, 91–93. https://doi.org/10.1038/sdata.2015.32.

Gillaspy, G.E., Senger, R.S., 2015. Designing metabolic engineering strategies with genome-scale metabolic flux modeling. Adv. Genomics Genet. https://doi.org/10.2147/AGG.S58494.

Gilman, J., Love, J., 2016. Synthetic promoter design for new microbial chassis. Synth. Biol., 731–737. https://doi.org/10.1042/BST20160042.

Goujon, M., Mcwilliam, H., Li, W., Valentin, F., Squizzato, S., Paern, J., et al., 2010. A new bioinformatics analysis tools framework at EMBL – EBI. Nucleic Acids Res. 38, 695–699. https://doi.org/10.1093/nar/gkq313.

Gu, C., Kim, G.B., Kim, W.J., Kim, H.U., Lee, S.Y., 2019. Current status and applications of genome-scale metabolic models. Genome Biol. 20, 1–18.

Guan, Z., Xue, D., Abdallah, I.I., Dijkshoorn, L., Setroikromo, R., Lv, G., et al., 2015. Metabolic engineering of *Bacillus subtilis* for terpenoid production. Appl. Microbiol. Biotechnol., 9395–9406. https://doi.org/10.1007/s00253-015-6950-1.

Guevara, G., Flores, Y.O., Ferna, L., Marı, J., 2019. Metabolic engineering of *Rhodococcus ruber* Chol-4: a cell factory for testosterone production. PLoS One 14 (7), 1–16.
Guldener, U., Munsterkotter, M., Kastnmuller, G., Strack, N., van Helden, J., Lemer, C., et al., 2005. CYGD: the comprehensive yeast genome database. Nucleic Acids Res. 33, 364–368. https://doi.org/10.1093/nar/gki053.
Gupta, R.M., Musunuru, K., 2014. Expanding the genetic editing tool kit: ZFNs, TALENs, and CRISPR-Cas9. J. Clin. Invest. 124. https://doi.org/10.1172/JCI72992.transcription.
Hajar, S., Azhar, M., Abdulla, R., Jambo, S.A., Marbawi, H., Azlan, J., et al., 2017. Yeasts in sustainable bioethanol production: a review. Biochem. Biophys. Rep. 10, 52–61. https://doi.org/10.1016/j.bbrep.2017.03.003.
Hartig, J.S., 2018. Aptamer-based control of gene expression utilizing endogenous miRNAs. Mol. Ther. 26, 1178–1180. https://doi.org/10.1016/j.ymthe.2018.04.010.
Haugen, S.P., Ross, W., Gourse, R.L., 2008. Advances in bacterial promoter recognition and its control by factors that do not bind DNA. Nat. Rev. Microbiol. https://doi.org/10.1038/nrmicro1912.
Hoops, S., Sahle, S., Gauges, R., Lee, C., Simus, N., Singhal, M., et al., 2006. Systems biology COPASI—a COmplex PAthway SImulator. Bioinformatics 22, 3067–3074. https://doi.org/10.1093/bioinformatics/btl485.
Howe, K.L., Achuthan, P., Allen, J., Allen, J., Alvarez-jarreta, J., Amode, M.R., et al., 2021. Ensembl 2021. Nucleic Acids Res. 49, 884–891. https://doi.org/10.1093/nar/gkaa942.
Hu, J., Mungall, C., Law, A., Papworth, R., Nelson, J.P., Brown, A., et al., 2001. The ARKdb: genome databases for farmed and other animals. Nucleic Acids Res. 29, 106–110.
Hughes, R.A., Ellington, A.D., 2017. Synthetic DNA synthesis and assembly: putting the synthetic in synthetic biology. Cold Spring Harb. Peerspect. Biol. 9, 1–17.
Ignea, C., Pontini, M., Motawia, M.S., Maffei, M.E., Makris, A.M., Kampranis, S.C., 2019. Synthesis of 11-carbon terpenoids in yeast using protein and metabolic engineering. Nat. Chem. Biol. https://doi.org/10.1038/s41589-018-0166-5.
Jain, M., Olsen, H.E., Paten, B., Akeson, M., 2016. The Oxford Nanopore MinION: delivery of nanopore sequencing to the genomics community. Genome Biol. 17, 1–11. https://doi.org/10.1186/s13059-016-1103-0.
Jassal, B., Matthews, L., Viteri, G., Gong, C., Lorente, P., Fabregat, A., et al., 2020. The reactome pathway knowledgebase. Nucleic Acids Res. 48, 498–503. https://doi.org/10.1093/nar/gkz1031.
Jeschek, M., Gerngross, D., Panke, S., 2017. Combinatorial pathway optimization for streamlined metabolic engineering. Curr. Opin. Biotechnol. 47, 142–151. https://doi.org/10.1016/j.copbio.2017.06.014.
Jung, S., Chun, J., Yim, S., Lee, S., Cheon, C., Song, E., et al., 2010. Transcriptional regulation of histidine biosynthesis genes in *Corynebacterium glutamicum*. Can. J. Microbiol. 187, 178–187. https://doi.org/10.1139/W09-115.
Kalvari, I., Nawrocki, E.P., Ontiveros-palacios, N., Argasinska, J., Lamkiewicz, K., Marz, M., et al., 2021. Rfam 14: expanded coverage of metagenomic, viral and microRNA families. Nucleic Acids Res. 49, 192–200. https://doi.org/10.1093/nar/gkaa1047.
Kanehisa, M., Furumichi, M., Sato, Y., Ishiguro-watanabe, M., Tanabe, M., 2021. KEGG: integrating viruses and cellular organisms. Nucleic Acids Res. 49, 545–551. https://doi.org/10.1093/nar/gkaa970.
Kautsar, S.A., Duran, H.G.S., Blin, K., Osbourn, A., Medema, H., 2017. plantiSMASH: automated identification, annotation and expression analysis of plant biosynthetic gene clusters. Nucleic Acids Res. 45, 55–63. https://doi.org/10.1093/nar/gkx305.

Kavšček, M., Stražar, M., Curk, T., Natter, K., Petrovič, U., 2015. Yeast as a cell factory: current state and perspectives. Microb. Cell Factories, 1–10. https://doi.org/10.1186/s12934-015-0281-x.

Keasling, J., Martin, H.G., Lee, T.S., Mukhopadhyay, A., Singer, S.W., Sundstrom, E., 2021. Microbial production of advanced biofuels. Nat. Rev. Microbiol. https://doi.org/10.1038/s41579-021-00577-w.

Khersonsky, O., Tawfik, D.S., 2010. Enzyme promiscuity: a mechanistic and evolutionary perspective. Annu. Rev. Biochem. https://doi.org/10.1146/annurev-biochem-030409-143718.

Kim, H., Kim, J., 2014. A guide to genome engineering with programmable nucleases. Nat. Rev. Genet. 15, 321–334. https://doi.org/10.1038/nrg3686.

Kimple, M.E., Brill, A.L., Pasker, R.L., 2013. Overview of affinity tags for protein purification. Curr. Protoc. Protein Sci., 1–23. https://doi.org/10.1002/0471140864.ps0909s73.

Knudsen, M., Wiuf, C., 2010. The CATH database. Hum. Genomics 4, 207–212.

Kochar, M., Upadhyay, A., Srivastava, S., 2011. Indole-3-acetic acid biosynthesis in the biocontrol strain *Pseudomonas fluorescens* Psd and plant growth regulation by hormone overexpression. Res. Microbiol. 162, 426–435. https://doi.org/10.1016/j.resmic.2011.03.006.

Kosuri, S., Church, G.M., 2014. Large-scale de novo DNA synthesis: technologies and applications. Nat. Methods 11, 499–507. https://doi.org/10.1038/nmeth.2918.

Kumar, V.S., Maranas, C.D., 2009. GrowMatch: an automated method for reconciling in silico/in vivo growth predictions. PLoS Comput. Biol. 5, 18–20. https://doi.org/10.1371/journal.pcbi.1000308.

Kumar, P., Kumar, D., Sharma, D., 2021. The identification of novel promoters and terminators for protein expression and metabolic engineering applications in Kluyveromyces marxianus. Metab. Eng. Commun. 12, e00160. https://doi.org/10.1016/j.mec.2020.e00160.

Kung, Y., Runguphan, W., Keasling, J.D., 2012. From fields to fuels: recent advances in the microbial production of biofuels. ACS Synth. Biol. 11, 498–513. https://doi.org/10.1021/sb300074k.

Kurosawa, K., Wewetzer, S.J., Sinskey, A.J., 2013. Engineering xylose metabolism in triacylglycerol-producing *Rhodococcus opacus* for lignocellulosic fuel production. Biotechnol. Biofuels, 1–13.

Kurosawa, K., Plassmeier, J., Kalinowski, J., Rückert, C., 2015. Engineering L-arabinose metabolism in triacylglycerol-producing *Rhodococcus opacus* for lignocellulosic fuel production. Metab. Eng., 1–7. https://doi.org/10.1016/j.ymben.2015.04.006.

Lai, Z., Tsugawa, H., Wohlgemuth, G., Mehta, S., Mueller, M., Zheng, Y., et al., 2017. Identifying metabolites by integrating metabolome databases with mass spectrometry cheminformatics. Nat. Methods. https://doi.org/10.1038/nmeth.4512.

Lang, X., Besada-lombana, P.B., Li, M., Da Sliva, N.A., Wheeldon, I., 2020. Developing a broad-range promoter set for metabolic engineering in the thermotolerant yeast Kluyveromyces marxianus. Metab. Eng. Commun. 11, e00145. https://doi.org/10.1016/j.mec.2020.e00145.

Lee, S.O., Fried, S.D., 2021. An error prone PCR method for small amplicons. Anal. Biochem. 628, 114266. https://doi.org/10.1016/j.ab.2021.114266.

Leggieri, P.A., Liu, Y., Hayes, M., Connors, B., Seppälä, S., Malley, M.A.O., et al., 2021. Integrating systems and synthetic biology to understand and engineer microbiomes. Annu. Rev. Biomed. Eng. 23, 169–201.

Li, D., Gaquerel, E., 2021. Next-generation mass spectrometry metabolomics revives the functional analysis of plant metabolic diversity. Annu. Rev. Plant Biol. 72, 867–891.

Lin, Z., Nielsen, J., Liu, Z., 2020. Bioprospecting through cloning of whole natural product biosynthetic gene clusters. Front. Bioeng. Biotechnol. 8. https://doi.org/10.3389/fbioe.2020.00526.

Liu, J., Li, J., Shin, H., Du, G., Chen, J., Liu, L., 2017. Metabolic engineering of Aspergillus oryzae for efficient production of l-malate directly from corn starch. J. Biotechnol. https://doi.org/10.1016/j.jbiotec.2017.09.021.

Liu, J., Wu, X., Yao, M., Xiao, W., 2020. Chassis engineering for microbial production of chemicals: from natural microbes to synthetic organisms. Curr. Opin. Biotechnol. 66, 105–112. https://doi.org/10.1016/j.copbio.2020.06.013.

Loenen, W.A.M., Dryden, D.T.F., Raleigh, E.A., Wilson, G.G., Murray, N.E., 2014. Highlights of the DNA cutters: a short history of the restriction enzymes. Nucleic Acids Res. 42. https://doi.org/10.1093/nar/gkt990.

Mannan, A.A., Bates, D., 2021. Designing an irreversible metabolic switch for scalable induction of microbial chemical production. Nat. Commun., 1–11. https://doi.org/10.1038/s41467-021-23606-x.

Mansy, S., 2010. Membrane transport in primitive cells. Cold Spring Harb. Perspect. Biol. 2 (8). https://doi.org/10.1101/cshperspect.a002188.

McNeil, B., Archer, D., Giavasis, I., Harvey, L., 2013. Microbial Production of Food Ingredients, Enzymes and Nutraceuticals. Elsevier.

Mehrshahi, P., Nguyen, G.T.D.T., Rovira, A.G., Sayer, A., Llavero-pasquina, M., Lim, M., et al., 2020. Development of novel riboswitches for synthetic biology in the green alga Chlamydomonas. ACS Synth. Biol. 9. https://doi.org/10.1021/acssynbio.0c00082.

Metzker, M.L., 2009. Sequencing technologies—the next generation. Nat. Rev. Genet. 11, 31–46. https://doi.org/10.1038/nrg2626.

Mi, H., Ebert, D., Muruganujan, A., Mills, C., Albou, L., Mushayamaha, T., et al., 2021. PANTHER version 16 : a revised family classification, tree-based classification tool, enhancer regions and extensive API. Nucleic Acids Res. 49, 394–403. https://doi.org/10.1093/nar/gkaa1106.

Mohammad, T., Hassan, M.I., 2018. Modern approaches in synthetic biology: genome editing, quorum sensing, and microbiome engineering. In: Singh, S. (Ed.), Synthetic Biology. Springer, pp. 189–205.

Montaño López, J., Durán, L., Avalos, J., 2021. Physiological limitations and opportunities in microbial metabolic engineering. Nat. Rev. Microbiol. https://doi.org/10.1038/s41579-021-00600-0.

Monzon, A.M., Palopoli, N., Aykac-fas, B., Bassot, C., Ben, G.I., Bevilacqua, M., et al., 2020. DisProt: intrinsic protein disorder annotation in 2020. Nucleic Acids Res. 48, 269–276. https://doi.org/10.1093/nar/gkz975.

Moradali, M.F., Rehm, B.H.A., 2020. Bacterial biopolymers: from pathogenesis to advanced materials. Nat. Rev. Microbiol. 18, 195–210. https://doi.org/10.1038/s41579-019-0313-3.

Nadine, T., Fuchs, L., Aharoni, A., 2017. The PhytoClust tool for metabolic gene clusters discovery in plant genomes. Nucleic Acids Res. 45, 7049–7063. https://doi.org/10.1093/nar/gkx404.

Napolitano-tabares, P.I., Negrín-santamaría, I., Gutiérrez-serpa, A., Pino, V., 2021. Recent efforts to increase greenness in chromatography. Curr. Opin. Green Sustain. Chem. 32, 100536. https://doi.org/10.1016/j.cogsc.2021.100536.

Nevoigt, E., Kohnke, J., Fischer, C.R., Alper, H., Stahl, U., Stephanopoulos, G., 2006. Engineering of promoter replacement cassettes for fine-tuning of gene expression in Saccharomyces cerevisiae. Appl. Environ. Microbiol. 72, 5266–5273. https://doi.org/10.1128/AEM.00530-06.

Nidhi, S., Anand, U., Oleksak, P., Tripathi, P., Lal, J.A., Thomas, G., et al., 2021. Novel CRISPR – Cas systems: an updated review of the current achievements, applications, and future research perspectives. Int. J. Mol. Sci. 22, 1–42.

Nishizaki, T., Tsuge, K., Itaya, M., Doi, N., Yanagawa, H., 2007. Metabolic engineering of carotenoid biosynthesis in *Escherichia coli* by ordered gene assembly in *Bacillus subtilis*. Appl. Environ. Microbiol. 73, 1355–1361. https://doi.org/10.1128/AEM.02268-06.

Ostergaard, S., Olsson, L., Nielsen, J., 2000. Metabolic engineering of *Saccharomyces cerevisiae*. Microbiol. Mol. Biol. Rev. 64, 34–50.

Park, J., Throop, A.L., Labaer, J., 2015. Site-specific recombinational cloning using gateway and in-fusion cloning schemes. Curr. Protoc. Mol. Biol., 1–23. https://doi.org/10.1002/0471142727.mb0320s110.

Peña, D.A., Gasser, B., Zanghellini, J., Steiger, M.G., 2018. Metabolic engineering of Pichia pastoris. Metab. Eng. 50, 2–15. https://doi.org/10.1016/j.ymben.2018.04.017.

Perez-riverol, Y., Csordas, A., Bai, J., Bernal-llinares, M., Hewapathirana, S., Kundu, D.J., et al., 2019. The PRIDE database and related tools and resources in 2019: improving support for quantification data. Nucleic Acids Res. 47, 442–450. https://doi.org/10.1093/nar/gky1106.

Pharkya, P., Burgard, A.P., Maranas, C.D., 2004. OptStrain: a computational framework for redesign of microbial production systems. Genome Res., 2367–2376. https://doi.org/10.1101/gr.2872004.14.

Poblete-castro, I., Becker, J., Dohnt, K., 2012. Industrial biotechnology of *Pseudomonas putida* and related species. Appl. Microbiol. Biotechnol., 2279–2290. https://doi.org/10.1007/s00253-012-3928-0.

Poblete-castro, I., Binger, D., Rodrigues, A., Becker, J., Martins, V.A.P., Wittmann, C., 2013. In-silico-driven metabolic engineering of *Pseudomonas putida* for enhanced production of poly-hydroxyalkanoates. Metab. Eng. 15, 113–123. https://doi.org/10.1016/j.ymben.2012.10.004.

Ramzi, A., Bahaudin, K., Baharum, S., Che Me, M., Hassan, M., Noor, N., 2018. Rapid assembly of yeast expression cassettes for phenylpropanoid biosynthesis in Saccharomyces cerevisiae. Sains Malays. 47, 2969–2974.

Redden, H., Alper, H.S., 2015. The development and characterization of synthetic minimal yeast promoters. Nat. Commun., 1–9. https://doi.org/10.1038/ncomms8810.

Redden, H., Morse, N., Alper, H.S., 2015. The synthetic biology toolbox for tuning gene expression in yeast. FEMS Yeast Res. 15, 1–10. https://doi.org/10.1111/1567-1364.12188.

Ro, D., Paradise, E.M., Ouellet, M., Fisher, K.J., Newman, K.L., Ndungu, J.M., et al., 2006. Production of the antimalarial drug precursor artemisinic acid in engineered yeast. Nature 440, 3–6. https://doi.org/10.1038/nature04640.

Romero, S., Merino, E., Bolı, F., Gosset, G., Martinez, A., 2007. Metabolic engineering of *Bacillus subtilis* for ethanol production: lactate dehydrogenase plays a key role in fermentative metabolism. Appl. Environ. Microbiol. 73, 5190–5198. https://doi.org/10.1128/AEM.00625-07.

Saha, R., Chowdhury, A., Maranas, C.D., 2014. Recent advances in the reconstruction of metabolic models and integration of omics data. Curr. Opin. Biotechnol. 29, 39–45. https://doi.org/10.1016/j.copbio.2014.02.011.

Sandberg, T.E., Salazar, M.J., Weng, L.L., Palsson, B.O., Feist, A.M., 2019. The emergence of adaptive laboratory evolution as an efficient tool for biological discovery and industrial biotechnology. Metab. Eng. 56, 1–16. https://doi.org/10.1016/j.ymben.2019.08.004.

Sanger, F., Nicklen, S., Coulson, A., 1977. DNA sequencing with chain-terminating inhibitors. Proc. Natl. Acad. Sci. U. S. A. 74, 5463–5467. https://doi.org/10.1073/PNAS.74.12.5463.

Schellenberger, J., Que, R., Fleming, R.M.T., Thiele, I., Orth, J.D., Feist, A.M., et al., 2011. Quantitative prediction of cellular metabolism with constraint-based models: the COBRA Toolbox v2.0. Nat. Protoc., 6. https://doi.org/10.1038/nprot.2011.308.

Segré, D., Vitkup, D., Church, G.M., 2002. Analysis of optimality in natural and perturbed metabolic networks. Proc. Natl. Acad. Sci. U. S. A. 99, 15112–15117.

Seiboth, B., Herold, S., Kubicek, C., 2012. Metabolic engineering of inducer formation for cellulase and hemicellulase gene expression in Trichoderma reesei. Subcell. Biochem. 64, 367–390.

Shetty, R.P., Endy, D., Jr, T.F.K., 2008. Engineering BioBrick vectors from BioBrick parts. J. Biol. Eng. 12, 1–12. https://doi.org/10.1186/1754-1611-2-5.

Shlomi, T., Berkman, O., Ruppin, E., 2005. Regulatory on ¯off minimization of metabolic flux changes after genetic perturbations. Proc. Natl. Acad. Sci. U. S. A. 102.

Singh, V., 2014. Recent advancements in synthetic biology: current status and challenges. Gene 535, 1–11. https://doi.org/10.1016/j.gene.2013.11.025.

Sleight, S.C., Sauro, H.M., 2013. BioBrick assembly using the in-fusion PCR cloning kit. Synth. Biol. 1073, 19–30. https://doi.org/10.1007/978-1-62703-625-2.

Smith, C., O'Maille, E., Quin, C., Trauger, S., Brandon, T., 2005. METLIN: a metabolite mass spectral database. Ther. Drug Monit. 27.

Smolke, C.D., 2009. Building outside of the box: iGEM and the BioBricks foundation. Nat. Biotechnol. 27, 1099–1102. https://doi.org/10.1038/nbt1209-1099.

Sohn, J., Park, H.J., Bae, J., Ko, H., Lee, S., Sung, H., et al., 2018. Low-pH production of D-lactic acid using newly isolated acid tolerant yeast Pichia kudriavzevii NG7. Biotechnol. Bioeng., 0–2. https://doi.org/10.1002/bit.26745.

Spohner, S.C., Schaum, V., Quitmann, H., Czermak, P., 2016. Kluyveromyces lactis : an emerging tool in biotechnology. J. Biotechnol. 222, 104–116. https://doi.org/10.1016/j.jbiotec.2016.02.023.

Stalidzans, A.P.E., Kokina, J.L.A., 2017. Model-based biotechnological potential analysis of Kluyveromyces marxianus central metabolism. J. Ind. Microbiol. Biotechnol. 44, 1177–1190. https://doi.org/10.1007/s10295-017-1946-8.

Starcevic, A., Zucko, J., Simunkovic, J., Long, P.F., Cullum, J., Hranueli, D., 2008. ClustScan: an integrated program package for the semi-automatic annotation of modular biosynthetic gene clusters and in silico prediction of novel chemical structures. Nucleic Acids Res. 36, 6882–6892. https://doi.org/10.1093/nar/gkn685.

Stark, C., Breitkreutz, B., Reguly, T., Boucher, L., Breitkreutz, A., Tyers, M., 2006. BioGRID: a general repository for interaction datasets. Nucleic Acids Res. 34, 535–539. https://doi.org/10.1093/nar/gkj109.

Stevenson, J., Krycer, J.R., Phan, L., Brown, A.J., 2013. A practical comparison of ligation-independent cloning techniques. PLoS One 8, 8–14. https://doi.org/10.1371/journal.pone.0083888.

Stovicek, V., Holkenbrink, C., Borodina, I., 2017. CRISPR/Cas system for yeast genome engineering : advances and applications. FEMS Yeast Res., 1–16. https://doi.org/10.1093/femsyr/fox030.

Su, L., Jia, W., Hou, C., Lei, Y., 2011. Microbial biosensors : a review. Biosens. Bioelectron. 26, 1788–1799. https://doi.org/10.1016/j.bios.2010.09.005.

Sun, J., Yu, H., Chen, J., Luo, H., Shen, Z., 2016. Ammonium acrylate biomanufacturing by an engineered *Rhodococcus ruber* with nitrilase overexpression and double - knockout of nitrile hydratase and amidase. J. Ind. Microbiol. Biotechnol. 43, 1631–1639. https://doi.org/10.1007/s10295-016-1840-9.

Sun, W., Vila-santa, A., Liu, N., Prozorov, T., Xie, D., Torres, N., et al., 2020. Metabolic engineering of an acid-tolerant yeast strain Pichia kudriavzevii for itaconic acid production. Metab. Eng. Commun. 10, e00124. https://doi.org/10.1016/j.mec.2020.e00124.

Sutherland, I.W., 1997. Microbial biopolymers from agricultural products: production and potential. Int. Biodeterior. Biodegradation 38, 246–261.

Thoma, F., Blombach, B., 2021. Metabolic engineering of *Vibrio natriegens*. Essays Biochem. 14048, 381–392.

Tillett, D., Neilan, B.A., 1999. Enzyme-free cloning: a rapid method to clone PCR products independent of vector restriction enzyme sites. Nucleic Acids Res. 27, 26–28.

Toivari, M., Vehkomäki, M., Nygård, Y., Penttilä, M., Ruohonen, L., Wiebe, M.G., 2013. Bioresource technology low pH D -xylonate production with Pichia kudriavzevii. Bioresour. Technol. 133, 555–562. https://doi.org/10.1016/j.biortech.2013.01.157.

Tseng, C., Lin, Y., Liu, W., Lin, H., Chou, H., Wu, J.H., et al., 2019. Metabolic engineering probiotic yeast produces 3S, 3′S-astaxanthin to inhibit B16F10 metastasis. Food Chem. Toxicol. 110993. https://doi.org/10.1016/j.fct.2019.110993.

Uchiyama, I., 2003. MBGD: microbial genome database for comparative analysis. Nucleic Acids Res. 31, 58–62. https://doi.org/10.1093/nar/gkg109.

Urnov, F.D., Rebar, E.J., Holmes, M.C., Zhang, H.S., Gregory, P.D., 2010. Genome editing with engineered zinc finger nucleases. Nat. Rev. Genet. 11, 636–646. https://doi.org/10.1038/nrg2842.

Vandemoortele, G., Eyckerman, S., Gevaert, K., 2019. Pick a tag and explore the functions of your pet protein. Trends Biotechnol. 37, 1078–1090. https://doi.org/10.1016/j.tibtech.2019.03.016.

Vanderstraeten, J., Briers, Y., 2020. Synthetic protein scaffolds for the colocalisation of co-acting enzymes. Biotechnol. Adv. 44, 107627. https://doi.org/10.1016/j.biotechadv.2020.107627.

Vázquez, E., Villaverde, A., 2013. Microbial biofabrication for nanomedicine: biomaterials, nanoparticles and beyond Editorial. Nanomedicine 8, 1895–1898.

Vento, J.M., Crook, N., Beisel, C.L., 2019. Barriers to genome editing with CRISPR in bacteria. J. Ind. Microbiol. Biotechnol. 46, 1327–1341. https://doi.org/10.1007/s10295-019-02195-1.

Vigouroux, A., Bikard, D., 2020. CRISPR tools to control gene expression in bacteria. Microbiol. Mol. Biol. Rev. 84.

Vu, H., Hirschbein, B., 1991. Internucleotide phosphite sulfurization with tetraethylthiuram disulfide. Phosphorothioate oligonucleotide synthesis via phosphoramidite chemistry. Tetrahedron Lett. 32.

Walker, A.W., Keasling, J.D., 2002. Metabolic engineering of *Pseudomonas putida* for the utilization of parathion as a carbon and energy source. Biotechnol. Bioeng. https://doi.org/10.1002/bit.10251.

Wang, Y., Hui, S., Wondisford, F.E., Su, X., 2020. Utilizing tandem mass spectrometry for metabolic flux analysis. Lab. Investig. https://doi.org/10.1038/s41374-020-00488-z.

Wang, B., Yang, H., Sun, J., Dou, C., Huang, J., Charles, T.C., 2021. BioMaster: an integrated database and analytic platform to provide comprehensive information about BioBrick parts. Front. Microbiol. 12, 1–6. https://doi.org/10.3389/fmicb.2021.593979.

Waugh, D.S., 2011. An overview of enzymatic reagents for the removal of affinity tags. Protein Expr. Purif. 80, 283–293. https://doi.org/10.1016/j.pep.2011.08.005.

Westbrook, A.W., Ren, X., Oh, J., Moo-young, M., Chou, C.P., 2018. Metabolic engineering to enhance heterologous production of hyaluronic acid in *Bacillus subtilis*. Metab. Eng. 47, 401–413. https://doi.org/10.1016/j.ymben.2018.04.016.

Wittig, U., Kania, R., Golebiewski, M., Rey, M., Shi, L., Jong, L., et al., 2012. SABIO-RK—database for biochemical reaction kinetics. Nucleic Acids Res. 40, 790–796. https://doi.org/10.1093/nar/gkr1046.

Wu, F., 2019. Mining plant genomes for rapid discovery of nature products. Mol. Plant Cell Press. 12, 13. https://doi.org/10.1016/j.molp.2018.12.009.

Xu, Z., Colosimo, A., Gruenert, D., 2003. Site-directed mutagenesis using the megaprimer method. *E. coli* plasmid vectors. Methods Mol. Biol. 235, 1–5.

Xu, P., Vansiri, A., Bhan, N., 2012. ePathBrick: a synthetic biology platform for engineering metabolic pathways in *E. coli*. ACS Synth. Biol. 7, 256–266.

Yang, K., Han, X., 2016. Lipidomics : techniques, applications, and outcomes related to biomedical sciences. Trends Biochem. Sci. https://doi.org/10.1016/j.tibs.2016.08.010.

Yang, Z., He, J., Wei, X., Ju, J., Ma, J., 2019a. Exploration and genome mining of natural products from marine Streptomyces. Appl. Microbiol. Biotechnol. 3.

Yang, D., Park, S.Y., Park, Y.S., Eun, H., Lee, S.Y., 2019b. Metabolic engineering of Escherichia coli for natural product biosynthesis. Trends Biotechnol. 38, 745–765. https://doi.org/10.1016/j.tibtech.2019.11.007.

Yang, D., Surya Prabowo, C., Eun, H., Young Park, S., Cho, I.J., Jiao, S., et al., 2021. *Escherichia coli* as a platform microbial host for systems metabolic engineering. Essays Biochem. 65, 225–246.

Yao, K., Xu, L., Wang, F., Wei, D., 2014. Characterization and engineering of 3-ketosteroid-△1-dehydrogenase and 3-ketosteroid-9α-hydroxylase in *Mycobacterium neoaurum* ATCC25795 to produce 9α-hydroxy-4-androstene-3,17-dione through the catabolism of sterols. Metab. Eng., 1–11. https://doi.org/10.1016/j.ymben.2014.05.005.

Young, R., Haines, M., Storch, M., Freemont, P.S., 2021. Combinatorial metabolic pathway assembly approaches and toolkits for modular assembly. Metab. Eng. 63, 81–101. https://doi.org/10.1016/j.ymben.2020.12.001.

Yu, K., Liu, C., Kim, B., Lee, D., 2014. Synthetic fusion protein design and applications. Biotechnol. Adv. https://doi.org/10.1016/j.biotechadv.2014.11.005.

Yuan, S., Guo, G., 2017. Ethanol production from dilute-acid steam exploded lignocellulosic feedstocks using an isolated multistress-tolerant Pichia kudriavzevii strain. Microb. Biotechnol. 10, 1581–1590. https://doi.org/10.1111/1751-7915.12712.

Zaparucha, A., Berardinis, B., Vaxelaire-Vergne, C., 2018. Genome mining for enzyme discovery. In: Zaparucha, A. (Ed.), Modern Biocatalysis: Advances Towards Synthetic Biological Systems. Royal Society of Chemistry, pp. 1–27.

Zhang, J., 2021. Molecular biology methods to construct recombinant fibrous protein. Methods Mol. Biol. 2347.

Zhang, X., Zhang, W., 2018. Antisense RNA: the new favorite in genetic research. J. Zhejiang Univ. B 19.

Zhang, H., Wang, L., Hunter, D., Voogd, C., Joyce, N., Davies, K., 2013. A Narcissus mosaic viral vector system for protein expression and flavonoid production. Plant Methods 9, 10–16.

Zhang, Q.W., Lin, L.G., Ye, W.C., 2018. Techniques for extraction and isolation of natural products: a comprehensive review. Chin. Med. 13, 1–26. https://doi.org/10.1186/s13020-018-0177-x.

Zhang, Y., Xhou, X., Yao, Y., Xu, Q., Shi, H., Wang, K., et al., 2021. Coexpression of VHb and MceG genes in Mycobacterium sp. Strain LZ2 enhances androstenone production via immobilized repeated batch fermentation. Bioresour. Technol. 342, 1–23.

Ziemert, N., Podell, S., Penn, K., Badger, J.H., Allen, E., Jensen, P.R., 2012. The natural product domain seeker NaPDoS : a phylogeny based bioinformatic tool to classify secondary metabolite gene diversity. PLoS One 7, 1–9. https://doi.org/10.1371/journal.pone.0034064.

Ziemert, N., Alanjary, M., Weber, T., 2016. Natural product reports the evolution of genome mining in microbes—a review. Nat. Prod. Rep. 33, 988–1005. https://doi.org/10.1039/C6NP00025H.

Chapter 3

Computational approaches for smart cell creation in the bioeconomy era

Sachiyo Aburatani[a,b], Koji Ishiya[b], Tomokazu Shirai[c], Yosuke Shida[d], Wataru Ogasawara[d], Hiroaki Takaku[e], and Tomohiro Tamura[a,b]
[a]*Computational Bio-Big Data Open Innovation Laboratory, National Institute of Advanced Industrial Science and Technology (AIST), Tokyo, Japan,* [b]*Bioproduction Research Institute, National Institute of Advanced Industrial Science and Technology (AIST), Sapporo, Japan,* [c]*Center for Sustainable Resource Science, RIKEN, Yokohama, Japan,* [d]*Department of Bioengineering, Nagaoka University of Technology, Nagaoka, Japan,* [e]*Department of Applied Life Science, Niigata University of Pharmacy and Applied Life Science, Niigata, Japan*

Newly bioproduction era

A Smart Cell is a host microorganism that is artificially designed using information analysis technology and is expected to accelerate the production of substances by microorganisms to realize the bioeconomy while contributing to the realization of the Sustainable Development Goals (SDGs). In Japan, a 5-year project called the NEDO Smart Cell Project was implemented from 2016 to 2021 (Nedo Smart Cell Project, 2020) to bring this idea to fruition. Smart Cells will enable a shift from fossil fuel-derived material production to microbial-derived material production, which will greatly contribute to the reduction of carbon dioxide. In addition, microorganisms created by Smart Cell technology will contribute to the use of renewable energy, as they utilize biomass of organic matter derived from plants and animals as an energy source, which has been discarded until now. In the future, from the viewpoint of carbon neutrality and the use of renewable energy, the use of microorganisms is expected to replace the use of fossil fuels for practical substance production, such as liquid fuels and many commodity chemicals, e.g., plastics, chemical fiber raw materials, and fragrances. However, in order to replace substances produced from fossil fuels with microbial-derived materials, several problems must be solved. Substance production capacity of host microorganisms must be greatly improved, production efficiency must be enhanced, and the purification cost must be reduced. The increased efficiency of the production process and the

Genomics and the Global Bioeconomy. http://doi.org/10.1016/B978-0-323-91601-1.00012-2

decrease in the cost of the purification process must be solved industrially. On the other hand, the improvement of useful substance production by microorganisms is expected to be solvable by biological approaches. But, how can we improve microorganisms to give them the ability to produce high levels of useful substances?

Microbial production has been used in the food industry for centuries. Fermented foods such as wine and cheese are good representatives, since wine production involves a system of alcoholic fermentation by the yeast *Saccharomyces cerevisiae* (Cox, 2015; Fugelsang and Edwards, 2007), and cheese is produced by lactic acid fermentation, using enzymes from lactic acid bacteria (Voet and Voet, 2007). Fermentation is a system in which pyruvic acid, obtained as an end substance in the process of microbial energy production using sugars such as glucose, is converted into various useful substances for humans by the microbial metabolic systems. In this context, many empirical advancements have been made in microorganisms. The improvement of microorganism abilities for bioproduction is an important theme in bioengineering fields, and several types of empirical breeding approaches have been developed and applied to many host microorganisms. The main breeding techniques include the selection of useful strains from random mutant strains (Sakaguchi and Okanishi, 1980; Zhang et al., 2014, 2015) and the improvement of key enzyme activities within the biosynthesis pathways (Liu et al., 2015; Zhong and Jousset, 2017). These conventional breeding approaches have effectively yielded useful host strains, but their development is quite time-consuming and costly. Furthermore, conventional approaches are aimed at enhancing the original functions of the host microorganism, and thus, it is difficult to modify microorganisms to produce alternative substances from fossil fuel-derived products, as currently required.

Over the past decade, synthetic biology has been used to address this problem by improving host organisms for the production of useful substances (Lorenzo et al., 2018). Synthetic biology is an interdisciplinary field that combines biotechnology and molecular biology with engineering and informatics sciences, including gene transfer into host microorganisms to redesign existing biological systems for useful purposes. One of the goals of synthetic biology in bioproduction is to develop biologically based alternative methods to reduce carbon dioxide emissions during the production of fossil fuel-derived substances. In particular, a lot of research is aimed at the production of biofuels and bioplastics, and some companies are putting them to practical use, e.g., Synthetic Genomics and Global Bioenergies (Synthetic Genomics, 2021; Global Bioenergies, n.d.). In the field of microbial production, a great deal of research and development combining synthetic biology and information science has been conducted (Nicoletti, 2009). Many approaches are aimed at reducing the time-consuming and high financial cost from conventional breeding methods, including the use of robots to obtain large amounts of data and machine learning methods to estimate the genes to be modified from the host.

However, the generation of such a large amount of data for each host microorganism and each substance is not practical from a cost perspective. With this challenge in mind, we created the "Smart Cell Project" in 2016 (Nedo Smart Cell Project, 2020) to test if it is possible to find candidate genes for modification to improve host microorganisms, without the need to acquire large amounts of data for each host microorganism or substance.

For the construction of Smart Cells, we applied the network modeling technology described in this chapter to clarify the mechanisms involved in the production of substances by the host microorganism and to identify the genes that are bottlenecks in the system. Computational design of novel metabolic pathways and optimized sequence design for gene transfer are important techniques in both the metabolic engineering and bioengineering fields (Hatzimanikatis et al., 2005; Guimaraes et al., 2014; Araki et al., 2015), and they are frequently used as effective methods for bioproduction with synthetic biology (Shirai et al., 2016; Vavricka et al., 2019; Saito et al., 2019; Mori et al., 2021). However, network modeling has not been attempted for the modification of host microorganisms so far. By using network modeling technology, we sought to improve host microorganisms with the same accuracy level as the current mainstream method for proposing modifications based on machine learning from large amounts of data (e.g., 10,000 samples), but in a shorter time than the conventional breeding method, and with one-hundredth less data (e.g., <100 samples). In the next section, we will discuss the network modeling approach in more detail.

Ability of network modeling

Network modeling is a method to visualize the complex interplay of intracellular systems as a graph structure. When an organism, especially a microorganism, is expressing a special function, such as the production of some useful substances, there is a specific set of processes happening inside the cell. Although microorganisms have thousands to tens of thousands of genes within a cell, only a limited number of genes are required to express certain special functions. During substance production, these genes are temporarily in an activated state and are used as a specific system in the cell. To improve the ability of the host microorganism to produce substances, this system must become more efficient and its overall ability should be enhanced. By using network modeling, the system can be visualized as a graph structure, which makes it possible to efficiently search for the necessary improvement points in the system.

The application of network modeling to biology has mainly been used to infer regulatory relationships between genes from gene expression data. The history of network modeling from gene expression data begins around the year 2000, and various algorithms have been developed to infer complex gene networks. Although improved algorithms are still being developed, most network modeling algorithms are derived from four models: the Boolean modeling developed by Akutsu et al. (1999), the Bayesian model developed by

Friedman et al. (2000), the differential equation model developed by Chen et al. (1999), and the graphical Gaussian model developed by Toh and Horimoto (2002). Since each of these algorithms has its own specific features, a different algorithm must be applied, depending on the type of data to be calculated and the shape of the network to be inferred. Boolean modeling is suitable for time series data. It expresses whether a change in the gene expression occurs, as measured at each time point as binarized data, and estimates the causal relationship between the genes as variables from the state in which the pattern of this binarized data changes at each time. Therefore, it is difficult for Boolean modeling to estimate the network structure from data when the host is randomly mutated, which is often measured in microbial breeding. Bayesian modeling is a powerful method that can be applied to any data where the variables are numerical. It is a probabilistic model that represents qualitative dependencies between multiple random variables in a directed acyclic graph structure, and the estimated causal relationship indicates the probability that when the expression of one gene changes, the expression of other genes will also change. Bayesian modeling is the most commonly utilized network modeling method, and many algorithms have been developed to estimate causal relationships between variables. Each of these algorithms is mathematically correct, but since the estimated network structure varies from algorithm to algorithm, it is not clear which structure should be chosen. Differential equation modeling is a method that is primarily applied to time series data, but it is slightly different from other algorithms. Although Boolean modeling, Bayesian modeling, and graphical Gaussian modeling have different features, they are commonly used methods for inferring the relationships between genes as variables from numerical data. However, in differential equation modeling, the causal relationships between variables must first be defined. Differential equation modeling is utilized to estimate the strength of the regulatory relationship between genes, based on the expression level of each gene that is changing over time within the known regulatory structure. Graphical Gaussian modeling is also applicable to any data where the variables are fluctuating. It is a conditional independence-based method to estimate direct relationships by calculating partial correlation coefficients between variables. Graphical Gaussian modeling requires the absence of multicollinearity among the variables, since the inverse matrix is calculated among them. Thus, to infer the network model among all genes, clustering or other methods must first be performed to classify the genes as a single cluster with high correlation (Toh and Horimoto, 2002; Aburatani et al., 2003). This method is not suitable for searching for system bottlenecks in host microorganisms because it can infer direct relationships between genes, but not causal relationships.

In our previous research, we developed a new network modeling method SENET (**S**tructural **E**quation modeling-based **NET**work inference) based on structural equation modeling (SEM) (Aburatani, 2011). From a mathematical viewpoint, SEM was established in the 1970s and summarized by Bollen (1989), including path analysis and factor analysis. The significant features of SEM are the inclusion of unobserved variables as latent variables within the constructed model and the ability to infer the network, including the cyclic structure such as feedback control. Additionally, one of the important aspects of network models is the strength of the relationship between variables (in this case genes), and in SEM, the covariance matrices of the model variables (genes) are calculated based on the assumed network model to be as close as possible to the covariance matrices calculated for the real measured numerical data. This means that the strength of the relationship between genes is estimated in a way that reflects the real data as closely as possible. In addition, fitting scores have been defined to quantify the degree to which the model represents the real data. By using these scores, it is possible to determine more precisely whether the assumed model corresponds to the real data or not. SEM has been successfully utilized to elucidate causal relationships in the econometrics, sociology, and psychology fields (Haavelmo, 1943; Duncan, 1975; Pearl, 2001). In addition, it has been applied to the quantitative analysis of quality trait loci (QTLs), which examine the relationship among DNA marker genes, the phenotypic data, and gene linkage (Liu et al., 2008; Aten et al., 2008), as well as to identify genetic networks from the expression level of genes or single-nucleotide polymorphism (SNP) information (Xiong et al., 2004; Lee et al., 2007; Shieh et al., 2008). However, these network analyses are based on the general use of SEM, which is to first assume the network structure and then verify whether the assumed network structure is correct or not from the measured data. In the case of host microorganisms' modification in Smart Cell creation, we are dealing with issues where the control structure of the production system is not known. For example, we know the metabolic pathway of oil biosynthesis in oil yeast. However, the activation of individual parts of the biosynthesis pathway does not give the desired results due to the robustness of the cell. We need a regulatory model that searches for regulators that improve the performance of the pathway as a whole, while balancing other systems in the cell.

In the recent Smart Cell Project, we developed ASENET system (**A**dvanced **S**tructural **E**quation modeling-based **NET**work inference system: https://nife.cbbd-oil-rinkai.aist.go.jp/asenet/), an improved version of SENET as a network modeling tool for host microbial modification. In order to create Smart Cells, it is necessary not only to infer a theoretically correct model, as in the case of conventional network models, but also to suggest a model that can actually improve the substance production capacity through empirical experiments. The main advantage of SEM network modeling as compared to other algorithms is that we can numerically evaluate the consistency of the estimated network model structure with the measured data. This is an important advantage for the creation of Smart Cells to experimentally validate candidate genes for modification. Our ASENET takes advantage of this feature and makes it possible to construct a network model of the most optimal form that can explain the measured data. This increases the confidence in the network model as a representation of what is actually going on in the cell and enables the identification of bottleneck genes that need to be modified. However, there is a restriction on the number of variables (genes) that can be included within the SEM network model, as compared to the Boolean modeling and graphical Gaussian modeling. While Boolean and graphical modeling can produce network models of thousands of genes, SEM is limited to producing network models of about 50 genes. This is because in the SEM calculation, the regulatory relationships for one gene are represented by a single equation, so for 50 genes you have 50 simultaneous equations to solve. It is difficult to obtain a unique solution if the number of variables is too large. However, the limitation on the number of variables, such as less than 50, might not be a disadvantage for network modeling of a host microbial improvement.

The main reason why the restriction that the number of variables must be less than 50 is not a disadvantage is that, as mentioned above, only a limited number of genes are required for a microorganism to express a particular function. Furthermore, constructing a network model that covers all or many genes is a major deviation from the goal of constructing a network model to improve the substance production capacity of host microorganisms. Fig. 3.1 shows that the construction of a network model of all genes is not necessary for the creation of a Smart Cell. Here, we have randomly generated a network model with 2000 relationships between 1000 genes, a network model with 200 relationships among 100 genes, and a network model with 100 relationships among 50 genes by computational simulation. The number of genes in a real cell is thousands, and their relationships among genes are much more complex than these models indicate. From this figure, even with a network model of only 2000 relationships with 1000 genes, which is simpler than the actual regulatory relationships in a cell, extracting meaning from this complicated model and understanding this network as a control system are not straightforward. Constructing this complex network model among all genes would not be appropriate for the purpose of controlling a particular function of microorganisms. If you want to control only

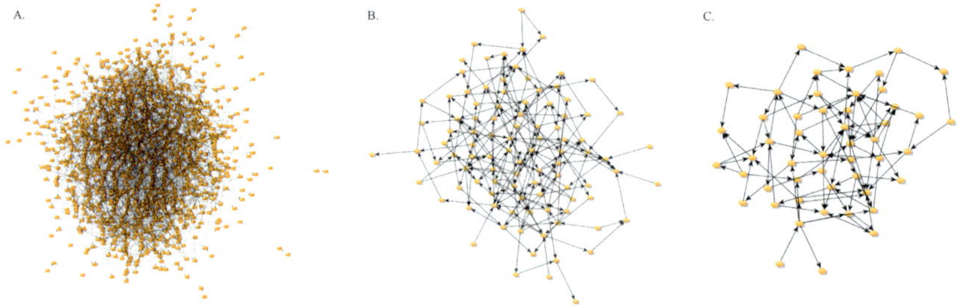

FIG. 3.1 Network complexity of many genes. *Rectangles* indicate genes, and *arrows* indicate relationships between genes. These networks were randomly generated by computer simulations with a set number of genes and relationships. (A) The network model with 2000 relations among 1000 genes. (B) The network model with 200 relations among 100 genes. (C) The network model with 100 relations among 50 genes. If some relationship exists between all of the genes, then the complexity of the model depends on the number of genes represented as the nodes and the directions of the relationships indicated by the edges.

one system in a cell, as in the creation of a Smart Cell, it is sufficient to know the regulatory relationships of only the important genes running that system, not of all the genes.

Network modeling in the creation of a Smart Cell aims to understand the functions of cells in terms of industrial systems and to search for bottlenecks and modifications to improve those systems. The most important aspect of using our network modeling to improve host microorganisms is the "selection of genes that are operating in a particular system." To understand the importance of this genetic selection, the correspondence between the process of manufacturing industrial systems, such as some kinds of machines, and the process of Smart Cell creation is shown in Fig. 3.2. When deciding to build a certain system, where would you start? You cannot use all of the various industrial components in the world to produce something. Therefore, you would probably start by selecting the components that are required for the system you are developing. Similarly, when reconstructing or elucidating a biological system, all genes and proteins are not required to be included in the system. We only need to extract the genes that are required for operating a certain system. This is the gene selection process. In the same way that an industrial system cannot be manufactured without sufficient amounts of the required components, it is

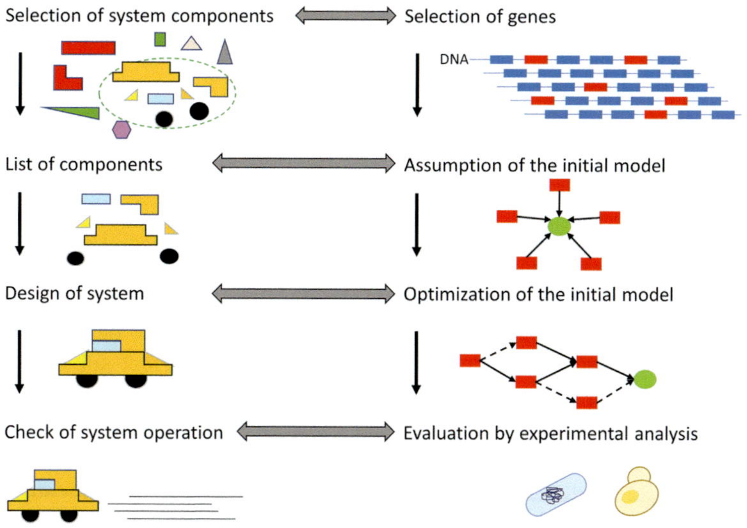

FIG. 3.2 Correspondence between the manufacturing process of industrial products and the Smart Cell creation process. The left side shows the general manufacturing process of an industrial product, and the right side shows the Smart Cell creation process, using the network modeling. The *rectangles* in the right side indicate genes, and circles indicate substance production. The *arrows* at "Assumption of the initial model" and "Optimization of the initial model" indicate the regulations between genes or genes to substance production. *Solid arrows* indicate positive regulation, and *dashed arrows* indicate negative regulation.

important to select the list of genes that function in the substance production in order to control the microbial system.

The next step is to construct the initial model, which involves gathering the collected parts in a list of genes at the former step and making a tentative plan for arranging them. Of course, if the function of each component is certainly known from the beginning, it can be arranged based on a blueprint, but in the case of biology, the functions of even single genes are not completely understood. Therefore, as an initial model, we arrange the selected genes in the form of "probably this" based on published literature. Based on our initially constructed model, we optimize the model structure from experimental data coming from real microorganisms producing the substance in question using our structural equation model so it is more suitable for the measured data. Our ASE-NET system is capable of estimating a model with a minimum of 10 data points and with a higher accuracy of about 100 data points. This is only one-hundredth of the number of data needed to suggest a candidate gene for modification with the same accuracy as is currently possible with commonly used machine learning techniques. By structural equation modeling, we can estimate the model structure best suited to represent the measured data. Assume that the estimated model structure is an intrinsic system that is excited in the cell during substance production. Once the excited intracellular system is revealed as a graph structure, we can estimate the modification points needed to maximize the outcome variables, i.e., substance production, cell amount, and so on. This modification point is a candidate for a modified gene to improve the substance production capacity of the host microorganism.

Effectiveness of SEM network modeling

Previous studies have confirmed that our method is effective for visualizing some systems controlled by gene expression in a cell or an organism, and we have been testing its effectiveness in various organisms. First, we used it to estimate the gene regulatory network model of the GAL system in the budding yeast, *Saccharomyces cerevisiae* (Aburatani, 2011). The GAL system is known to be a continuous system of transcription factor expression and was the perfect subject to test whether SEM could be used to infer a hierarchical regulatory system in the cell. In the GAL system, a specific transcription factor protein regulates the expression of a group of genes at the next level, but the gene expression data used in the analysis does not contain any information on the protein. However, the estimation of the network model by SEM made it possible to incorporate information on transcription factor proteins into the model, which is not actually observed. We found that the SEM network model can represent the gene regulations by transcription factor proteins that actually occur during the operation of the GAL system (Aburatani, 2011). Next, to test whether the network model that can be estimated with this method is only a simple hierarchical control system, we applied it to the yeast cell cycle control system. Since the cell cycle is regulated by several factors, and there is also a feedback control to prevent the cell cycle from reversing, it is assumed that the cell cycle has a more complex regulatory structure, but it was found that each cell cycle is strictly regulated by different factors (Aburatani and Toh, 2013). We successfully visualized the regulatory system involving gene expression as a graph using SEM network modeling, and we further checked whether this was possible for various systems in other organisms besides yeast. In concrete terms, we applied our methods to pal-1 lineage-specific regulation in *Caenorhabditis elegans* embryo development (Aburatani, 2012), anteroposterior (AP) pattern formation in *Drosophila melanogaster* embryogenesis (Aburatani and Toh, 2014), and transcriptional control for pluripotency in mouse embryonic stem (ES) cells (Aburatani, 2015).

The study of *C. elegans* and *D. melanogaster* was undertaken as a task to model spatial and temporal regulation during developmental stages. A variety of intracellular factors are known to regulate specific gene expression during development, but the activity information of these regulators is not directly reflected in gene expression profiles. We confirmed whether our optimal SEM modeling approach could be used to model the regulatory system that determines the fate of the C lineage in *C. elegans* development, which is controlled by nontranscription factor proteins. The inferred network model showed that 13 genes involved in the C lineage were sequentially regulated by factors such as protein accumulation and localization in addition to pal-1 expression, indicating that the regulation of known C lineage genes was dependent on the sequential regulation of these regulatory factors (Aburatani, 2012). Next, we modeled the regulatory system of anteroposterior (AP) pattern formation in *D. melanogaster* embryogenesis: unlike *C. elegans*, it is difficult to infer the initial model of *D. melanogaster* from the relevant genes. Therefore, we developed a method to build an initial model using only gene expression data and optimized it by modeling with SEM to see how far it could explain the known AP pattern formation. The estimated network model shows that the 18 transcription factor genes, which are related to AP pattern formation, are regulated by factors involved in the expression of other genes, maternal effects, and spatial morphological distribution. We also found that feedback control is applied as the stages progress to ensure that each stage is strictly followed in morphogenesis (Aburatani and Toh, 2014). The *C. elegans* and *D. melanogaster* models confirm that SEM can also be used to model gene expression control systems that are not caused by proteins but by other factors in the cell.

We also modeled a more complex biological system, the system that controls pluripotency and self-renewal in mouse embryonic stem (ES) cells. In the inferred model, the cell-initiating TF genes (Pou5f1, Sox2, and Nanog) were regulated by different factors. It was speculated that the respective switches for pluripotency and self-renewal of mouse ES cells are inputs of different signaling pathways (Aburatani, 2015). This shows that SEM modeling can also neatly represent two different phenotypes that can occur simultaneously in a cell as two different systems.

One of the significant features of SEM is that unobserved elements can be incorporated within the network model as latent variables, but when understanding the shape of the system structure, it is often better to clarify the control relationships between known variables (genes) without latent variables. Therefore, we have improved our network modeling technique with SEM to be able to build network model without latent variables. In order to test the effectiveness of modeling without latent variables, we applied this method to detect the toxicity-specific effects of environmental chemicals on several functional genes in human embryonic stem cells (Aburatani et al., 2013; Yamane et al., 2016). It is believed that environmental chemicals have a negative impact on pregnant women and can cause problems for the development of the unborn child. In this study, we investigated which genes were affected when ES cells from early human embryos were exposed to various environmental chemicals. The estimated network model clearly showed that the network structure affected by the toxicity of environmental chemicals was different. This shows that it is possible to visualize the regulatory system in a cell even with a SEM network model without latent variables. Therefore, we thought that a modeling technique that does not include latent variables would be suitable as a network model for Smart Cell creation.

Applying structural equation modeling (SEM) methods to microbial bioproduction

Gene selection

Gene selection is an important aspect of network modeling to improve host microbes and utilize them as Smart Cells. In the past, we have applied SEM network modeling to problems where the genes involved in the regulatory system are already known. However, for most of the industrial hosts that Smart Cells are aimed at, the genes involved in substance production system are not known. Even if the metabolic pathways are known, the expression levels of the enzyme genes in the metabolic pathways do not directly contribute to the amount of substance produced. The important thing is to extract a set of genes that actually contribute to the substance production system in the cell from the measured data alone, regardless of any known metabolic pathways or other known information. To identify the genes that are directly related to the output in terms of substance production, we then performed a correlation analysis between gene expression data and substance production data.

The first step was to standardize the data for the analysis of multiple samples in an integrated manner. Gene expression data is generally measured by RNA sequence technology (RNA-Seq) or DNA microarray methods. RNA-Seq is a sequencing technique, which uses next-generation sequencing (NGS) to measure the presence and quantity of RNA in biological samples at any given moment (Wang et al., 2009; Chu and Corey, 2012). With RNA-Seq, we first extract the mRNA produced by gene transcription in living cells, copy these fragments, and sequence them by NGS short-read sequencing. By comparing the determined sequence with the reference genome sequence (the whole-genome sequence of the host microorganism) and reconstructing which genomic regions are transcribed, the expressed genes and their relative expression levels can be expressed numerically and at a global level. DNA microarrays are a more targeted technique for measuring the expression levels of a large number of genes simultaneously rather than the whole transcriptome (Pollack et al., 1999). DNA microarrays are used to quantify the expression levels of genes by detecting the labeled targets with fluorescent dyes or chemiluminescence, which are then expressed as numerical data.

Gene expression data measured by RNA-Seq or DNA microarrays can vary depending on experimental conditions. In order to perform network analysis, these variations must be kept constant. Therefore, we used logarithmic transformation and *Z*-scoring to quantitatively detect the increase or decrease in gene expression as a way to keep the variation of the measured data constant. Here is a simple example to explain the problem when there is variation in the measurement data. For example, if all the genes are highly expressed at one experimental data, a gene expression value of 10 would be an average gene expression value, but if many genes are not expressed well in another experimental data, the same gene expression value of 10 would mean that the gene is expressed at a very high level. In other words, even with the same numerical information of 10, the meaning of that number changes for each experiment.

The first step is to calculate the logarithm of all the gene expression data measured by RNA-Seq or DNA microarray techniques. The logarithmic transformation makes it possible to show the increase or decrease in gene expression in positive or negative values. The next step is to convert the disparate measurements for each experimental data into similarly distributed numerical data (Z-scores). To calculate the *Z*-scores, we first calculate the mean and standard deviation values of each experimental measured data from the log-transformed gene expression data. Next, we subtract the mean value from the log-transformed expression level of each gene and divide it by the value of the standard deviation. In this way, gene expression measured under different conditions can be integrated and analyzed in a uniform way, with a mean value of zero and a standard deviation of one. Here, the mean and standard deviation values were calculated for each experimental data, but if there is a sufficient number of experimental data, for example, more than 100, the mean and standard deviation for each gene in 100 experimental data can also be calculated. However, if the number of experimental data is less than 50, it would be better not to calculate the mean and standard deviation for each gene. The reason for this is that if the number of data is small, the variation in expression of each gene is likely to be biased. If the number of experimental data is small, it is recommended to calculate and use the mean and standard deviation for each experimental data.

In order to select the genes that are related to the amount of substance production from all genes, correlation analysis between the Z-scored gene expression and the measured values of substance production was performed. First, to calculate the similarity between changes in gene expression and changes in substance production, the commonly used Pearson correlation coefficient was calculated between all genes and material production. The Pearson correlation coefficient can be used to evaluate how similar the changes in numerical data are on a scale from -1 to $+1$. Therefore, genes with correlation coefficients close to -1 are likely to inhibit substance production, while genes with correlation coefficients close to $+1$ are likely to induce substance production

When improving the substance production capacity of microorganisms, a single target is not always the only outcome variable. For example, when a microorganism produces a certain useful substance, or when we want to control the production balance of multiple useful substances, unnecessary substances can be created as by-products. Therefore, to select genes that can control such multiple outcome variables at the same time, we performed the above correlation analysis for each outcome variable and then narrowed down the genes by logical operations.

Network modeling by ASENET

The fact that a change in gene expression correlates with a change in substance production does not necessarily mean that all selected genes are actually regulating the substance production of the host microbe. The expression of other genes may also be altered as a by-product of the intracellular behavior of changing the amount of a substance. Therefore, it is necessary to search for the genes that control the material productivity by the network model by ASENET. The first step in the construction of the ASENET network model is to assume that all selected genes are involved in the production of substances in the host microbe. Specifically, we set up a network model of the form shown in Fig. 3.3 as an initial model.

This initial model has a very simple structure. The substance production is arranged in the center, and the selected genes are arranged around the substance production. If you want to control several substances at the same time, you can use an arrangement like B. This model represents a situation in which each selected gene controls the production of a substance independently. This "independently" means that the initial model does not take into account the regulatory relationships between the selected genes. This is because regulatory relationships among genes will be properly estimated during the optimization of ASENET modeling and it is not necessary to take them into account at the beginning.

After assuming the initial model, ASENET uses SEM calculations to estimate the weights of the regulatory relationships between genes and substance production in the model. Since network modeling for Smart Cell creation does

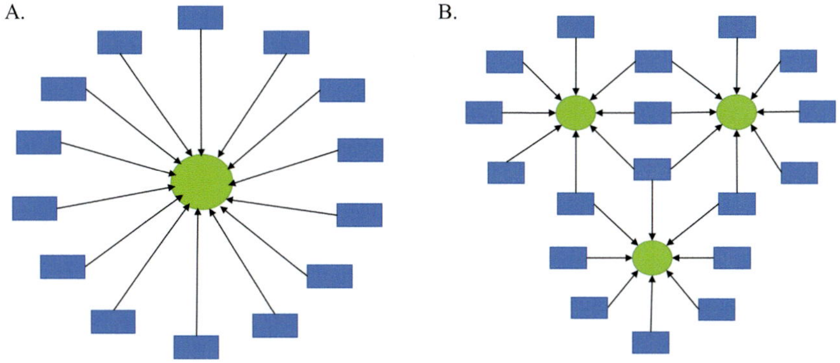

FIG. 3.3 Shapes of initial models. *Rectangles* indicate the selected genes, *circles* indicate substance production, and *arrows* indicate positive regulation between nodes on the graph. (A) Star network shape when there is only one type of target substance. All of the selected genes are assumed as equally effective in substance production. (B) Star network shape when there are several target substances. The initial model does not take into account the relationships between substances or genes. Therefore, the model is constructed while assuming only the regulatory relationship of each gene to the target substance.

not include latent variables, all variables in the model (genes and substance production) can be defined by very simple linear equations, as follows:

$$p = \alpha_{1p}g_1 + \alpha_{2p}g_2 \cdots \alpha_{np}g_n + \varepsilon_p \tag{3.1}$$

$$\begin{aligned} g_1 &= \alpha_{21}g_2 + \alpha_{31}g_3 + \cdots \alpha_{n1}g_n + \varepsilon_{g1} \\ g_2 &= \alpha_{12}g_1 + \alpha_{32}g_3 + \cdots \alpha_{n2}g_n + \varepsilon_{g2} \\ g_3 &= \alpha_{13}g_1 + \alpha_{23}g_2 + \cdots \alpha_{n3}g_n + \varepsilon_{g3} \\ &\vdots \end{aligned} \tag{3.2}$$

Here, p is the amount of substance production; g_i is the gene expression level of gene i; n is the number of selected genes; α_{kl} is the weight of regulatory relationships from k to l. Thus, α_{1p} in Eq. (3.1) shows the strength of the influence of gene 1 on the amount of substance production, and α_{21} in Eq. (3.2) shows the strength of the regulatory relation from gene 2 to gene 1. The errors that affect the genes and substance productions are ε. Next, we will solve the values of the weight coefficients α, from these equations. First, we will make these equations as simple as possible. These linear equations can be expressed by a single equation using matrix form. In this matrix equation, three matrices are used. The first matrix contains genes and substances, the second matrix contains the weights of linear equations, and the third matrix contains error terms. By using these three matrices, we can think of any number of linear equations as a single equation. No matter how many linear equations are presented in Eqs. (3.1) and (3.2), by using the three matrices, we can think of them as a very simple equation:

$$x = Ax + e \tag{3.3}$$

where x indicates substance production and gene expression, A indicates the weight coefficients described in the linear Eqs. (3.1) and (3.2), and e indicates error terms.

The next step is to solve for x from Eq. (3.3). This is very easy. If we follow the same steps as in the general method, we can express x in terms of A and e as follows:

$$x = (1 - A)^{-1}e \tag{3.4}$$

In this case, since A is a matrix, we need to use the unit matrix I as 1, but this is merely a matter of form. In this way, gene expression and substance production can be expressed as mathematical equations in a network model.

The next step is to solve the value of the most important weight coefficient α in the network model. We will use x, which we solved for in Eq. (3.4). First, we use x to express the variance-covariance matrix that represents the relationship between the values of substance production and gene expression. The variance-covariance matrix calculated here is a matrix in the network model and is represented by equations. The next step is to calculate the variance-covariance matrix of substance production and gene expression from the actual measured data. Since this is calculated from numerical values, it is determined as a unique value. In SEM calculation, the weight coefficient α is calculated so that the variance-covariance matrix expressed in the equations and the variance-covariance matrix determined numerically are as close as possible. In this way, the value of each weight coefficient α indicated on the network model is calculated.

Finally, we have to evaluate how adaptively the estimated weight coefficients α and the network structure fit the measured substance production and gene expression data. This evaluation of model adaptability t is one of the features of SEM. In SEM, the network structure is first assumed and then calculated, so if the structure is significantly wrong, the network structure needs to be re-examined. There are several indicators of goodness-of-fit scores. In our method, seven major indices were used to evaluate the model: CMIN, GFI, AGFI, CFI, RMSEA, AIC, and BIC. We will leave the calculation method of each indicator to technical books, and only the values that we judged to be good for each indicator will be described here. If the value of CMIN (Wheaton et al., 1977) is higher than 0.05, the model is considered to be well-fitted. GFI stands for goodness-of-fit index, and AGFI is a modified GFI according to the degrees of freedom (Joreskog and Sorbom, 1983). When the values of GFI and AGFI are above 0.9, the model is judged to fit the measurement data. Furthermore, comparative fit index (CFI) and root mean square error of approximation (RMSEA) are both independent of the huge number of samples to determine if the model is suitable (Bentler, 1990; Browne and Cudeck, 1993). A value of CFI higher than 0.9 or a value of RMSEA less than or equal to 0.05 indicates that the model fits the measured data; a value of RMSEA higher than or equal to 0.10 suggests that model is far from the measurement data. To compare two or more different models, the Akaike information criterion (AIC) (Akaike, 1974) and the Bayesian information criterion (BIC) (Findley, 1991) are suitable. In AIC and BIC, models with lower scores are considered to be well-fitted with the measurement data.

Model optimization by ASENET

In general, SEM is utilized to evaluate the fitness of an assumed model with the measured data. To infer the regulation relationships for the creation of Smart Cells by SEM, we improved our iteration algorithm that optimizes the initial model to the measurement data in ASENET (Aburatani et al., 2019). The basic flow of our model optimization algorithm is to calculate the probabilities of all the control relationships between genes and substance production assumed in the initial model from the inverse of the Fisher information matrix. Regulatory relationships for which the calculated probabilities are not significant are considered unnecessary in the substance production system and are removed from the model. In addition, we use the modification index (MI) scores, which measure how much the chi-square statistic is expected to decrease if a particular parameter setting is constrained, to add new regulatory relationships between genes (Aburatani, 2011; Aburatani and Toh, 2013).

To apply this iteration algorithm for model optimization to Smart Cell creation, ASENET has been modified to optimize the model by including the correlation of error terms along the way. By first optimizing the outline of the substance production system among variables and then optimizing the error term at an early stage, we can construct a network model that includes the effects of other regulatory systems in the cell on the substance production system. In addition, to avoid local minima during the optimization phase, we combined genetic algorithms (GAs) to add random regulatory relations in the network model or to randomly remove estimated control relations. A schematic of the improved iteration algorithm is displayed in Fig. 3.4.

Application examples of ASENET modeling

Ex. 1 Cellulase biosynthesis balance control in *Trichoderma reesei*

The use of biomass is being considered as one solution toward curbing the use of fossil fuels for carbon neutrality. As for biomass utilization, the production of bioethanol from sugar biomass such as sugarcane and corn has been widely used, but it conflicts with the food problem and the environmental implications. Therefore, the use of cellulosic biomass as a nonfood source is attracting attention. In order to put cellulosic biomass to practical use, a large amount of cellulase is required for the enzymatic saccharification of cellulosic biomass, but the problem is the high costs and the fact that the balance of cellulase must be adjusted according to the cellulose to be saccharified.

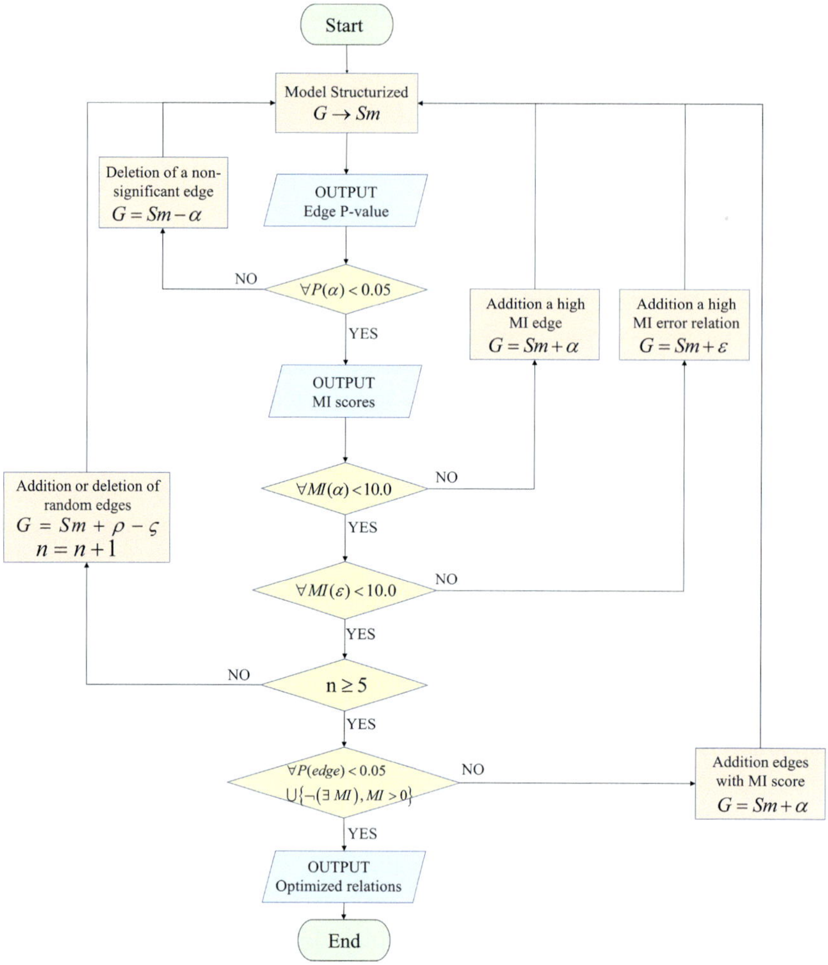

FIG. 3.4 The improved model optimization algorithm for bioproduction. The genetic algorithm is applied by randomly adding and subtracting edges as point mutations. This algorithm is stopped when all of the edges in the graph are significant and there are no more estimated relationships for which a positive MI score can be calculated.

The ascomycete *Trichoderma reesei* is known to produce a variety of cellulases and hemicellulases, and the hypercellulolytic mutants of this fungus are useful as industrial cellulase producers (Nogawa et al., 2001; Bischof et al., 2016). Since the composition of secreted enzymes greatly influences biomass saccharification, we performed network modeling to artificially regulate the enzyme balance. Many cellulases are produced by *T. reesei*, but there are four major types: CBH, BGL, EGL, and XYN. Although Japanese breeding techniques have successfully constructed host strains of *T. reesei* with high cellulase production capacity (Hirasawa et al., 2019), it has been difficult to change this enzyme balance. Therefore, in the Smart Cell Project, we have developed a network model to artificially regulate this enzyme balance.

In this network modeling, multiple enzyme genes were modeled as outcome variables from over 400 expression profiles. Representative results are shown in Fig. 3.5.

It is clear from the inferred network that the enzyme genes are intricately intertwined, and it is difficult to increase or decrease any single enzyme by itself. However, it has been experimentally confirmed that host microorganisms with modified genes, which were inferred from our network model to affect enzyme levels, were able to significantly alter the enzyme balance, which had previously remained completely unchanged. Although many of the regulatory factors estimated by network modeling have unknown functions, by estimating the system based on numerical data, namely, gene expression data, it is now possible to discover new regulatory factors that have not been described previously.

Ex. 2 Improvement of oil accumulation in *Lipomyces starkeyi*

There are 17 goals in the SDGs, and the palm oil substitute is required as an issue related to the 12th goal, "Responsible consumption, production" and the 15th goal, "Life on land." Palm oil is widely used not only in food products such as fried oil for cup noodles, margarine, chocolates, and cookies, but also as one of the raw materials for detergents and cosmetics. However, oil palm plantations have long been considered to have environmental impacts such as deforestation and loss of biodiversity due to the conversion of thousands to tens of thousands of hectares of land for monoculture cultivation for efficient production. In Indonesia and Malaysia, which currently account for 85% of the world's palm oil production, oil palm plantation development is the most significant cause of deforestation. In addition, a large amount of carbon is fixed in peatlands where plants cannot decompose because they are always submerged in water. The development of oil palm plantations will cause the carbon fixed in peat bogs to be emitted as greenhouse gases. It is estimated that the amount of carbon dioxide equivalent greenhouse gas emissions per ton of palm oil produced is more than 2.4 tons of coal. One solution to this problem is the use of microorganisms that biosynthesize oils with a composition ratio similar to that of palm oil. Among the oleaginous microorganisms that produce oil intracellularly, *Lipomyces starkeyi* has an excellent oil production capacity (Angerbauer et al., 2008) and the composition of the lipids it produces is similar to palm oil (Béligon et al., 2016), so *L. starkeyi* is expected to have industrial applications. Fig. 3.6 shows how *L. starkeyi* accumulates oils in its cells.

To realize the industrial use of *L. starkeyi*, a Smart Cell with further improved oil productivity must be created. Therefore, we applied network

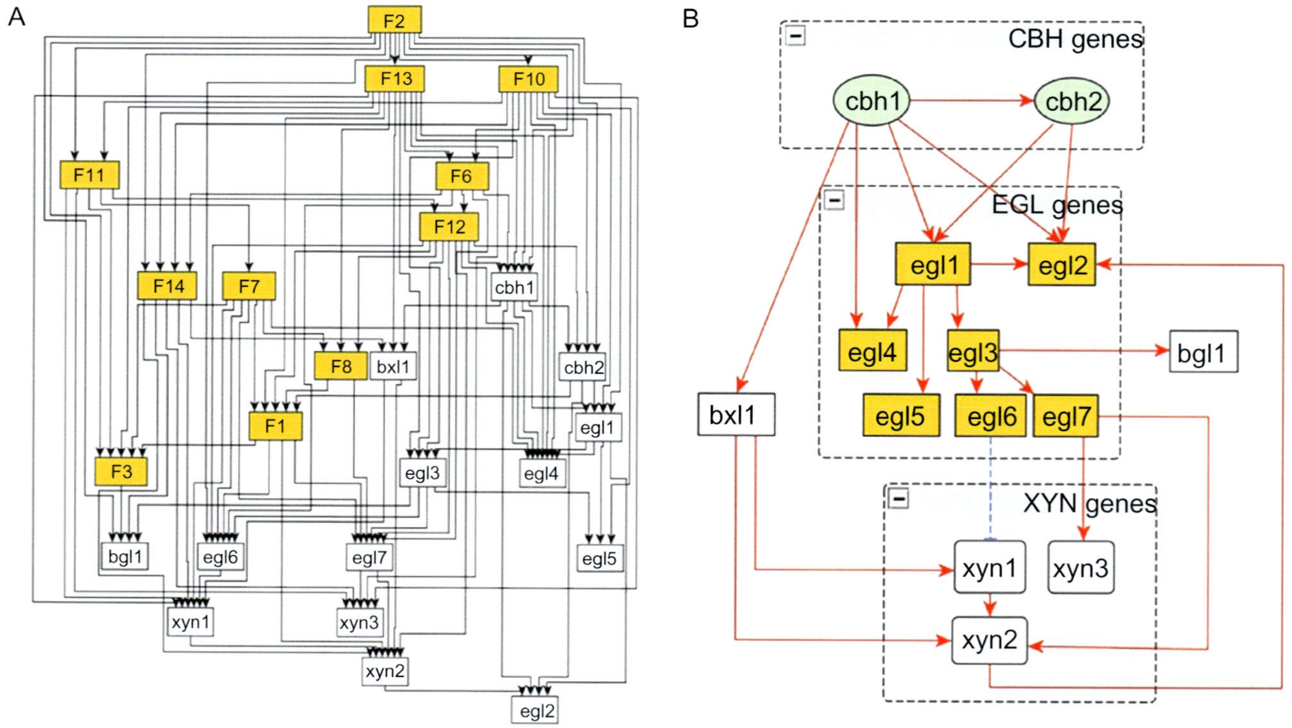

FIG. 3.5 The inferred network for controlling constitutive cellulase balance. The *rectangles* marked with the letter "F" and a number indicate regulator genes for cellulase genes, *white rectangles* indicate cellulase genes such as CBH family genes, EGL family genes, BGL family genes, and XYN family genes. (A) The whole network between all genes. *Arrows* indicate regulations between genes. For the group of regulatory genes indicated by the *black rectangles*, specific names are not provided because they are related to patent applications. (B) The extracted model among cellulase and hemicellulase genes. *Solid arrows* indicate positive regulations, and the *dashed line* indicates negative regulation. The estimated graph structures were well-fitted with the measured data with high goodness-of-fit scores.

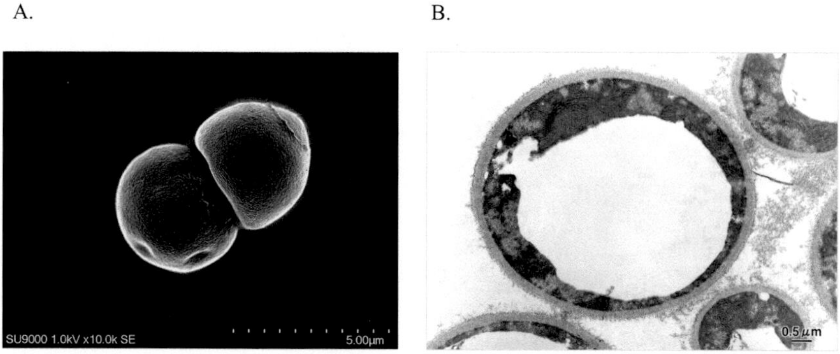

FIG. 3.6 The oil accumulation in *Lipomyces starkeyi*. These are electron micrographs of *L. starkeyi* accumulating lipids in its cells. (A) The shape of *L. starkeyi*. This shape is rounder and more swollen than normal cells. (B) Inside a cell of *L. starkeyi*. The white area inside the cell is accumulated lipids.

modeling to find candidate genes that can be modified to significantly affect oil production. The estimated network model is shown in Fig. 3.7.

Although we cannot mention the names of specific genes due to pending patent applications, we have successfully found several genes that contribute to oil production from about 200 expression profiles. We are currently constructing a genetically modified industrial host and optimizing the cultivation conditions and oil purification process for a practical production process in the ongoing project. Some of these genes had unknown functions, and others had functions that were difficult to predict even from sequence information. The fact that we did not include known biological knowledge, and instead estimated the regulatory relationships completely from numerical data alone, led to this novel discovery.

Future direction of network modeling for smart cells

In this Smart Cell Project, we improved network modeling technologies for constructing gene expression regulatory network models and have contributed to the creation of Smart Cells for new bioproducts in partnership with our industrial and academic collaborators' robot-based data accumulation, combined with information analysis technology to optimize metabolic pathways, and enzyme optimization using MD simulations. Through this project, we have found that the network modeling is applicable to a variety of host organisms and is versatile enough to be used with any host organism. Although many of these factors are newly discovered and have not yet been published due to patent applications, we were able to discover factor genes that increased measured yields by factors of 2- to 10-fold.

In the past, microbial production of substances involved optimizing metabolic pathways and increasing enzyme activity. However, in the cell, many

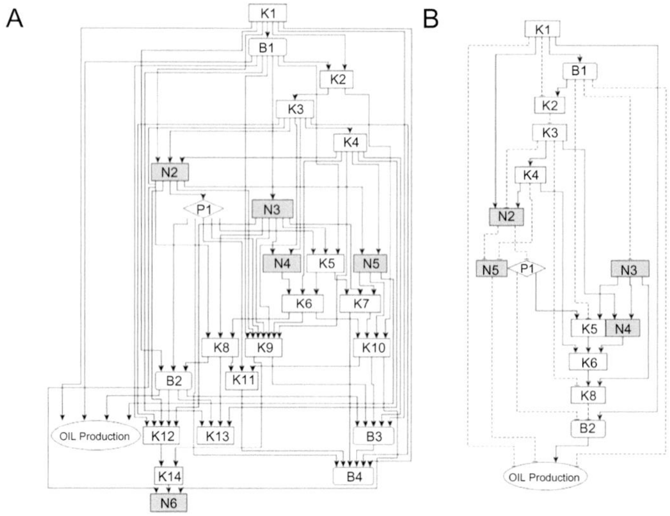

FIG. 3.7 The inferred network for oil production in *L. starkeyi*. *White rectangles* with a capital K indicate enzyme-related genes, *rounded rectangles* with a capital B indicate transcription factor genes, *gray rectangles* with a capital N indicate function unknown genes, and a *circle* indicate the amount of oil production. Gene names are not provided as they are patent pending. (A) The whole network structure of selected genes and oil production. *Arrows* indicate the regulation between variables. (B) Extracted subgraph related to oil production. *Solid lines* indicate positive regulations and *dashed lines* indicate negative regulations. Oil synthesis was found to be regulated by genes with unknown functions.

factors are intricately intertwined and behave as a system to constitute a single phenomenon. The network modeling is a method of visualizing a specific system that is activated in a cell, based solely on measured numerical data. Therefore, we have often found that genes that were difficult to detect by metabolic engineering approaches were actually contributing to the bioproduction system. Furthermore, by clarifying the causal relationship in a network model, we were able to propose many candidate genes for modification and host improvement.

Until now, network modeling has been regarded as more of a theoretical exercise. In particular, network modeling of gene expression regulation is difficult to confirm in living cells, and thus, establishing the biological plausibility of the inferred network model has been problematic. Even now, the network modeling of phenomena in various model organisms is difficult to confirm experimentally. However, within the scope of microbial production systems, we have found that network modeling of biological systems allows us to estimate candidate genes for modification for host improvement with smaller amounts of data than methods using machine learning.

It is difficult to substitute microbial production for all fossil fuel-derived materials. However, once the Smart Cells are fully realized, some materials can be converted to microbial production. Network modeling is expected to significantly reduce the time required to breed microorganisms for Smart Cell creation and be a useful technology for discovering new ways to improve host microorganisms. We are still working on the practical application of microbial production of useful compounds using Smart Cells in a new project to be applied to real production processes by companies. We hope that in the near future, the materials produced by Smart Cells will be used around the world and thereby contribute to the SDGs.

Acknowledgment

This article is based on results obtained from a project, JPNP16009, subsidized by the New Energy and Industrial Technology Development Organization (NEDO).

References

Aburatani, S., 2011. Application of structure equation modeling for inferring a serial transcriptional regulation in yeast. Gene Regul. Syst. Bio. 5, 75–88.

Aburatani, S., 2012. Network inference of pal-1 lineage-specific regulation in the C. elegans embryo by structural equation modeling. Bioinformation 8 (14), 652–657.

Aburatani S., 2015. Inference of transcriptional network for pluripotency in mouse embryonic stem cells. J. Phys. Conf. Ser.; 574: 012138. https://iopscience.iop.org/article/10.1088/1742-6596/574/1/012138. https://doi.org/10.1088/1742-6596/574/1/012138 [Accessed 28 July 2021].

Aburatani, S., Toh, H., 2013. Estimation of physical transcriptional control in yeast cell cycle by structure equation modeling. In: Iconcept Press (Ed.), Protein Purification and Analysis: Methods and Applications. Createspace Independent Pub, Scotts Valley, CA (Chapter 5).

Aburatani S., Toh H., 2014. Network inference of AP pattern formation system in D. melanogaster by structural equation modeling. J. Phys. Conf. Ser.; 490: 012145. https://iopscience.iop.org/article/10.1088/1742-6596/490/1/012145. https://doi.org/10.1088/1742-6596/490/1/012145 [Accessed 28 July 2021].

Aburatani, S., Kuhara, S., Toh, H., Horimoto, K., 2003. Deduction of a gene regulatory relationship framework from gene expression data by the application of graphical Gaussian modeling. Signal Process. 83 (4), 777–788.

Aburatani, S., Nagano, R., Sone, H., Fujibuchi, W., Yamane, J., Imanishi, S., Ohsako, S., 2013. Inference of gene regulatory networks to detect toxicity-specific effects in human embryonic stem cells. Int. J. Adv. Life Sci. 5 (1&2), 103–114.

Aburatani S., Shida Y., Ogasawara W., Yamazaki H., Takaku H., 2019. Application of structural equation modelling for oil accumulation system control in oleaginous yeast. J. Phys. Conf. Ser.; 1391: 012043. https://iopscience.iop.org/article/10.1088/1742-6596/1391/1/012043/meta. https://doi.org/10.1088/1742-6596/1391/1/012043 [Accessed 28 July 2021].

Akaike, H., 1974. A new look at the statistical model identification. IEEE Trans. Autom. Control 19 (6), 716–723.

Akutsu, T., Miyano, S., Kuhara, S., 1999. Identification of genetic networks from a small number of gene expression patterns under the Boolean network model. Pac. Symp. Biocomput., 17–28.

Angerbauer, C., Siebenhofer, M., Mittelbach, M., Guebitz, G.M., 2008. Conversion of sewage sludge into lipids by Lipomyces starkeyi for biodiesel production. Bioresour. Technol. 99 (8), 3051–3056.

Araki, M., Cox III, R.S., Makiguchi, H., Ogawa, T., Taniguchi, T., Miyaoku, K., Nakatsui, M., Hara, K.Y., Kondo, A., 2015. M-path: a compass for navigating potential metabolic pathways. Bioinformatics 31 (6), 905–911.

Aten J.E., Fuller T.F., Lusis A.J., Horvath S., 2008. Using genetic markers to orient the edges in quantitative trait networks: the NEO software. BMC Syst. Biol. [Accessed 28 July 2021]; 2: 34. https://bmcsystbiol.biomedcentral.com/articles/10.1186/1752-0509-2-34. https://doi.org/10.1186/1752-0509-2-34.

Béligon, V., Christophe, G., Fontanille, P., Larroche, C., 2016. Microbial lipids as potential source to food supplements. Curr. Opin. Food Sci. 7, 35–42.

Bentler, P.M., 1990. Comparative fit indexes in structural models. Psychol. Bull. 107 (2), 238–246.

Bischof R.H., Ramoni J., Seiboth B. 2016. Cellulases and beyond: the first 70 years of the enzyme producer *Trichoderma reesei*. Microb. Cell Factories; 15: 106. https://microbialcellfactories.biomedcentral.com/articles/10.1186/s12934-016-0507-6. https://doi.org/10.1186/s12934-016-0507-6. [Accessed 28 July 2021].

Bollen, K.A., 1989. Structural Equations With Latent Variables. Wiley-Interscience, New York, NY.

Browne, M.W., Cudeck, R., Bollen, K.A., 1993. Alternative ways of assessing model fit. In: Long, J. S. (Ed.), Testing Structural Equation Models. Sage, Newbury Park, CA, pp. 136–162.

Chen, T., He, H.L., Church, G.M., 1999. Modeling gene expression with differential equations. Pac. Symp. Biocomput., 29–40.

Chu, Y., Corey, D.R., 2012. RNA sequencing: platform selection, experimental design, and data interpretation. Nucleic Acid Ther. 22 (4), 271–274.

Cox, J., 2015. From Vines to Wines: The Complete Guide to Growing Grapes and Making Your Own Wine, fifth ed. Storey Publishing, LLC, North Adams, MA.

Duncan, O., 1975. Introduction to Structural Equation Models. Academic Press, New York, NY.

Findley, D.F., 1991. Counterexamples to parsimony and BIC. Ann. Inst. Stat. Math. 43 (3), 505–514.

Friedman, N., Linial, M., Nachman, I., Pe'er, D., 2000. Using Bayesian networks to analyze expression data. J. Comput. Biol. 7, 601–620.

Fugelsang, K.C., Edwards, C.G., 2007. Wine Microbiology, second ed. Springer, Berlin Heidelberg (DE).

Global Bioenergies. n.d. Évry-Courcouronnes (FR). Global Bioenergies. https://www.global-bioenergies.com/. [Accessed 28 July 2021].

Guimaraes, J.C., Rocha, M., Arkin, A.P., Cambray, G., 2014. D-tailor: automated analysis and design of DNA sequences. Bioinformatics 30 (8), 1087–1094.

Haavelmo, T., 1943. The statistical implications of a system of simultaneous equations. Econometrica 11, 1–12.

Hatzimanikatis, V., Li, C., Ionita, J.A., Henry, C.S., Jankowski, M.D., Broadbelt, L.J., 2005. Exploring the diversity of complex metabolic networks. Bioinformatics 21 (8), 1603–1609.

Hirasawa, H., Shioya, K., Mori, K., Tashiro, K., Aburatani, S., Shida, Y., Kuhara, S., Ogasawara, W., 2019. Cellulase productivity of *Trichoderma reesei* mutants developed in Japan varies with varying pH conditions. J. Biosci. Bioeng. 128 (3), 264–273.

Joreskog, K.G., Sorbom, D., 1983. Lisrel VI Analysis of Linear Structural Relationships by the Method of Likelihood. Scientific Software, Lincolnwood, IL.

Lee S., Jhun M., Lee E.K., Park T., 2007. Application of structural equation models to construct genetic networks using differentially expressed genes and single-nucleotide polymorphisms.

BMC Proc. [Accessed 28 July 2021]; 1: S76. https://bmcproc.biomedcentral.com/articles/10.1186/1753-6561-1-s1-s76. https://doi.org/10.1186/1753-6561-1-S1-S76.

Liu, B., de la Fuente, A., Hoeshele, I., 2008. Gene network inference via structural equation modeling in genomics experiments. Genetics 178 (3), 1763–1776.

Liu Y., Li Q., Zheng P., Zhang Z., Liu Y., Sun C., Cao G., Zhou W., Wang X., Zhang D., Zhang T., Sun J., Ma Y., 2015. Developing a high-throughput screening method for threonine overproduction based on an artificial promoter. Microb. Cell Factories; 14: 121. https://microbialcellfactories.biomedcentral.com/articles/10.1186/s12934-015-0311-8. https://doi.org/10.1186/s12934-015-0311-8. [Accessed 28 July 2021].

Lorenzo V., Prather K.L., Chen G.Q., O'Day E., Kameke C., Oyarzún D.A., Hosta-Rigau L., Alsafar H., Cao C., Ji W., Okano H., Roberts R.J., Ronaghi M., Yeung K., Zhang F., Lee S.Y., 2018. The power of synthetic biology for bioproduction, remediation and pollution control: the UN's sustainable development goals will inevitably require the application of molecular biology and biotechnology on a global scale. EMBO Rep.; 19(4): e45658. https://www.embopress.org/doi/full/10.15252/embr.201745658. https://doi.org/10.15252/embr.201745658 [Accessed 28 July 2021].

Mori Y., Noda S., Shirai T., Kondo A., 2021. Direct 1,3-butadiene biosynthesis in *Escherichia coli* via a tailored ferulic acid decarboxylase mutant. Nat. Commun.; 12: 2195. https://www.nature.com/articles/s41467-021-22504-6. https://doi.org/10.1038/s41467-021-22504-6 [Accessed 28 July 2021].

Nedo Smart Cell Project, 2020. Tokyo (JP). Nedo Smart Cell Project. https://www.jba.or.jp/nedo_smartcell/en/. (Accessed 28 July 2021).

Nicoletti, M.C., 2009. Computational Intelligence Techniques for Bioprocess Modelling, Supervision and Control. Springer, Berlin Heidelberg (DE).

Nogawa, M., Goto, M., Okada, H., Morikawa, Y., 2001. L-sorbose induces cellulase gene transcription in the cellulolytic fungus *Trichoderma reesei*. Curr. Genet. 38 (6), 329–334.

Pearl, J., 2001. Causality: Models, Reasoning, and Inference, second ed. Cambridge University Press, Cambridge, MA.

Pollack, J.R., Perou, C.M., Alizadeh, A.A., Eisen, M.B., Pergamenschikov, A., Williams, C.F., Jeffrey, S.S., Botstein, D., Brown, P.O., 1999. Genome-wide analysis of DNA copy-number changes using cDNA microarrays. Nat. Genet. 23 (1), 41–46.

Saito Y., Kitagawa W., Kumagai T., Tajima N., Nishimiya Y., Tamano K., Yasutake Y., Tamura T., Kameda T., 2019. Developing a codon optimization method for improved expression of recombinant proteins in actinobacteria. Sci. Rep.; 9: 8338. https://www.nature.com/articles/s41598-019-44500-z. https://doi.org/10.1038/s41598-019-44500-z [Accessed 28 July 2021].

Sakaguchi, K., Okanishi, M., 1980. Molecular Breeding and Genetics of Applied Microorganisms. Academic Press, Cambridge, MA.

Shieh G.S., Chen C.M., Yu C.Y., Huang J., Wang W.F., Lo Y.C., 2008. Inferring transcriptional compensation interactions in yeast via stepwise structure equation modeling. BMC Bioinf.; 9: 134. https://bmcbioinformatics.biomedcentral.com/articles/10.1186/1471-2105-9-134. https://doi.org/10.1186/1471-2105-9-134. [Accessed 28 July 2021].

Shirai T., Osanai T., Kondo A., 2016. Designing intracellular metabolism for production of target compounds by introducing a heterologous metabolic reaction based on a *Synechosystis* sp. 6803 genome-scale model. Microb. Cell Factories; 15: 13. https://microbialcellfactories.biomedcentral.com/articles/10.1186/s12934-016-0416-8. https://doi.org/10.1186/s12934-016-0416-8. [Accessed 28 July 2021].

Synthetic Genomics, 2021. La Jolla (CA). Advancing Genomics to Better Life. https://syntheticgenomics.com/. (Accessed 28 July 2021).

Toh, H., Horimoto, K., 2002. Inference of a genetic network by a combined approach of cluster analysis and graphical Gaussian modeling. Bioinformatics 18 (2), 287–297.

Vavricka C.J., Yoshida T., Kuriya Y., Takahashi S., Ogawa T., Ono F., Agari K., Kiyota H., Li J., Ishii J., Tsuge K., Minami H., Araki M., Hasunuma T., Kondo A., 2019. Mechanism-based tuning of insect 3,4-dihydroxyphenylacetaldehyde synthase for synthetic bioproduction of benzylisoquinoline alkaloids. Nat. Commun.; 10(1): 2015. https://www.nature.com/articles/s41467-019-09610-2. https://doi.org/10.1038/s41467-019-09610-2 [Accessed 28 July 2021].

Voet, D., Voet, J.G., 2007. Biochemistry, fourth ed. Wiley, Hoboken, NJ.

Wang, Z., Gerstein, M., Snyder, M., 2009. RNA-Seq: a revolutionary tool for transcriptomics. Nat. Rev. Genet. 10 (1), 57–63.

Wheaton, B., Muthen, B., Alwin, D.F., Summers, G., 1977. Assessing reliability and stability in panel models. Sociol. Methodol. 8, 84–136.

Xiong, M., Li, J., Fang, X., 2004. Identification of genetic networks. Genetics 166 (2), 1037–1052.

Yamane, J., Aburatani, S., Imanishi, S., Akanuma, H., Nagano, R., Kato, T., Sone, H., Ohsako, S., Fujibuchi, W., 2016. Prediction of developmental chemical toxicity based on gene networks of human embryonic stem cells. Nucleic Acids Res. 44 (12), 5515–5528.

Zhang, X., Zhang, X.F., Li, H.P., Wang, L.Y., Zhang, C., Xing, X.H., Bao, C.Y., 2014. Atmospheric and room temperature plasma (ARTP) as a new powerful mutagenesis tool. Appl. Microbiol. Biotechnol. 98 (12), 5387–5396.

Zhang G., Lin Y.P., Qi X., Wang L., He P., Wang Q., Ma Y., 2015. Genome shuffling of the nonconventional yeast *Pichia anomala* for improved sugar alcohol production. Microb. Cell Factories; 14: 112. https://microbialcellfactories.biomedcentral.com/articles/10.1186/s12934-015-0303-8. https://doi.org/10.1186/s12934-015-0303-8. [Accessed 28 July 2021].

Zhong, W., Jousset, A., 2017. Plant breeding goes microbial. Trends Plant Sci. 22 (7), 555–558.

Further reading

Mulaik, S.A., Millasap, R.E., 2000. Doing the four-step right. Struct. Equ. Model. 7 (1), 36–73.

Spirtes, P., Glymour, C., Scheines, R., 2001. Causation, Prediction, and Search, second ed. The MIT Press, Cambridge, MA.

Stuart, A., Ord, K., 2010. Kendall's Advanced Theory of Statistics, Distribution Theory, sixth ed. Wiley, Hoboken, NJ.

Chapter 4

Young innovators and the bioeconomy

Xinyi E. Chen[a,b], Samuel King[c,d], Sarah W.S. Ng[a,b], Paarsa Salman[c,d], Janella C. Schwab[e], and Parneet Sekhon[a]
[a]*Department of Microbiology and Immunology, University of British Columbia, Vancouver, BC, Canada,* [b]*Department of Computer Science, University of British Columbia, Vancouver, BC, Canada,* [c]*Department of Botany, University of British Columbia, Vancouver, BC, Canada,* [d]*Department of Zoology, University of British Columbia, Vancouver, BC, Canada,* [e]*Faculty of Land and Food Systems, University of British Columbia, Vancouver, BC, Canada*

On July 20, 1969, millions of children sat with their eyes glued to boxy televisions as the world held its breath. The first footsteps on the surface of the once distant moon suddenly changed the definition of "impossible" forever. The generation of young people who witnessed humankind's most daring adventure to date became the ones who moved mountains through the aeronautical frontiers that built today's world. History tends to repeat itself, and now, the young people who grew up amidst revolutions in the genome sciences will be the same ones to fundamentally change society through feats of genomic and biotechnological innovation.

Biological innovation is happening at a rate faster than ever before. The things that are commonplace today were thought impossible just decades ago. For example, the first draft of the human genome sequence was a colossal global project in 1999, estimated to have cost 300 million USD. Now, a high-quality human genome can be generated for under 1000 USD (Wetterstrand, 2021). Young people are being raised in a time where the natural world is no longer a clouded mystery, but rather an increasingly transparent collection of billions of systems ranging from the nanoscale to the ecological scale. At the very core of these levels of life is the genome. Illuminating genomics is helping us to understand life as a living, programmable language, where it is possible to engineer and build living systems from scratch. It is these kinds of possibilities that are inspiring young people to become changemakers in tomorrow's bioeconomy.

The effects of the current surge in genomic innovation on those entering the early stages of their career, including high school students, undergraduates, and

Genomics and the Global Bioeconomy. http://doi.org/10.1016/B978-0-323-91601-1.00005-5

graduate students, have not been directly measured. However, it can be paralleled against previous technological explosions. In her book *The Age of Living Machines: How Biology Will Build the Next Technology Revolution*, Susan Hockfield (2019) discusses the historical technology phenomenon that took place in the 20th century. She calls it "convergence 1.0," describing the incredible combination of physics and chemistry that produced the atomic bomb, space travel, the Internet, and more. Hockfield argues that "convergence 2.0" is on the horizon, where biology and engineering will revolutionize the world as we know it. New methods in genomics, ranging from base editing, to conservation genotyping, to DNA computing, are painting the backdrop for young innovators as they create their own ideas that trickle back into the bioeconomy.

Students have been the cornerstone of many great discoveries in genomics, playing a seminal role in executing ideas and testing hypotheses. For example, at age 27, Raymond Gosling, a student under Rosalind Franklin (aged 32), took "Photo 51," the first photograph capturing the structure of DNA. Another instance took place in a course named SEA-PHAGES, implemented across 73 institutions, where first-year undergraduate students isolated, sequenced, and genomically characterized more than 3000 completely novel bacteriophages from the environment from 2008 to 2014 (Jordan et al., 2014). These young scientists contributed hundreds of new sequences and annotations to the GenBank database. Albert Einstein once postulated that "a person who has not made his great contribution to science before the age of thirty will never do so." Although this statement is quite contradictory and untrue in many cases (including the Nobel Prize), his message is clear: the potential for brilliance ripens when you are young. This emphasizes the need to support and nurture early career innovators to maximize their chances of success.

With enough accessibility to resources and knowledge, early innovators are poised to impact genomics and the bioeconomy more than ever before. As undergraduate students from the University of British Columbia (UBC), we will explore a variety of case studies that have taken place in British Columbia, Canada. Throughout this chapter, we will see how young people already play key roles in the bioeconomy, where they fuel innovation, regenerate talent, and provide new perspectives.

Changing curricula spark early interest in genomics

As genomic technology advances, the curricula taught to budding scientists must also keep pace. Before 2008, the average high school student in British Columbia was not introduced to the concept of DNA until the 11th grade, instead spending the majority of their science classes on physics, chemistry, and the biology of ecosystems. Now, science students in the 10th grade are already learning the processes and applications of genetic engineering, cloning, gene therapy, and even the ethical considerations involved in using these technologies.

This trend in curricula continues for those who pursue postsecondary education. For example, at UBC, many core life science courses now teach the science behind DNA technologies and its possible applications to the topic of interest. Entirely new graduate programs have sprung up around the field, including genome sciences and technology, biomedical engineering, and bioinformatics. And because synthetic biology and genomic sciences are inherently multidisciplinary studies, we are seeing an increased focus on interdisciplinary learning in both undergraduate and graduate studies.

In addition to enhanced focus on genomics within education programs, opportunities outside of the classroom have evolved as well. Traditionally, the undergraduate experience in the life sciences was predominantly lecture-based with limited exposure to self-directed hands-on research. But now, there are increasing opportunities for students to apply their textbook knowledge to cutting-edge research. This includes organizations that provide research opportunities by connecting students to research mentors, or by helping students secure positions in university laboratories. Other organizations encourage the formation of competitive student-led teams, which fund and conduct independent research with the guidance of a host laboratory.

One such organization, which includes more than 350 student research teams around the world, is the MIT-founded International Genetically Engineered Machine (iGEM) Foundation, which runs an annual synthetic biology research competition. In addition to engaging both undergraduate and graduate researchers to tackle real-world problems with synthetic biology, they also manage an evergrowing registry of genetic parts, which can be assembled into biological systems like wires in a circuit.

We are students from UBC's iGEM team and have selected three past projects by UBC iGEM to highlight the myriad ways in which young innovators can contribute to the field of synthetic biology: Paralyte, a whole-cell biosensor for shellfish toxin; Probeeotics, an engineered metabolic pathway for bee microbiota; and VPRE, a machine learning model to predict the evolution of SARS-CoV-2. For each project, the teams moved through the ideation, proof-of-concept, early prototyping, and market validation stages within 10–12 months.

Paralyte: Mining the metagenome to prevent shellfish poisoning

On the southwestern coast of Canada lies British Columbia, a province known for its natural beauty and traditionally inhabited by Indigenous and rural communities. These communities have gathered resources from the surrounding forests and oceans for generations, with shellfish being one important source of food and income. During algal blooms, explosions of marine microorganism populations produce toxins that can bioaccumulate in shellfish. While these toxins are harmless to shellfish, they can cause paralytic shellfish poisoning

(PSP) in humans (Cusick & Sayler, 2013). PSP may lead to nausea, numbness, paralysis, and, in severe cases, death (CFIA, 2019). According to the Canadian Food Inspection Agency (CFIA), as little as 0.8 parts per million (ppm) of saxitoxin can cause serious harm (CFIA, 2019). Worryingly, the frequency of harmful algal blooms has increased due to climate change (Moore et al., 2008).

There is no visual difference between toxic and nontoxic shellfish, making it challenging to distinguish between them. Upon interview by the UBC iGEM team, first nation community representatives said that samples of their shellfish are sent to the Canadian Food Inspection Agency to be tested. In the days to weeks before test results return, they have no way to guarantee that the shellfish they sell is safe. Traditional methods of toxin detection, which involve testing for discoloration of silver coins and numbness when placed on the lips, are neither reliable nor scalable. Outside of the CFIA, rapid test kits are available but still require expensive equipment and training to use and are inaccessible to remote communities.

Having defined the scope of the problem through conversations with shareholders, the iGEM team set out to create a synthetic biology biosensor that would produce visual signals when exposed to unsafe levels of saxitoxin, in order to provide an easily usable, portable device for rapid testing.

Very quickly, the team bumped up against their first challenge—no known genetic regulatory mechanism is responsive to saxitoxin. The team reasoned that because many species have coevolved with the toxin for millions of years, such a mechanism must exist. And what better place to look for a detection mechanism than in ecosystems where saxitoxin naturally exists?

The first step was to select a screening method. Substrate-induced gene expression (SIGEX) was first developed by Uchiyama and Watanabe (2008) to screen metagenomic DNA from groundwater for benzoate-sensitive operon fragments. The team hypothesized that they could use SIGEX to isolate environmental DNA fragments from saxitoxin-contaminated water and screen for genetic regions regulated by saxitoxin (Fig. 4.1).

SIGEX functions by ligating environmental DNA upstream of a reporter protein. If the ligated fragment contains a genetic region regulated by the compound of interest, the compound will trigger the production of the reporter protein. However, if the regulatory region of the ligated fragment happened to be oriented in an opposite direction to the reporter protein, no reporter signal would be generated. To address this, the team added a reverse reporter unit to the screening vector. Now, a reporter protein would be produced if a saxitoxin-responsive genetic unit existed in the ligated fragment, regardless of fragment orientation.

Then, the team had to find saxitoxin-contaminated water. The Hallam Lab at UBC regularly samples water around the Saanich Inlet, where according to the CFIA, a high concentration of saxitoxin was present in June 2016. In collaboration with the CFIA and the Hallam Lab, the team obtained and extracted bacterial DNA fragments from water samples collected around that time.

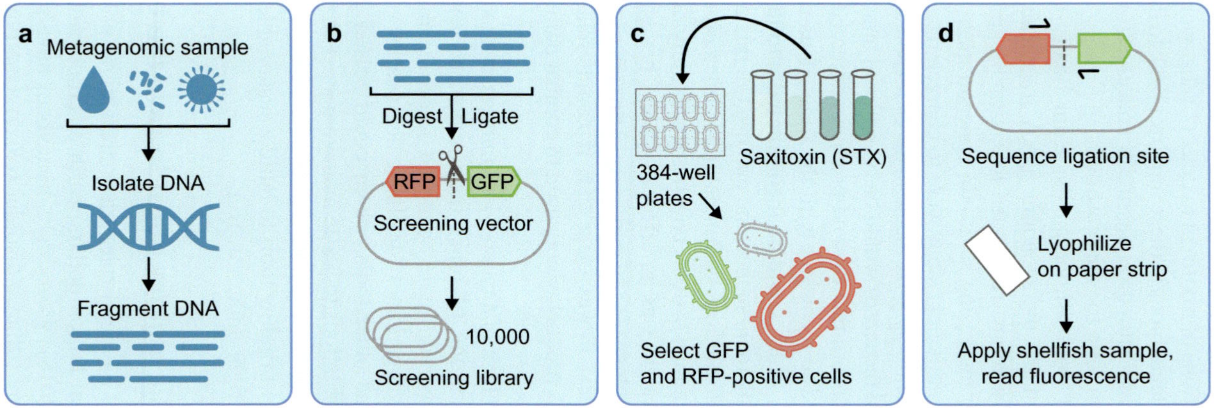

FIG. 4.1 Paralyte project workflow. (A) Bacterial DNA from metagenomic samples collected from algal bloom-affected waters is isolated and fragmented. (B) Each fragment of DNA is ligated into a digest site between a forward reporter (green fluorescent protein, GFP) and a reverse reporter (red fluorescent protein, RFP) in the screening vector. A screening library of 10,000 different RFP-insert-GFP vectors is created to search for inserts containing saxitoxin-inducible regulatory units that may ligate into the vector in the forward or reverse direction. (C) *Escherichia coli* cells separately transformed with screening vectors are treated with increasing concentrations of saxitoxin to screen for a GFP or RFP response. (D) The screening vector with a saxitoxin-inducible regulatory unit is identified by sequencing before being applied to a paper strip as an accessible paralytic shellfish toxin detection device.

With their environmental DNA fragments obtained, the team was ready to harness the power of high-throughput technology with the mentorship and expertise of the Hallam Lab. However, important questions remained. What would be relevant concentrations of saxitoxin to detect? What would be a desirable test response time? The team brought these inquiries back to the Indigenous community and the CFIA and settled on saxitoxin concentrations from 0 to 2 ppm and a 24 hour reaction time for the biosensor.

After multiple weeks of troubleshooting, the team finally constructed a library containing 10,000 unique metagenomic DNA fragments. In the first round of screening, seven fragments responded to saxitoxin, and one showed an almost linear response to increasing saxitoxin concentrations. Encouraged by the positive results, the team's next steps were to characterize and optimize the biosensor to understand and apply its mechanism of response to saxitoxin. Following that, the optimized biosensor would be lyophilized onto paper stripes for convenience of use. To ensure that the end product addressed the needs of the target users, they planned to reconnect with the Indigenous community to learn about desired features, which could be incorporated into the biosensor.

High-throughput screening tools such as the one used in this project enable researchers to find biological solutions within nature—a method of innovation used since the beginning of human history, now made infinitely more powerful by genomic technology. Paralyte was an effort that went beyond research to directly confirm and address the needs of end users. Through discussions with local Indigenous communities, it was clear that if commercially available, Paralyte would be a genomics product widely used to combat paralytic shellfish poisoning.

Paralyte was just a component of a much wider movement in synthetic biology as high-throughput screening becomes more advanced and accessible. Apart from biosensor engineering, more advanced screening methods could be used to mine environmental genomes for plastic degradation enzymes, bioremediation mechanisms, and other tools, which could bolster sustainability efforts and offer alternative solutions to waste management. The results of this and similar projects hint at the massive potential of yet undiscovered genomic elements, and the possibility of finding tools to build a sustainable future in the microbes on our shores and right under our feet.

Probeeotics: A synthetic biology quest to save the bees

In the web of life, bees play a crucial role—they are natural pollinators, and pollinate an estimated several billion dollars worth of crops each year in the United States alone (FDA, 2018). Worryingly, honeybees have suffered from mass deaths, known as colony collapse disorder (CCD). One potential culprit is the use of neonicotinoid pesticides (European Commission, 2013). Neonicotinoids are banned in the European Union, but without these pesticides, farmers are facing economic losses. Using traditional pesticides could be even more harmful since neonicotinoids are comparatively safe for mammals (European Commission, 2013).

An idea sparked within the UBC iGEM team—what if a probiotic could be introduced into the honeybee's digestive tract to neutralize the neonicotinoid's harmful effects? The team named their project Probeeotics and set out to engineer a bee gut microbe, which could degrade neonicotinoids into biologically useful byproducts (Fig. 4.2).

They first attempted to find already existing degradation pathways for neonicotinoids. Imidacloprid (IMI) is a common neonicotinoid pesticide which certain bacteria are able to transform (Elbert et al., 1991; Hu et al., 2013). Although the mechanism was unknown, the team collaborated with a research laboratory at UBC to screen for bacteria with the capability to use IMI as a sole nutrient source. This approach came to a dead end when they were unable to find fast-growing bacteria that fit the criteria in their relatively small genetic library.

This was an opportunity for innovation: to formulate a novel degradation pathway. Previous research shows that certain proteins called cytochromes can transform IMI into 6-chloronicotinic acid (6-CNA) (Joußen et al., 2008; Schulz-Jander & Casida, 2002; Karunker et al., 2009). While 6-CNA is significantly less toxic, it is still a persistent environmental contaminant (Nauen et al., 2001; Rouchaud et al., 1996). Encouragingly, the degradation of 6-CNA into harmless 6-hydroxynicotinic acid (6-HNA) has been found in bacteria in soil (Shettigar et al., 2012). More excitingly, 6-HNA is degraded by the nicotinate degradation pathway (nic pathway) in *Pseudomonas* bacteria into compounds that can be used in central metabolism (Tang et al., 2012). Motivated by these findings, the team planned to construct a metabolic pathway to convert IMI to 6-CNA and then to 6-HNA to be finally rendered harmless by the nic pathway.

As a proof of concept, the team used a divide-and-conquer approach to show that each step of the proposed degradation pathway is achievable in the model organism *Escherichia coli*. First, to test IMI degradation to 6-CNA, three cytochrome proteins from insects and humans were chosen (Joußen et al., 2008; Schulz-Jander & Casida 2002; Karunker et al., 2009; Pritchard et al., 2006; Nikou et al., 2003). With the help of graduate student mentors, they created a plan for producing cytochrome proteins in *E. coli* and targeting them to the correct compartment within the bacteria. However, it was unclear whether human or insect proteins would be functional in bacteria, and preliminary testing was inconclusive on whether this section of the pathway was successfully implemented.

To convert 6-CNA to 6-HNA in the next step, the team used the cch2 enzyme from the soil bacteria *Bradyrhizobiaceae* (Shettigar et al., 2012). They developed and optimized a strain of *E. coli*, which could produce the cch2 enzyme, and demonstrated the strain's ability to completely consume 6-CNA within 60 min, with a transformation rate of 43.3 + 11.2 μM/min. Lastly, they incorporated all six proteins involved in the nic pathway into their engineered strain of *E. coli*. This pathway would allow the bacteria to convert 6-HNA into compounds that feed into the cells' main metabolic pathway and thus be consumed.

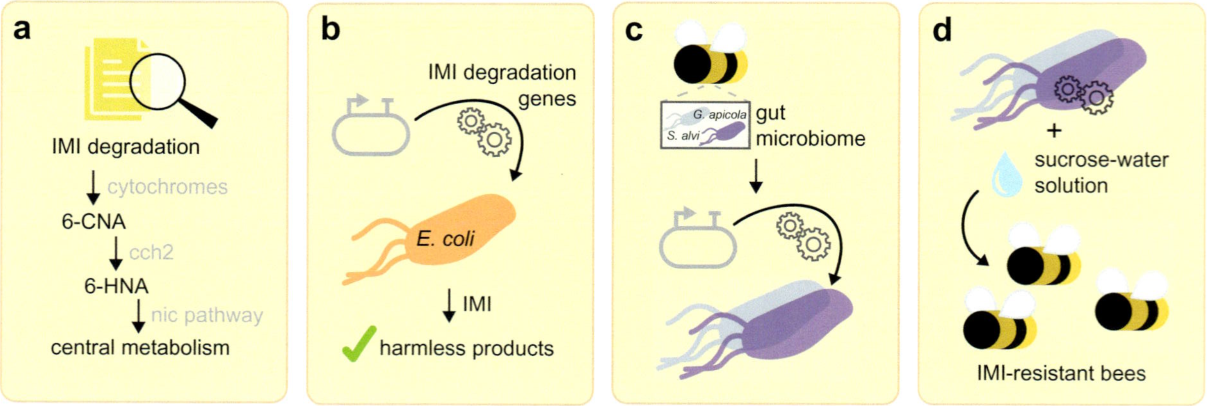

FIG. 4.2 Probeeotics project workflow. (A) A novel degradation pathway for IMI is researched and designed. (B) A series of IMI degradation genes chosen based on the pathway are transformed into *E. coli*, allowing it to degrade IMI through the nic pathway into harmless byproducts. (C) The same genes are transformed into two bacteria native to bees, *Snodgrassella alvi* and *Gilliamella apicola*, which are subsequently (D) Fed to bees in sugar water. The engineered bacteria repopulate the bee's gut microbiome, rendering them resistant to IMI.

Every step in the IMI degradation pathway was now implemented. Excited and nervous, the team decided it was time to test Probeeotics in real honeybees. They sought guidance from Dr. Leonard Foster at UBC, who studies the honeybee gut microbiome. Suited up in protective clothing, members retrieved honeybees from beehives and brought them into the laboratory. The bees were then fed IMI and 6-CNA with or without Probeeotics, and it seemed that bees fed with Probeeotics were more resistant to IMI—an amazing result. Seeing their hypothesis confirmed in living organisms, the students were filled with excitement.

However, it was not enough to show that the degradation pathway worked in *E. coli*—Probeeotics should ideally be implemented in bacteria native to the bee gut microbiome. *Gilliamella apicola* and *Snodgrassella alvi* are the most abundant microbes in bees' guts (Kwong et al., 2014). But at the time, there was limited research on culturing and genetically modifying those organisms. Figuring out how to transfer the designed pathway into bees' natural gut microbiome was one of the most challenging aspects of the project. To tackle this knowledge gap, the team collected all of the approaches used to culture bacterial systems and applied them to *G. apicola* and *S. alvi* to characterize their effects on growth rate and genetic transformation efficiency. Future work would include engineering the IMI degradation pathway into bee-specific microbiota and implementing the product into commercial beekeeping practices.

With 10 months of hard work, collaboration, and mentorship from multiple experts, the building blocks of Probeeotics had been shown to be possible. Colony collapse disorder has been a hotly discussed topic, and most proposed solutions at the time involved treating the bees with fungicides or other medicines, or avoiding the use of neonicotinoids. This project highlights the unique, elegant solutions that synthetic biology could offer to global issues of sustainability, and the progress that can be made by creative young researchers in less than a year when more senior scientists are open to collaboration and generous with their mentorship.

The Viral Predictor for mRNA Evolution (VPRE): Foreseeing genetic evolution with machine learning

As genome sequencing and characterization become more high-throughput, bioinformatics methods must also keep pace. The new wealth of genomic data is an incredibly powerful resource, but only if computational methods can handle the sheer amount of data. This necessitates a growing intersection between biology and the computational sciences, enabling us to model the systems of life and evolution like never before.

When the SARS-CoV-2 virus halted society in a global pandemic, research unrelated to COVID-19 slowed and came to a standstill. With limited access to the laboratory, the UBC iGEM team was unable to carry out their annual project. The mRNA vaccines had just begun to be developed, and nobody was sure when normalcy would return. Many team members studied virology and immunology as their specialty and felt frustrated by their inability to contribute to the research efforts. But as the pandemic continued, advanced genomic technologies allowed a global explosion of SARS-CoV-2 genome sequencing to flood public repositories with genomic data. The influx of data opened the door to critical research pathways, including evolutionary tracing, protein characterization, and genomic surveillance. The glimmerings of an idea began to form—there were many members of the team who were skilled in computer science and engineering. So as the world's scientists raced to complete the first vaccines, the UBC iGEM team began a computational project to ensure that future vaccines would remain effective for as long as possible.

A critical barrier to long-term vaccine efficacy is that the targeted pathogenic protein can mutate and thus lower vaccine-induced immunity against new strains. Although vaccine developers attempt to target conserved viral protein regions to reduce this risk, mutations still sometimes occur. Single-stranded RNA viruses, such as SARS-CoV-2, have one of the highest viral rates of mutation (Duffy et al., 2008). As the virus becomes widespread, more viral particles exist to undergo mutations, which can give rise to new strains. Additionally, areas with lower vaccination rates promote the emergence of vaccine-resistant strains (Rella et al., 2021). Accordingly, SARS-CoV-2 variants have caused global concern over the continuing efficacy of available vaccines.

To address this problem, efforts have been made to computationally model the evolution of pathogens by simulating mechanisms of evolution such as genetic drift. However, these approaches impose theorized, conceptual rules of evolution onto the data and may not accurately capture the influence of environmental factors (Lee et al., 2016).

One solution is to use deep learning—a machine learning technique that uses neural networks to extrapolate patterns from available data. Early in the pandemic, researchers tried modeling SARS-CoV-2 evolution with deep learning, using discrete viral DNA sequences as input (Crossman, 2020). This requires tremendous computational power—just one protein sequence is thousands of base pairs long, with hundreds of new SARS-CoV-2 genomes being collected each day. The iGEM team realized that if genomic sequences could be accurately encoded into a small set of numbers, this would at once dramatically reduce the required computation for analysis and enable the use of mathematical analyses, which would normally be impossible with discrete variables. The team explored this route with the hypothesis that evolutionary trends invisible at the discrete, text level might become apparent at the continuous, numerical level (Fig. 4.3) (King et al., 2021).

One neural network is especially suited for the purposes of encoding: the variational autoencoder (VAE), an unsupervised deep learning method that can glean patterns from data with minimal human bias. A VAE can be trained to compress any single protein sequence into just several representative decimal numbers, which can then be analyzed.

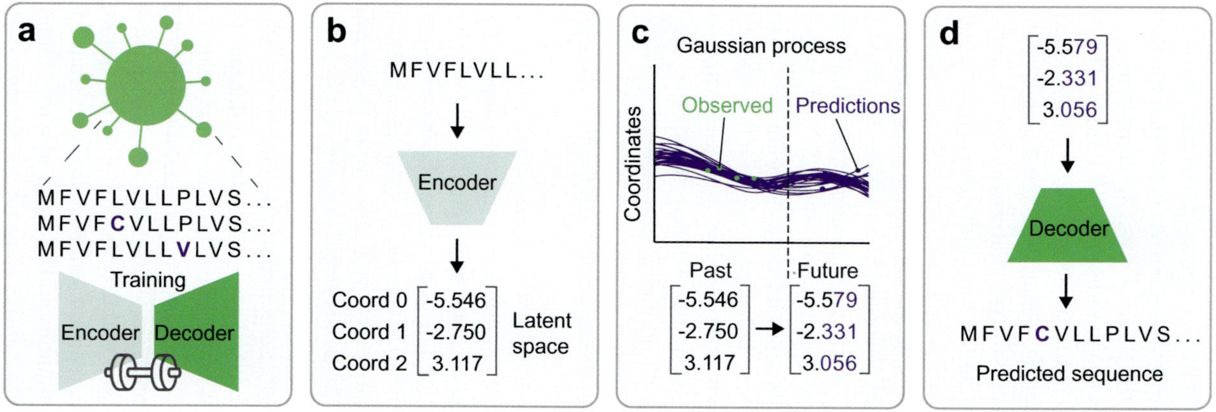

FIG. 4.3 VPRE project workflow. (A) The VAE is trained by repeatedly (1) encoding viral protein sequences into continuous numerical values and (2) decoding the numerical values back into protein sequences, until the neural network can reliably encode and decode sequences with minimal loss of information. (B) A protein dataset of interest is encoded into a three-dimensional, continuous latent space, where every protein sequence is individually represented by three numerical coordinates. (C) The observed coordinates enter separate Gaussian processes, where the best-fitting functions to the coordinates, along with their dates of appearance in the dataset, are modeled and projected into the future. (D) The predicted coordinates are decoded by the VAE back into protein sequences, which can be analyzed further for biological viability or compared with existing viral variants.

At the start of the iGEM team's project, Moderna was leading the race for a vaccine, with runners-up close behind with all-new mRNA vaccines that target the SARS-CoV-2 spike protein, the protein that allows the coronavirus to enter human cells. It was crucial that the first round of vaccines remained effective for as long as possible against the mutating virus. Therefore, the team honed in on the evolutionary patterns of the SARS-CoV-2 spike protein.

At the time, genomic data on the virus was still limited. They gathered 9534 spike protein sequences from online databases that were uploaded between January 25 and June 30, 2020, and simulated 20,000 more spike proteins to form a training dataset based on variants from real-world data. Using this, the VAE learned to accurately encode and decode individual spike sequences into just three numbers.

To forecast novel spike sequences, the team used the trained VAE to encode 104 sequences collected from Australia before May 30, 2020. Australia was chosen with the assumption that its relative geographic isolation would allow the simulation of a simpler phylogenetic pathway for SARS-CoV-2 spike proteins with less external influence on the emergence of novel variants and the fact that it had more spike sequences collected at the time than other island nations.

They passed the encoded coordinates and their associated collection dates through a Gaussian process (GP), a probability model that can perform time series regression. The GP model fits functions to the input dataset with associated time points and projects these functions into the future. Using the GP, the team projected 1000 possible evolutionary trajectories 1, 2, and 5 months into the future.

Because the GP predictions are continuous coordinates that are decoded into discrete amino acid sequences, multiple coordinate triplets can represent the same amino acid sequence. As a result, the frequency with which each sequence is predicted by the GP can be used to estimate the likelihood of their occurrence.

The VAE decoded the GP's 1000 predicted coordinates into 17 amino acid sequences. The two topmost frequent predictions were existing spike proteins, and the other 15 were novel. In the predicted sequences, there were nine amino acid variations (five substitutions and four deletions), which were not seen in the training data, suggesting that the model can generate novel variants.

Comparing the top three predictions to a validation dataset of 81 sequences collected in Australia within 1 month after the training data cutoff date (May 30, 2020), they found that the top prediction was identical to 70% of the validation sequences. After a similar analysis with another validation dataset of 8407 sequences collected in Australia within 5 months after the training data cutoff date, they found that the top predicted sequence differed from 85% of the validation sequences by only one amino acid (N477S). For the second and third top predictions, the most similar sequences from the validation cohort differed by two amino acids.

The GP's top prediction was only off by a single amino acid when predicting the most dominant spike protein 5 months into the future, despite being trained on a small dataset. This shows incredible promise for future iterations of similar models—if these results were possible with such limited data and a relatively unrefined model, imagine what might be possible with the vast amounts of genomic data that is now available. We would have been able to predict the emergence of the more virulent strains and adjusted our vaccines ahead of time, and prevented widespread infection and the need for subsequent boosters of the vaccine.

With the increasing risk of pandemic as a result of the changing climate and human interference with wildlife, it has become obvious that genomic analysis and predictive vaccinology are crucial parts of pandemic preparedness. It is a field poised to save millions of lives in the future and save global economies billions of dollars in health care. With further development, it is possible that a similar model could inform predictive vaccine development to ensure that future vaccines remain maximally effective for as long as possible.

The positive feedback loop of expertise in the field of synthetic biology

From the various research avenues explored earlier in this chapter, we can see that synthetic biology has profound potential. This potential is positively reinforced by young innovators such as undergraduate students, who eventually transform into valuable researchers and industry experts. Such transformation enables an exchange of knowledge, methods, and resources—all of which encourage new research initiatives to grow from existing techniques. Through the lens of iGEM, this section explores tangible ways in which the field of synthetic biology has developed, grown, and matured with the students who contribute to this discipline.

iGEM projects are typically completed over the course of a year, with students driving every aspect from project design to presentation at the annual competition. However, some projects go on to form the foundations for theses, graduate work, and the creation of 150 successful start-ups.

One of these start-ups, Ginkgo Bioworks, is now an innovative biotechnology company that uses genetic engineering to make advancements in enzyme discovery, strain improvement, therapeutics, and biosecurity. Their journey began as an iGEM team representing MIT in 2004, in which they engineered *E. coli* to make the scent of bananas using endogenous metabolites. Today, the company is worth more than $719 million in funding and provides an excellent example of the potential for synthetic biology to succeed within the business market. Similarly, the University of Washington iGEM team of 2011 designed a recombinant candidate enzyme therapeutic for celiac disease. Their project eventually developed into a product and received significant funding, which not only bolstered the company but also inspired other iGEM teams to pursue similar projects with the intention of entering the commercial market.

In addition to inspiring younger researchers, iGEM team alumni also make concrete contributions to training the new teams of budding synthetic biologists. It is not uncommon to see graduated members returning as advisors and mentors—passing on years of insight and expertise. Evan Gibbard, a former UBC iGEM researcher who is now a team advisor, describes how his experiences with UBC iGEM have given him the ability to understand and guide the enthusiasm of young researchers into ambitious and tangible projects. As advice for future researchers, he reminds us that iGEM is a great opportunity to take risks and think outside the box when it comes to research. Student-directed projects offer a large amount of flexibility to pursue areas of interest and collaborate between different disciplines. iGEM teams also typically consist of multiple subteam components including wet laboratory, software modeling, and public outreach. Mr. Gibbard suggests that students explore these different fields to experience new opportunities, make connections, and dip their fingers into potential future careers. Students should also remember to celebrate their achievements and enjoy learning through the process. He and other alumni who return to iGEM or serve elsewhere as mentors create a positive feedback loop of knowledge, accumulated in the young innovators of the next generation (Fig. 4.4).

In completing their projects, iGEM teams contribute to a collection of valuable, accessible resources. For example, the Synthetic Biology Open Language (SBOL) is a language for the description and exchange of standard biological part designs (McLaughlin et al., 2020). This enables researchers to access an entire repository of open-source software tools while simultaneously creating community-driven data exchange standards.

Similarly, the BioBricks Foundation (BBF) is a public-benefit organization that aims to promote the standardization of BioBricks parts. BioBricks are essentially the "lego blocks" that enable easy assembly of synthetic biological circuits. They act as DNA standards that assemble a specific restriction-enzyme assembly standard (Shetty et al., 2008). Promoters, enhancers, terminators, and ribosomal binding sites are all examples of BioBricks. Another similar but important resource is the Registry of Standard Biological Parts, which is a continuously growing collection of genetic parts that can be mixed and matched to build synthetic biology systems. Similar to the BioBricks Foundation, it also provides a resource of available genetic parts to teams and academic laboratories. Resources such as these help to standardize the characterization of genetic parts and enable quick progress for future researchers who can build off of previous work.

It is no secret that the knowledge, tools, and research imparted by each generation of innovators set up incoming students and synthetic biologists with a strong foundation. By learning from those before us and playing a role in the development of the younger generation, we can shape the field of synthetic biology just as the work of previous researchers simultaneously shapes us.

Supporting the next generation of innovators

As the frontiers of biology are pushed forward with each passing day, potential successors in the form of young scientists, entrepreneurs, and policymakers have both the drive and ability to fill the growing space of opportunity that is created. Institutional changes in curriculum, when coupled with student-centered research opportunities, kindle both a passion and a sense of ownership that empowers students to pursue significant advancements in biological research. Given the right tools, these efforts can be translated to substantial contributions to the growing bioeconomy.

Equal access to knowledge is increasing along with this momentum in biological research, through advances in open documentation and initiatives for access to scientific knowledge. Despite this, firsthand expertise is unparalleled between young and senior scientists, which can only be conferred through direct guidance. Additionally, considering that the current momentum in biotechnology also includes a more collaborative nature, which has allowed for initiatives like the Mouse Genome Encyclopedia Project and the Human Genome Project, the common factor between them is the need for a supportive mentorship system.

Scientists with more senior roles are in an optimal position to provide guidance and endow potential to younger scientists and, as such, create future professional partners. For example, UBC's iGEM team was founded unwaveringly by a group of undergraduate students supported by a faculty brimming with excitement at the cusp of synthetic biology research. Seeing undergraduates as fully capable equals, they bestowed the responsibility of building a hub for student-led research that could explore new ideas like biological logic gates and bacterial reporter systems, which were in their infancy at the time. Support was both direct and indirect; Eric Ma, cofounder of the team, described their experience as frictionless only due to the sponsorship of molecular biology professor Dr. Alice Mui, and guidance of Dr. Joanne Fox, Principal of UBC's Vantage College. This allowed the team to go on to obtain multiple Gold Medals in the competition and win Best Model out of hundreds of international teams. Whether it be through indirect support such as sponsorship or referral to another contact, or more direct support such as taking a student in as a trainee, or offering laboratory space to a student team, every extended hand leaves a lasting impression on a young scientist's success (Fig. 4.4).

In Réné Magritte's 1936 painting *La clairvoyance*, he illustrates himself using an unhatched egg as a reference to paint a bird opening its wings. The artwork makes a compelling point of visualizing the future yet to happen, or perhaps a more subtle reference to Magritte envisioning a beautiful possibility of what might be. Similarly, young creators are envisioning a world of possibility—a world that will accelerate the propulsion of biology into the global economy, to tackle issues both known and yet to be identified. The current generations of thinkers in high school, undergraduate, graduate, and early career

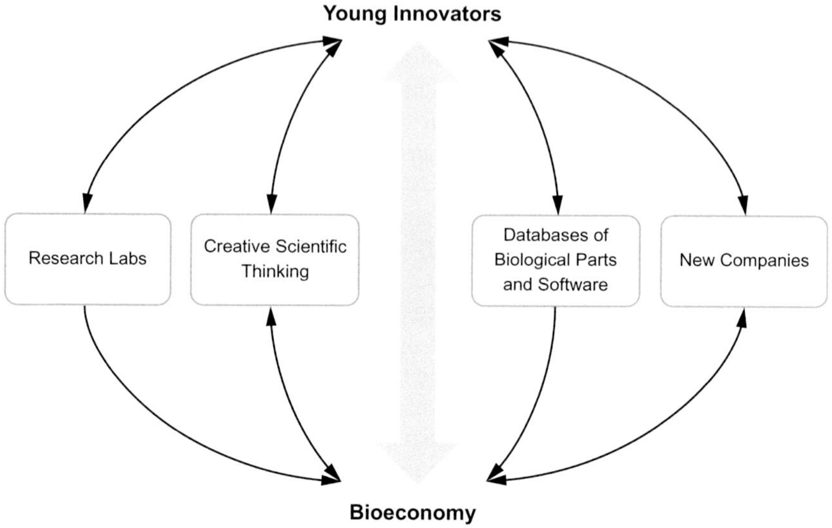

FIG. 4.4 The positive feedback loop between young innovators and the scientific community.

levels have their brushes on the canvas as they illustrate creative innovation in ways previously unimaginable. Magritte's *La clairvoyance* depicts the foresight that is needed to see what the incoming generation of pioneers, entwined with the bioeconomy, is set to become.

References

CFIA, 2019. Marine Biotoxins in Bivalve Shellfish: Paralytic Shellfish Poisoning, Amnesic Shellfish Poisoning and Diarrhetic Shellfish Poisoning. Retrieved 28 July 2021, from http://www.bccdc.ca/health-info/diseases-conditions/paralytic-shellfish-poisoning.

Crossman, L.C., 2020. Leveraging deep learning to simulate coronavirus spike proteins has the potential to predict future zoonotic sequences. bioRxiv.

Cusick, K.D., Sayler, G.S., 2013. An overview on the marine neurotoxin, saxitoxin: genetics, molecular targets, methods of detection and ecological functions. Mar. Drugs 11 (4), 991–1018.

Duffy, S., Shackelton, L.A., Holmes, E.C., 2008. Rates of evolutionary change in viruses: patterns and determinants. Nat. Rev. Genet. 9 (4), 267–276.

Elbert, A., Becker, B., Hartwig, J., Erdelen, C., 1991. Imidacloprid—a new systemic insecticide. Pflanzenschutz-Nachr. Bayer 44, 113–135.

European Commission, 2013 July 16. Press Release: European Commission. Bee Health: EU Takes Additional Measures on Pesticides to Better Protect Europe's Bees. (Accessed 18 September 2021). https://ec.europa.eu/commission/presscorner/detail/en/IP_13_708.

Hockfield, S., 2019. The Age of Living Machines: How Biology Will Build the Next Technology Revolution. WW Norton & Company.

Hu, G., Zhao, Y., Liu, B., Song, F., You, M., 2013. Isolation of an indigenous imidacloprid-degrading bacterium and imidacloprid bioremediation under simulated in situ and ex situ conditions. J. Microbiol. Biotechnol. 23 (11), 1617–1626.

Jordan, T.C., Burnett, S.H., Carson, S., Caruso, S.M., Clase, K., Dejong, R.J., Hatfull, G.F., 2014. A broadly implementable research course in phage discovery and genomics for first-year undergraduate students. MBio 5 (1).

Joußen, N., Heckel, D.G., Haas, M., Schuphan, I., Schmidt, B., 2008. Metabolism of imidacloprid and DDT by P450 CYP6G1 expressed in cell cultures of Nicotiana tabacum suggests detoxification of these insecticides in Cyp6g1-overexpressing strains of Drosophila melanogaster, leading to resistance. Pest Manag. Sci. 64, 65–73.

Karunker, I., Morou, E., Nikou, D., Nauen, R., Sertchook, R., Stevenson, B.J., Vontas, J., 2009. Structural model and functional characterization of the Bemisia tabaci CYP6CM1vQ, a cytochrome P450 associated with high levels of imidacloprid resistance. Insect Biochem. Mol. Biol. 39 (10), 697–706.

King, S., Chen, X.E., Ng, S.W., Rostin, K., Roberts, T., Hahn, S.V., Schwab, J.C., Sekhon, P., Kagieva, M., Reilly, T., Qi, R.C., Salman, P., Hong, R.J., Ma, E.J., Hallam, S.J., 2021. Modeling the trajectory of SARS-CoV-2 spike protein evolution in continuous latent space using a neural network and Gaussian process. bioRxiv.

Kwong, W., Engel, P., Koch, H., Moran, N., 2014. Genomics and host specialization of honey bee and bumble bee gut symbionts. Proc. Natl. Acad. Sci. 111, 11509–11514.

Lee, E.K., Nakaya, H.I., Yuan, F., Querec, T.D., Burel, G., Pietz, F.H., Pulendran, B., 2016. Machine learning for predicting vaccine immunogenicity. Interfaces 46 (5), 368–390.

McLaughlin, J., Beal, J., Mısırlı, G., Grünberg, R., et al., 2020. The Synthetic Biology Open Language (SBOL) Version 3: simplified data exchange for bioengineering. Front. Bioeng. Biotechnol. 8 (1009). https://doi.org/10.3389/fbioe.2020.01009. In this issue.

Moore, S., Trainer, V., Mantua, N., Parker, M., Laws, E., Backer, L., Fleming, L., 2008. Impacts of climate variability and future climate change on harmful algal blooms and human health. Environ. Health 7 (Suppl. 2), S4.

Nauen, R., Ebbinghaus-Kintscher, U., Schmuck, R., 2001. Toxicity and nicotinic acetylcholine receptor interaction of imidacloprid and its metabolites in Apis mellifera (Hymenoptera: Apidae). Pest Manag. Sci. 57 (7).

Nikou, D., Ranson, H., Hemingway, J., 2003. An adult-specific CYP6 P450 gene is overexpressed in a pyrethroid-resistant strain of the malaria vector, Anopheles gambiae. Gene 318, 91–102.

Pritchard, M.P., McLaughlin, L., Friedberg, T., 2006. Establishment of functional human cytochrome P450 monooxygenase systems in Escherichia Coli. In: Cytochrome P450 Protocols. Humana Press, Totowa, NJ, pp. 19–29.

Rella, S.A., Kulikova, Y.A., Dermitzakis, E.T., Kondrashov, F.A., 2021. Rates of SARS-CoV-2 transmission and vaccination impact the fate of vaccine-resistant strains. Sci. Rep. 11 (1), 1–10.

Rouchaud, J., Gustin, F., Wauters, A., 1996. Imidacloprid insecticide soil metabolism in sugar beet field crops. Bull. Environ. Contam. Toxicol. 56 (1), 29–36.

Schulz-Jander, D.A., Casida, J.E., 2002. Imidacloprid insecticide metabolism: human cytochrome P450 isozymes differ in selectivity for imidazolidine oxidation versus nitroimine reduction. Toxicol. Lett. 132, 65–70.

Shettigar, M., Pearce, S., Pandey, R., Khan, F., Dorrian, S.J., et al., 2012. Cloning of a novel 6-chloronicotinic acid chlorohydrolase from the newly isolated 6-chloronicotinic acid mineralizing bradyrhizobiaceae strain SG-6C. PLoS One 7 (11), e51162.

Shetty, R.P., Endy, D., Knight, T.F., 2008. Engineering BioBrick vectors from BioBrick parts. J. Biol. Eng. 2 (1), 1–12.

Tang, H., Yao, Y., Wang, L., Yu, H., Ren, Y., et al., 2012. Genomic analysis of Pseudomonas putida: genes in a genome island are crucial for nicotine degradation. Sci. Rep. 2 (1), 1–8.

The United States Food and Drug Administration, 2018 July 30. Animal & Veterinary: The United States Food and Drug Administration. Helping Agriculture's Helpful Honey Bees. (Accessed 18 September 2021). https://www.fda.gov/animal-veterinary/animal-health-literacy/helping-agricultures-helpful-honey-bees.

Uchiyama, T., Watanabe, K., 2008. Substrate-induced gene expression (SIGEX) screening of metagenome libraries. Nat. Protoc. 3 (7), 1202–1212.

Wetterstrand, K.A., 2021 Nov 1. The Cost of Sequencing a Human Genome. National Human Genome Research Institute. https://www.genome.gov/about-genomics/fact-sheets/Sequencing-Human-Genome-cost. (Accessed 20 November 2021).

Part II

Genomic monitoring is revolutionizing our understanding of biodiversity and ecological services

Chapter 5

Environmental DNA: Revolutionizing ecological assessments with genomics

Neha Acharya-Patel, Michael J. Allison, and Caren C. Helbing
Department of Biochemistry and Microbiology, University of Victoria, Victoria, BC, Canada

Overview

The diverse methods of environmental DNA (eDNA) hold great promise for informing environmental assessments, natural resource operations, and industrial practices and span fisheries, agriculture, forestry, and mining among other economic sectors. Having already demonstrated suitability for ecological assessment and monitoring as a complementary approach to traditional methods, several facets of eDNA are being refined to address a larger range of management and decision-making applications.

Introduction

Human impact intensity on the environment is increasing. Monitoring the effects of these impacts on species and whole ecosystems is vital to mitigate damages in the future. Environmental DNA (eDNA) shows potential as a powerful monitoring tool. Referring to genetic material shed by organisms into their environment, eDNA can be isolated from either micro- or macroorganisms (Fig. 5.1). For years, microbiologists have been extracting and reading the DNA of microorganisms such as bacteria and viruses from environmental samples, but this technology has been applied to macroscopic organisms (e.g., animals and plants) for just over a decade (Ficetola et al., 2008) (Fig. 5.1). Through natural processes like shedding skin, excreting waste, or through reproductive fluids, organismal DNA is constantly being released into the environment. Extracting eDNA from easily obtained environmental samples has opened the door to asking myriad biological and ecological questions regarding species presence, diversity, distribution, and abundance.

Genomics and the Global Bioeconomy. https://doi.org/10.1016/B978-0-323-91601-1.00004-3

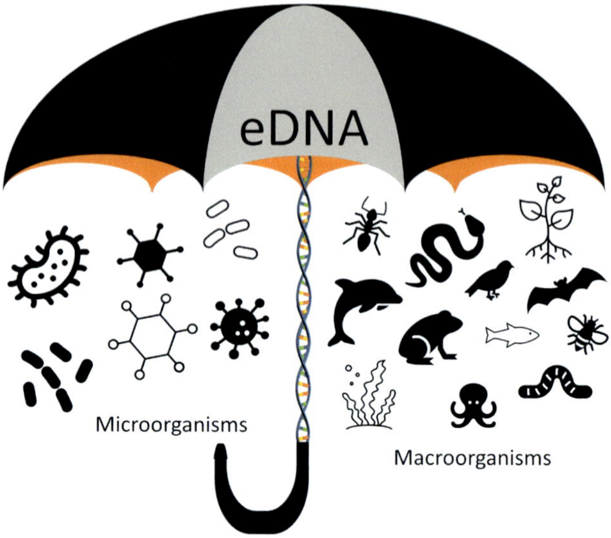

FIG. 5.1 eDNA is an "umbrella" term that includes DNA isolated from microorganisms in environmental samples or DNA shed from macroorganisms that is released into the environment.

As advancements in environmental genomics continue to progress rapidly, eDNA tools are becoming increasingly powerful and applicable. In environments that are difficult to sample, or in finding elusive or cryptic species, the utility of eDNA is apparent. Additionally, eDNA sampling can often be more cost- and resource-effective in comparison with traditional methods, while offering greater species identification power (Bakker et al., 2017; Boussarie et al., 2018). Because of this, eDNA sampling is being integrated into large-scale biomonitoring programs globally. As the sampling protocols, data sharing practices, and results become more standardized, eDNA data will also be useful for informing policy and management decisions.

eDNA approaches

Currently, three broad approaches are applied to detect eDNA and infer species distribution: targeted analysis and two community-based approaches—metabarcoding and metagenomics (Fig. 5.2). A summary of the advantages and disadvantages of each method is presented in Table 5.1. Global applications of these methods are growing and addressing a broad range of biological questions. The present chapter provides an overview of the current state of the field, its opportunities and challenges, and projected outlook.

Targeted eDNA analysis

Targeted studies focus on questions revolving around certain target species. These types of studies use species-specific primers and probes that have been

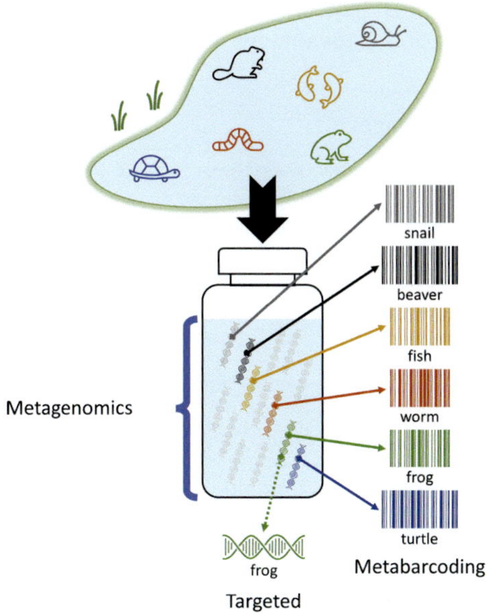

FIG. 5.2 Depiction of three common approaches to the analysis of eDNA from environmental samples. Targeted approaches are derived from previous knowledge regarding specificity of a DNA fragment to identify the presence of the target species within a complex mixture. Metabarcoding uses "universal" primers to amplify specific regions of an organism's genome and then determine the genetic sequence of that region as a "barcode." Metagenomics involves sequencing and assembling the much larger portions of genomes or entire genomes found in an environmental sample.

specially designed to detect gene sequences unique to the target species from environmental samples (Langlois et al., 2020). Targeted eDNA tests are powerful tools to find rare and elusive species and are relatively inexpensive to perform and interpret. For example, swabbing eDNA from artificial cover objects can be used to detect the elusive, endangered sharp-tailed snake (Matthias et al., 2021). The most common targeted eDNA analytical method uses quantitative real-time polymerase chain reaction (qPCR), which is an established, accessible, and robust technology when properly validated (Langlois et al., 2020).

Due to their sensitivity, targeted eDNA tests have also begun to be used to quantify the amount of target species DNA found in the environmental sample that can subsequently be used to infer organismal abundance. Statistical methods for quantifying copy number in low eDNA concentration samples using qPCR (Lesperance et al., 2021) and digital PCR (dPCR) methods (Doi et al., 2015; Nathan et al., 2014) facilitate this application. Targeted approaches excel in sensitivity and quantitation, with the disadvantages that they require prior knowledge of DNA sequence information and will miss unexpected species presence.

TABLE 5.1 Summary of the advantages and disadvantages of the three main eDNA approaches.

Method	Advantages	Disadvantages
Targeted analysis	• Reliably detects very low eDNA concentrations • Estimates target species abundance • Relatively inexpensive • Uses accessible technologies	• Relies on available sequence information for eDNA test design • Detects a limited number of species at a time
Metabarcoding	• Detects many species at once • Assesses community biodiversity • Detects new species	• Relies on reference sequence availability for barcode identification • Is amplification biased and therefore cannot estimate abundance • High false-negative detection rates
Metagenomics	• Elucidates differences within species to understand population • Detects new species	• Relies on next-generation sequencing and extensive bioinformatics • Expensive and inaccessible to many

Community metabarcoding

In contrast, metabarcoding focuses on characterizing a community from an environmental sample by using "universal" primer sets that can amplify across species to examine a specific region of DNA that leads to the reading of its sequence or "barcode." This method uses high-throughput sequencing techniques and bioinformatics analysis (for further details, see Helbing and Hobbs, 2019). Metabarcoding can be used to simultaneously determine species presence and diversity while enabling the detection of novel species. For example, a 2021 study used universal vertebrate primers to assess vertebrate diversity in the rivers of northern Colombia. They detected over 200 native taxa in multiple different phyla including endangered species like the Antillean manatee (Lozano Mojica and Caballero, 2021). Despite its power, metabarcoding is prone to amplification bias causing high false-negative detection rates. This also broadly limits metabarcoding to diversity questions as abundance data may not be accurately and reliably inferred. Furthermore, confident species identification relies on published genetic sequence information and the uniqueness of the barcode. Today, most organisms do not have published sequence information to allow for identification and most metabarcoding studies can only confidently characterize communities to the genus or family level. The ability to survey entire groups of species from a sample is extremely powerful but requires intensive verification of sequence databases and survey results to avoid misleading conclusions.

Metagenomics

The term "metagenomics" refers to the study of the collective genetic material from many organisms contained in each environment. Like metabarcoding, metagenomics methods use high-throughput sequencing and bioinformatics. However, this approach is more demanding as it aims to assemble large genome segments or whole genomes. Multiple recent studies have demonstrated the emerging utility of this approach in addressing complex population-level questions by identifying multiple mitochondrial haplotypes of the same species from environmental samples (Adams et al., 2019; Gehri et al., 2021; Nichols et al., 2021; Parsons et al., 2018; Sigsgaard et al., 2016; Székely et al., 2021; Truelove et al., 2019). One example is a laboratory-based study that compared the allelic frequencies of the round goby detected from a water sample from a mesocosm to those detected from tissues of the animals within that mesocosm. Their results showed that intraspecific nuclear gene information could reliably reflect population differences (Andres et al., 2021). As with metabarcoding, however, elevated scrutiny of the genetic resources is needed to mitigate erroneous results.

Current applications of eDNA approaches

Due to the numerous advantages offered by eDNA methods compared to conventional survey approaches, eDNA has been used in a wide range of applications including environmental assessment and monitoring, ecological recovery, species inventories in challenging environments, and conservation and resource management (Fig. 5.3). Below are selected examples from around the world illustrating proof of concept for influencing decision-making and, eventually, policy.

Environmental assessment and monitoring

In many countries when a proposed project or development is being considered, an environmental impact assessment that can quantify its ramifications on the surrounding environment and ecosystem must be completed. For example, extensive evaluation of the impacts on seabed ecosystems was required before an environmental license was granted to a seafloor sulfide deposit mining project in New Zealand (Durden et al., 2018). These surveys are usually highly multidisciplinary, considering interrelated socioeconomic, cultural, ecological, and human health impacts, both beneficial and adverse. eDNA has multiple

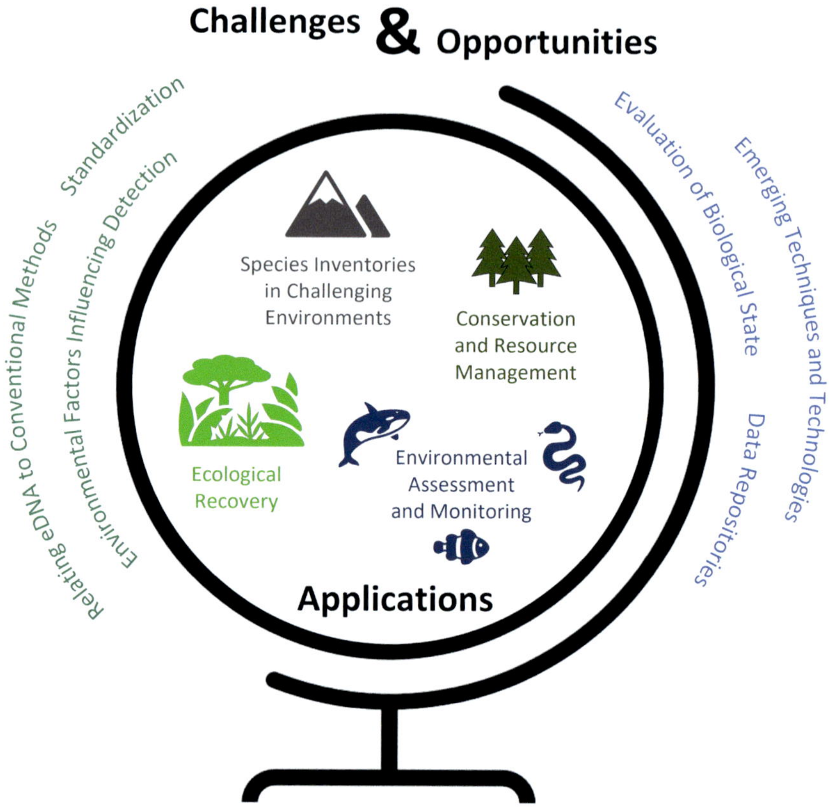

FIG. 5.3 Each challenge encountered in the development of eDNA techniques provides an opportunity for the advancement of the field. While the applications of eDNA methods are diverse, all stand to benefit from improvements in standardization, data integration and management, and our understanding of the biotic and abiotic influences on genetic material throughout the eDNA workflow. In addressing these challenges, new possibilities arise in sampling technologies, organism health data, and increased accessibility of the methods to a wider range of users.

applications to help achieve this: from detecting protected species triggering changes in development plans to identifying key habitats needing protection. For instance, the deep sea is under increasing anthropogenic stress as interest in deep-sea mining grows. Being extremely inaccessible, characterizing what factors are important in maintaining biodiversity in the deep sea is very difficult. In one metabarcoding study, researchers collected eDNA using remotely operated vehicles (ROVs) in deep-sea plains and seamounts near proposed mining areas. They found that, though seamounts were very important in terms of species endemism and richness as expected, the vast deep-sea plains in between them also held equally distinct communities dominated by nematodes, arthropods, mollusks, and cnidaria (Laroche et al., 2020a,b). Therefore, deep-sea mining must have these considerations in mind as we are only beginning to understand how the survival of these deep-sea animals has very real effects upon the surface. This study clearly demonstrates the utility of eDNA as a biomonitoring tool, an environmental impact assessment tool, and a tool to reach these highly inaccessible environments.

eDNA methods can also prove highly useful in increasing the efficiency of long-term site monitoring, which can be achieved using multiple approaches. Ecosystem health is commonly determined by monitoring the presence and abundance of indicator species whose life history is closely associated with healthy habitat functioning (Czechowski et al., 2020). In the Swiss National Park, six spring-dwelling invertebrate indicator species were detected using targeted eDNA tests paired with conventional morphological identification (Blattner et al., 2021). eDNA analysis produced comparable results to conventional methods, while removing the invasive aspects and improving both cost and time efficiency of long-term monitoring.

The utility of metabarcoding has been demonstrated for tracking pollution in fragile ecosystems to their sources and accentuated the degree of interconnectedness between wetland environments. For example, in Australia, a proof-of-concept assessment of the eDNA of specific crops in the water of sensitive wetland areas near farmlands revealed remarkable mobility of agricultural species and associated organic pollution (Adame and Reef, 2020).

Ecological recovery

Pathogenic species monitoring typically relies on morphological or genetic diagnosis of vector animals, which is at best invasive and at worst lethal to the vector depending on the species (Anh et al., 2008; Coutinho et al., 2015). Viable alternatives to these approaches have been established in studies targeting the necessary habitats of pathogens and vectors. For example, targeted eDNA analysis of water in Norway resulted in the detection of *Batrachochytrium dendrobatidis* infecting native amphibian species, providing the first known occurrence of the pervasive pathogen in the country (Taugbøl et al., 2021).

The utility of monitoring insect presence on forest plants using eDNA can be easily applied to a multitude of ecosystem recovery and protection scenarios, potentially providing an early warning against invasive species, and reducing the severity of pest outbreaks (Valentin et al., 2020). Similarly, these methods can be applied to determine the range and approximate biomass of invasive species, which has direct applications for informing ecological damage limitation and remediation efforts (Everts et al., 2021; Kamoroff et al., 2019; Spear et al., 2020). In 2021, eDNA assays were employed using digital PCR (dPCR) to monitor the occurrence and abundance of American bullfrog invading Belgium (Everts et al., 2021). The abundance and biomass measured by these assays were strongly correlated with concurrent conventional method results over a 2-year period.

The effectiveness of a metagenomics approach to indicator species monitoring was demonstrated in a study that tracked the change in fungal communities over a 10-year period of Australian woodland recovery after clearing and extensive use as grazing pasture (Yan et al., 2018). As the woodland regained its native plant composition, the eDNA analysis of soil from the area reflected a shift toward native fungal diversity. In addition, metagenomic analysis of soil composition can elucidate the history of plant growth in fields (Foucher et al., 2020), soil health (Wakelin et al., 2016), and impact on surrounding areas (Adame and Reef, 2020).

Many studies have successfully detected pathogens in water including trematodes and viruses (Huver et al., 2015; Jones et al., 2018; Kaganer et al., 2021; Mulero et al., 2020; Vilaça et al., 2020). The potential for eDNA approaches for pandemic monitoring has been recognized and applied around the world (Amarasiri et al., 2021; Farrell et al., 2021). For example, the SARS-CoV-2 virus can be detected in wastewater from cities providing advanced warning of surges in infection rate (Randazzo et al., 2020; Street et al., 2020; Wu et al., 2020).

Species inventories in challenging environments

Conventional monitoring of transient and pelagic species is frequently conducted using highly invasive catch methods (Colombano et al., 2020; Enders et al., 2021) or hydroacoustic methods that lack resolution in species identification (Lin et al., 2021). Application of eDNA tools presents an excellent opportunity as both stand-alone and complementary detection methods (Berger et al., 2020; Gold et al., 2021; McClenaghan et al., 2020). The effectiveness of metabarcoding for pelagic species monitoring at multiple ocean water column depths was assessed in a study that employed several universal fish primer sets, demonstrating that strong assays were able to detect fish up to at least 2500 m below the ocean surface (McClenaghan et al., 2020).

A particular challenge in researching pelagic or transient fauna in the open ocean is their elusiveness. It is very difficult to define distinct populations and even more difficult to establish their connectivity. A 2016 metagenomic eDNA study of a whale shark aggregation in the Arabian Gulf enabled researchers to elucidate population-level information, providing a snapshot of population diversity not possible with traditional biopsy methods (Sigsgaard et al., 2016). This information was invaluable to whale shark researchers as it indicated that populations were more interconnected than previously thought, and it was invaluable to eDNA researchers as it highlighted the benefits of designing studies to take advantage of natural aggregation events.

Conservation and resource management

A common aim of biomonitoring research is to provide compelling evidence in order to inform policy and management decisions for the conservation of species or of key habitats. eDNA has proven itself to be a very valuable tool that can fill in the gaps left by other biomonitoring techniques. For instance, a metabarcoding study aiming to quantify shark biodiversity in areas of differing anthropogenic impact neatly demonstrates that eDNA could detect more species than traditional underwater visual censuses or baited cameras. Additionally, they showed that decreased anthropogenic influence (either because of low human population in the area or because of mandated protection) corresponded with higher biodiversity (Bakker et al., 2017; Boussarie et al., 2018). This evidence is frequently used to make decisions regarding monitoring protocols and the delineation of protected areas.

Another group based in Alaska took eDNA samples from close proximity to the harbor porpoise—an apex species whose healthy presence is a critical indicator for the health of the entire ecosystem. This study elucidated population-level data from the eDNA samples that demonstrated far more genetic differentiation in the population than was previously thought. This suggests that it may be better to manage the harbor porpoise population as two distinct populations rather than one. This type of indirect stock assessment has the potential to revolutionize population evaluation and subsequent management for inaccessible marine taxa (Parsons et al., 2018).

In studies that seek to describe species range, the ability to detect threatened and cryptic species presence without visual observation or habitat destruction is extremely desirable for limiting survey invasiveness and reducing the requirement for sampling permits (Hobbs et al., 2020; Parsons et al., 2018; Spence et al., 2020). Correlation between eDNA signal strength and visual counts of threatened endemic species has established the viability of not only detecting threatened species, but also extrapolating context-dependent abundance estimates (Baldigo et al., 2016; Maruyama et al., 2018; Skinner et al., 2020). For example, population recovery of white sharks in California was effectively monitored using targeted dPCR assays (Lafferty et al., 2018). This approach provided a nonintrusive solution to the difficulty of monitoring a rare, protected marine species and demonstrated the effectiveness of eDNA in an ocean environment. In British Columbia, Canada, targeted qPCR approaches have also been successfully applied to threatened mountain stream-dwelling amphibians, whose conventional surveillance usually requires intensive time-constrained searches by expert biologists to properly identify the species (Hobbs et al., 2020, 2019).

Challenges and opportunities

While eDNA applications and methodologies are broadening and progressing at a breakneck speed, specific challenges must be addressed to facilitate widespread acceptance and uptake to fully embrace their potential. Review of these challenges highlights the need for collection of comprehensive metadata alongside eDNA samples, the integration of ecosystem-specific hydrological models into eDNA analysis and eDNA study design, standardized reporting and performance criteria, and other recommendations (Fig. 5.3) so that eDNA results can be assessed in a robust, comparative, meaningful way by experts and amateurs alike (Collins et al., 2018; Harrison et al., 2019; Levi et al., 2019; Sigsgaard et al., 2016). Certain technology-based approaches, discussed below, seek to address these limitations with creative implementation of advances from other scientific fields.

Environmental factors influencing eDNA detection

Overarchingly, there is a need to better understand how environmental factors, both biotic and abiotic, affect eDNA sources, dispersion, and degradation in differing environments. Abiotic factors such as salinity, acidity, temperature, and UV exposure, and biotic factors such as microbial activity, or source animal metabolism or size can affect how much eDNA is shed from source organisms and how long eDNA can persist in the environment before degrading into undetectable components. Additionally, factors like wind, currents, tides, and rainfall can all affect how far and how fast eDNA spreads from its source. Though there have been many mesocosm studies and some in situ studies done to address these factors, overwhelmingly the extent to which each of these factors affects eDNA is ecosystem-dependent (Allison et al., 2021; Barnes and Turner, 2015; Barnes et al., 2014; Collins et al., 2018; Holman et al., 2021; Jo et al., 2020; Moushomi et al., 2019; Murakami et al., 2019; Robinson et al., 2021; Saito and Doi, 2021; Stewart, 2019; Weltz et al., 2017; Wood et al., 2020; Zulkefli et al., 2019). These factors influence the ability to link eDNA data to other pertinent factors like species abundance or range expansion.

Relating eDNA results to conventional methods

Data pairing from different methodologies is the most promising way to establish these links, and for eDNA to be reliably integrated into global monitoring programs. However, while ground-truthing with conventional methods is desired for independent use of eDNA methods, it is important to acknowledge the limitations of such comparisons, particularly with respect to their relative sensitivities (Darling et al., 2021). For example, conventional monitoring can have extremely low detection rates (e.g., 0.4% in sharp-tailed snake surveys) that are far below eDNA detection rates (9%; Matthias et al., 2021). Nevertheless, using eDNA methods for addressing species abundance questions is a highly desired outcome.

Undoubtedly, eDNA has its advantages over other methods in its ability to detect the genetic presence of organisms that would have been missed by methods which require physical presence of the organisms. There is ample positive evidence that supports the reliability of quantitative data from eDNA methods, but variation around this positive correlation may be attributed to varying biotic and abiotic factors (Rourke et al., 2021), poorly performing eDNA tests (Langlois et al., 2020), or lack of standardized performance criteria (Hobbs et al., 2019). Currently, the majority of eDNA studies focused on abundance compare eDNA results with known biomass data collected with conventional methods like beach seining, bottom trawls, electrofishing, or camera trapping (Leempoel et al., 2020). These comparisons are necessary at this stage as the factors that impact strong correlations between eDNA surveys and biomass estimates determined by other survey methods are still being elucidated. For example, there is strong evidence that the size distribution of a population will change how much DNA is released into the environment and subsequently detected (Yates et al., 2020).

Variation in technical methodologies—field and technical replication, sample volume, filtration method, extraction method, eDNA assay specificity and sensitivity—will also impact these correlations and the overall quality of eDNA research. This variation highlights the need for systematic experimentation to determine best practices and subsequently methodological standardization (Baillie et al., 2019; Hobbs et al., 2020; Loeza-Quintana et al., 2020; Nicholson et al., 2020).

Data standards, methods, and harmonization

There is currently an international push for standardization in field and laboratory techniques for eDNA surveys. Several groups have made notable steps in the publication of methods guidance documents, and the creation of eDNA resource repositories (Table 5.2). In Canada, a national standard for eDNA reporting requirements and terminology has recently been released (Table 5.2).

As standards are established, eDNA application will be more globally accessible and used to inform and influence policy. A particular strength of eDNA is its amenability to citizen science. Citizen science entails the recruitment of volunteers from the public to sample environments. Public interest in science and demands for community consultation and involvement in development and remediation projects is expanding. Citizen science initiatives increase ecological survey capacity without a large increase in project costs, and the incorporation of citizen science into tourism promotes more sustainable consumption and better public understanding of the environment and the challenges it is facing (Meschini et al., 2021). Larger sampling capacity subsequently leads to more data that is helpful for academic and management initiatives. Effective citizen science projects have been completed by groups such as EnviroDNA (2021), Nature Metrics (2021), Living Lakes Canada (2021), Invasive Species Centre (2021), and the Hakai Institute (2021). However, global standardization of best practices is needed to make these community-based initiatives more reliable and comparable (Buxton et al., 2018). As the Anthropocene continues, harmonized research is vital to identify global patterns of change in biodiversity and species distribution.

TABLE 5.2 Examples of eDNA standards, guidance documents, and repositories.

Category	Type	Location	Reference
National standard	Environmental DNA (eDNA) reporting requirement and terminology	Canada	Canadian Standards Association (2021)
Guidance documents	Guidance on the use of targeted environmental DNA (eDNA) analysis for the management of aquatic invasive species and species at risk	Canada	Abbott et al. (2021)
	Environmental DNA protocol for freshwater aquatic ecosystems	Canada	Hobbs et al. (2017)
	A validation scale to determine the readiness of environmental DNA assays for routine species monitoring	International	Thalinger et al. (2021)
	Implementation option for DNA-based identification into ecological status assessment under the European Water Framework Directive	Europe	Hering et al. (2018)
	Environmental DNA sampling and experiment manual	Japan	Minamoto et al. (2020)
	A protocol for collecting environmental DNA samples from streams	United States	Carim et al. (2016)
	Quality assurance project plan eDNA monitoring of bighead and silver carps	United States	Woldt et al. (2020)
Repositories	Atlas of Living Australia	Australia	Atlas of Living Australia (2021)
	Barcode of life data systems	Canada	Barcode of Life Database (2007)
	Mitochondrial genomes of Danish vertebrate species generated for the national DNA reference database, DNAmark	Denmark	Margaryan et al. (2020)
	Environmental DNA information and resources	United States	eDNA Resources (2021)
	The aquatic eDNAtlas project	United States	The Aquatic eDNAtlas Project (2021)

Globalized research also generates immense amounts of data. In this age of information, having comprehensive, accessible, and well-organized data repositories is necessary to facilitate meaningful worldwide collaboration. For example, a single metabarcoding study may generate millions of DNA barcodes and identify many different taxa. This information is often published in the scientific literature in a way that is not comprehensive or comparable to other similar work. There are many examples of open-access repositories aiming to address this problem to provide the data depth that is needed for eDNA studies (Table 5.2). The barcode of life database (BOLD) is working to create reference genetic libraries of barcode regions for as many species as possible, and the Atlas of Living Australia contains occurrence data from thousands of different contributors (Berry et al., 2020). There are also repositories of different primer sets that have been designed and used, and repositories of longer gene sequences like the mitogenome repository of Danish vertebrates (Table 5.2). For targeted eDNA studies, these long gene sequence repositories are immensely valuable because they provide ample sequence information for powerful assay design. Though these are all important steps, the value of eDNA to industry, government, and decision-makers would be greatly improved through standardized reporting, and incorporation of eDNA data with pertinent metadata into widely used, accessible, and curated repositories.

Data repositories facilitate global sequence availability. Increasing the accessibility of genetic sequences will aid the creation of targeted assays and improve the ability of metabarcoding and metagenomics approaches to recognize species diversity. As the relative costs of some molecular biology technologies, including sequencing, decrease, the global demand and accessibility will increase. However, financial and technical barriers continue to exist, particularly impacting widespread access to new promising sequencing and molecular technologies. For example, nanopore sequencers are small portable sequencers that use third-generation sequencing methods to produce information very rapidly. They can also sequence multiple species in the same run and are not as error-prone in comparison with Illumina sequencing platforms as was previously thought. Metagenomic studies rely on the ability to sequence thousands of reads per study often on short timescales. These sequencers have shown utility in field studies providing results in a matter of days (Egeter et al., 2020; Truelove et al., 2019), but they require long DNA strands and are still relatively expensive to acquire and run, thereby limiting metagenomics to those who have the technology.

Emerging techniques and technologies

Beyond qPCR and next-generation sequencing approaches in common use, other methods are gaining traction for eDNA applications. dPCR can be more reliable and resistant to inhibitor presence for detecting and quantifying eDNA signals at low eDNA concentrations compared to conventional qPCR methods

(Doi et al., 2015), but it is more expensive per sample, requires specialized consumables, and many institutions do not have access to dPCR machines. Other researchers are testing CRISPR-Cas9 with isothermal recombinase polymerase amplification (CRISPR/RPA) technology to amplify eDNA (Phelps, 2019; Williams et al., 2020) instead of qPCR. This innovative approach is compatible for field use as a rapid bioassessment tool and has promise as an invaluable tool in species detection in remote environments without access to modern molecular equipment. However, CRISPR/RPA is still in its infancy in terms of eDNA applications with sensitivity issues (Phelps, 2019; Williams et al., 2020). While the above technologies show a lot of promise, qPCR remains the most promising, accessible technology for eDNA detection. It is most conducive to widespread global adoption and standardization due to its accessibility and cost-effectiveness (Goldberg et al., 2016; Hobbs et al., 2019; Langlois et al., 2020). With well-designed, robust assays, qPCR can be very sensitive and quantitate very low eDNA copy numbers from environmental samples (Langlois et al., 2020; Lesperance et al., 2021).

eDNA holds great promise as a revolutionary tool in molecular ecology because it enables the sampling of difficult-to-access organisms and ecosystems. However, the majority of eDNA studies today still require the presence of humans to collect and process samples. Engineering advancements with eDNA in mind may change this. Robotic samplers, once tried and tested, have the potential to overcome the temporal and spatial limitations faced by humans (Yamahara et al., 2019). For years now, ROVs, unmanned aerial vehicles (UAVs) such as drones, and autonomous underwater vehicles (AUVs) have been used to collect remote data. Some researchers have modified these tools to collect eDNA samples (Laroche et al., 2020a,b; Truelove et al., 2019). Others have even created samplers that are able to collect and preserve samples through carefully engineered instruments that capture and filter water, followed by injection of preservation fluid into individual capsules to maintain eDNA sample integrity for up to 3 weeks (Yamahara et al., 2019). Temporal and spatial constraints of ship- or shore-based sampling are alleviated through the higher-resolution sampling enabled by ROVs, UAVs, and AUVs. Through their use, time and spatial series data can be obtained at a frequency and resolution that is sufficient to reflect natural eDNA signal variation caused by the tide, spawning events, and vertical migrations of marine organisms. It is possible that one day, this type of remote sampling collection tightly paired to oceanographic and geographic data will be widely used and that even processes like extraction, amplification, or even sequencing could be done remotely with creatively designed instrumentation in areas inaccessible to human sampling.

Beyond taxa detection—Evaluation of biological state

Another eDNA challenge is elucidating the biological state of the source organism. Is it alive or dead? Is it a juvenile or an adult? Is it healthy or sick? RNA that is extracted from an environmental sample, referred to as "eRNA," can potentially address these issues. Unlike DNA, RNA is actively expressed by living organisms and differentially expressed based on health and environmental conditions, life history stage of the source organism, etc. Since RNA typically has a shorter half-life than DNA, it has until recently been largely assumed that accessing RNA in an environmental sample would be very unreliable. However, recent research has shown that eRNA may be able to persist in the environment for longer time periods and in larger quantities than previously thought (Cristescu, 2019; von Ammon et al., 2019). While the mechanisms underlying these findings are still being explored (e.g., protein capsids, extracellular vesicles, double-stranded RNAs), the fact remains that metabolically active organisms can release sufficient RNA into their environment for high-resolution population status determination—a very attractive prospect.

Outlook

There is vast potential in the use of eDNA methods, limited only by creativity and the necessity of experimental rigor. As new applications are explored, careful attention must be paid to each aspect of the methodological pipeline and collaboration encouraged between experts to maintain the highest quality of work and data interpretation. Best practice guidelines and adherence to standardized methods will reduce much of the influence of experimental unknowns, improve comparability between projects, and enable integration into policy. Virtually no area of environmental surveying lacks the potential to benefit from the many forms of eDNA analysis, provided the methods are appropriately applied. The rapid pace of globalization and climate change emphasizes the immediate need for rapid monitoring methods. Reliable and accessible eDNA methods have that ability to assess current biodiversity enabling effective estimation of future trajectories and negative impact mitigation. Through attentive implementation of these techniques, we add an entirely new set of tools to supplement conventional methods, providing a high caliber of resolution and confidence in all fields of environmental assessment.

References

Abbott, C., Coulson, M., Gagné, N., Lacoursière-Roussel, A., Parent, G.J., Bajno, R., Dietrich, C., May-McNally, S., 2021. Guidance on the use of targeted environmental DNA (eDNA) analysis for the management of aquatic invasive species and species at risk. In: DFO Can. Sci. Advis. Sec. Res. Doc. 2021/019, p. 42.

Adame, M.F., Reef, R., 2020. Potential pollution sources from agricultural activities on tropical forested floodplain wetlands revealed by soil eDNA. Forests 11 (8), 892.

Adams, C., Knapp, M., Gemmell, N.J., Jeunen, G.-J., Bunce, M., Lamare, M., Taylor, H.R., 2019. Beyond biodiversity: can environmental DNA (eDNA) cut it as a population genetics tool? Genes 10 (192), 20.

Allison, M.J., Round, J.M., Bergman, L.C., Mirabzadeh, A., Allen, H., Weir, A., Helbing, C.C., 2021. The effect of silica desiccation under different storage conditions on filter-immobilized environmental DNA. BMC Res. Notes 14 (1), 106.

Amarasiri, M., Furukawa, T., Nakajima, F., Sei, K., 2021. Pathogens and disease vectors/hosts monitoring in aquatic environments: potential of using eDNA/eRNA based approach. Sci. Total Environ. 796, 148810.

von Ammon, U., Wood, S.A., Laroche, O., Zaiko, A., Lavery, S.D., Inglis, G.J., Pochon, X., 2019. Linking environmental DNA and rna for improved detection of the marine invasive fanworm *Sabella spallanzanii*. Front. Mar. Sci. 6, 621.

Andres, K.J., Sethi, S.A., Lodge, D.M., Andres, J., 2021. Nuclear eDNA estimates population allele frequencies and abundance in experimental mesocosms and field samples. Mol. Ecol. 30 (3), 685–697.

Anh, N.T., Phuong, N.T., Ha, G.H., Thu, L.T., Johansen, M.V., Murrell, D.K., Thamsborg, S.M., 2008. Evaluation of techniques for detection of small trematode eggs in faeces of domestic animals. Vet. Parasitol. 156 (3–4), 346–349.

Atlas of Living Australia, 2021. Atlas of Living Australia, Canberra, Australia. https://www.ala.org.au/blogs-news/edna-records-now-available-on-ala/. (Accessed 26 July 2021).

Baillie, S.M., McGowan, C., May-McNally, S., Leggatt, R., Sutherland, B.J.G., Robinson, S., 2019. Environmental DNA and Its Applications to Fisheries and Oceans Canada: National Need and Priorities. Department of Fisheries and Oceans, Ottawa, ON, p. 100.

Bakker, J., Wangensteen, O.S., Chapman, D.D., Boussarie, G., Buddo, D., Guttridge, T.L., Hertler, H., Mouillot, D., Vigliola, L., Mariani, S., 2017. Environmental DNA reveals tropical shark diversity in contrasting levels of anthropogenic impact. Sci. Rep. 7 (1), 16886.

Baldigo, B.P., Sporn, L.A., George, S.D., Ball, J.A., 2016. Efficacy of environmental DNA to detect and quantify brook trout populations in headwater streams of the adirondack mountains, New York. Trans. Am. Fish. Soc. 146 (1), 99–111.

Barcode of Life Database, 2007. Guelph, Ontario. http://v4.boldsystems.org/. (Accessed 26 July 2021).

Barnes, M.A., Turner, C.R., 2015. The ecology of environmental DNA and implications for conservation genetics. Conserv. Genet. 17 (1), 1–17.

Barnes, M.A., Turner, C.R., Jerde, C.L., Renshaw, M.A., Chadderton, W.L., Lodge, D.M., 2014. Environmental conditions influence eDNA persistence in aquatic systems. Environ. Sci. Technol. 48 (3), 1819–1827.

Berger, C.S., Bougas, B., Turgeon, S., Ferchiou, S., Ménard, N., Bernatchez, L., 2020. Groundtruthing of pelagic forage fish detected by hydroacoustics in a whale feeding area using environmental DNA. Environ. DNA 2 (4), 477–492.

Berry, O., Jarman, S., Bissett, A., Hope, M., Paeper, C., Bessey, C., Schwartz, M.K., Hale, J., Bunce, M., 2020. Making environmental DNA (eDNA) biodiversity records globally accessible. Environ. DNA 3, 699–705.

Blattner, L., Ebner, J.N., Zopfi, J., von Fumetti, S., 2021. Targeted non-invasive bioindicator species detection in eDNA water samples to assess and monitor the integrity of vulnerable alpine freshwater environments. Ecol. Indic. 129, 107916.

Boussarie, G., Bakker, J., Wangensteen, O.S., Mariani, S., Bonnin, L., Juhel, J.B., Kiszka, J.J., Kulbicki, M., Manel, S., Robbins, W.D., et al., 2018. Environmental DNA illuminates the dark diversity of sharks. Sci. Adv. 4 (5), eaap9661.

Buxton, A., Groombridge, J., Griffiths, R., 2018. Comparison of two citizen scientist methods for collecting pond water samples for environmental DNA studies. Cit. Sci.: Theory Pract. 3 (2), 1–9.

Canadian Standards Association, 2021. Environmental DNA (eDNA) Reporting Requirement and Terminology. CSA Group, Ottawa, ON, p. 25.

Carim, K.J., McKelvey, K.S., Young, M.K., Wilcox, T.M., Schwartz, M.K., 2016. A Protocol for Collecting Environmental DNA Samples from Streams. United States Department of Agriculture, Fort Collins, CO, p. 26.

Collins, R.A., Wangensteen, O.S., O'Gorman, E.J., Mariani, S., Sims, D.W., Genner, M.J., 2018. Persistence of environmental DNA in marine systems. Commun. Biol. 1, 185.

Colombano, D.D., Manfree, A.D., O'Rear, T.A., Durand, J.R., Moyle, P.B., 2020. Estuarine-terrestrial habitat gradients enhance nursery function for resident and transient fishes in the San Francisco estuary. Mar. Ecol. Prog. Ser. 637, 141–157.

Coutinho, S.D., Burke, J.C., de Paula, C.D., Rodrigues, M.T., Catao-Dias, J.L., 2015. The use of singleplex and nested pcr to detect *Batrachochytrium dendrobatidis* in free-living frogs. Braz. J. Microbiol. 46 (2), 551–555.

Cristescu, M.E., 2019. Can environmental RNA revolutionize biodiversity science? Trends Ecol. Evol. 34 (8), 694–697.

Czechowski, P., Stevens, M.I., Madden, C., Weinstein, P., 2020. Steps towards a more efficient use of chironomids as bioindicators for freshwater bioassessment: exploiting eDNA and other genetic tools. Ecol. Indic. 110, 105868.

Darling, J.A., Jerde, C.L., Sepulveda, A.J., 2021. What do you mean by false positive? Environ. DNA 3 (5), 879–883.

Doi, H., Uchii, K., Takahara, T., Matsuhashi, S., Yamanaka, H., Minamoto, T., 2015. Use of droplet digital PCR for estimation of fish abundance and biomass in environmental DNA surveys. PLoS One 10 (3), e0122763.

Durden, J.M., Lallier, L.E., Murphy, K., Jaeckel, A., Gjerde, K., Jones, D.O.B., 2018. Environmental impact assessment process for deep-sea mining in 'the area'. Mar. Policy 87, 194–202.

eDNA Resources, 2021. Washington State University. https://ednaresources.science/. (Accessed 26 July 2021).

Egeter, B., Veríssimo, J., Lopes-Lima, M., Chaves, C., Pinto, J., Riccardi, N., Beja, P., Fonseca, N. A., 2020. Speeding up the detection of invasive aquatic species using environmental DNA and nanopore sequencing. bioRxiv. https://doi.org/10.1101/2020.06.09.142521.

Enders, E.C., Charles, C., van der Lee, A.S., Lumb, C.E., 2021. Temporal variations in the pelagic fish community of Lake Winnipeg from 2002 to 2019. J. Great Lakes Res. 47 (3), 626–634.

EnviroDNA, 2021. Victoria, Australia. https://www.envirodna.com/. (Accessed 11 October 2021).

Everts, T., Halfmaerten, D., Neyrinck, S., De Regge, N., Jacquemyn, H., Brys, R., 2021. Accurate detection and quantification of seasonal abundance of American bullfrog (*Lithobates catesbeianus*) using ddPCR eDNA assays. Sci. Rep. 11 (1), 11282.

Farrell, J.A., Whitmore, L., Duffy, D.J., 2021. The promise and pitfalls of environmental DNA and rna approaches for the monitoring of human and animal pathogens from aquatic sources. Bioscience 71 (6), 609–625.

Ficetola, G.F., Miaud, C., Pompanon, F., Taberlet, P., 2008. Species detection using environmental DNA from water samples. Biol. Lett. 4 (4), 423–425.

Foucher, A., Evrard, O., Ficetola, G.F., Gielly, L., Poulain, J., Giguet-Covex, C., Laceby, J.P., Salvador-Blanes, S., Cerdan, O., Poulenard, J., 2020. Persistence of environmental DNA in cultivated soils: implication of this memory effect for reconstructing the dynamics of land use and cover changes. Sci. Rep. 10 (1), 10502.

Gehri, R.R., Larson, W.A., Gruenthal, K., Sard, N.M., Shi, Y., 2021. eDNA metabarcoding outperforms traditional fisheries sampling and reveals fine-scale heterogeneity in a temperate freshwater lake. Environ. DNA 3 (5), 912–929.

Gold, Z., Sprague, J., Kushner, D.J., Zerecero Marin, E., Barber, P.H., 2021. eDNA metabarcoding as a biomonitoring tool for marine protected areas. PLoS One 16 (2), e0238557.

Goldberg, C.S., Turner, C.R., Deiner, K., Klymus, K.E., Thomsen, P.F., Murphy, M.A., Spear, S.F., McKee, A., Oyler-McCance, S.J., Cornman, R.S., et al., 2016. Critical considerations for the application of environmental DNA methods to detect aquatic species. Methods Ecol. Evol. 7 (11), 1299–1307.

Hakai Institute, 2021. Campbell River, British Columbia, Canada. https://www.hakai.org/. (Accessed 11 October 2021).

Harrison, J.B., Sunday, J.M., Rogers, S.M., 2019. Predicting the fate of eDNA in the environment and implications for studying biodiversity. Proc. Biol. Sci. 286 (1915), 20191409.

Helbing, C.C., Hobbs, J., 2019. Environmental DNA Standardization Needs for Fish and Wildlife Population Assessments and Monitoring. Canadian Standards Association, Calgary, AB, p. 41.

Hering, D., Borja, A., Jones, J.I., Pont, D., Boets, P., Bouchez, A., Bruce, K., Drakare, S., Hanfling, B., Kahlert, M., et al., 2018. Implementation options for DNA-based identification into ecological status assessment under the European water framework directive. Water Res. 138, 192–205.

Hobbs, J., Goldberg, C.S., Helbing, C.C., Veldhoen, N., 2017. Environmental DNA Protocol for Freshwater Aquatic Ecosystems Version 2.2. BC Ministry of Environment and Climate Change Strategy, Victoria, BC, p. 48.

Hobbs, J., Round, J.M., Allison, M.J., Helbing, C.C., 2019. Expansion of the known distribution of the coastal tailed frog, *Ascaphus truei*, in British Columbia, Canada, using robust eDNA detection methods. PLoS One 14 (3), 16.

Hobbs, J., Adams, I.T., Round, J.M., Goldberg, C.S., Allison, M.J., Bergman, L.C., Mirabzadeh, A., Allen, H., Helbing, C.C., 2020. Revising the range of rocky mountain tailed frog, *Ascaphus montanus*, in British Columbia, Canada, using environmental DNA methods. Environ. DNA 2 (3), 350–361.

Holman, L.E., Chng, Y., Rius, M., 2021. How does eDNA decay affect metabarcoding experiments? Environ. DNA 00, 1–9.

Huver, J., Koprivnikar, J., Johnson, P., Whyard, S., 2015. Development and application of an eDNA method to detectand quantify a pathogenic parasite in aquatic ecosystems. Ecol. Appl. 25 (4), 991–1002.

Invasive Species Centre, 2021. Sault Ste. Marie, Ontario, Canada. https://www.invasivespeciescentre.ca/. (Accessed 11 October 2021).

Jo, T., Arimoto, M., Murakami, H., Masuda, R., Minamoto, T., 2020. Estimating shedding and decay rates of environmental nuclear DNA with relation to water temperature and biomass. Environ. DNA 2 (2), 140–151.

Jones, R.A., Brophy, P.M., Davis, C.N., Davies, T.E., Emberson, H., Rees Stevens, P., Williams, H. W., 2018. Detection of *Galba truncatula*, *Fasciola hepatica* and *Calicophoron daubneyi* environmental DNA within water sources on pasture land, a future tool for fluke control? Parasit. Vectors 11 (1), 342.

Kaganer, A.W., Nagel, L.D., Youker-Smith, T.E., Bunting, E.M., Hare, M.P., 2021. Environmental DNA-derived pathogen gene sequences can expand surveillance when pathogen titers are decoupled in eDNA and hosts. Environ. DNA 00, 1–6.

Kamoroff, C., Daniele, N., Grasso, R.L., Rising, R., Espinoza, T., Goldberg, C.S., 2019. Effective removal of the American bullfrog (*Lithobates catesbeianus*) on a landscape level: long term monitoring and removal efforts in Yosemite Valley, Yosemite National Park. Biol. Invasions 22 (2), 617–626.

Lafferty, K.D., Benesh, K.C., Mahon, A.R., Jerde, C.L., Lowe, C.G., 2018. Detecting Southern California's white sharks with environmental DNA. Front. Mar. Sci. 5 (355), 6.

Langlois, V.S., Allison, M.J., Bergman, L.C., To, T.A., Helbing, C.C., 2020. The need for robust qPCR-based eDNA detection assays in environmental monitoring and species inventories. Environ. DNA 3 (3), 519–527.

Laroche, O., Kersten, O., Smith, C.R., Goetze, E., 2020a. Environmental DNA surveys detect distinct metazoan communities across abyssal plains and seamounts in the western Clarion Clipperton zone. Mol. Ecol. 29, 4588–4604.

Laroche, O., Kersten, O., Smith, C.R., Goetze, E., 2020b. From sea surface to seafloor: a benthic allochthonous eDNA survey for the abyssal ocean. Front. Mar. Sci. 7, 682.

Leempoel, K., Hebert, T., Hadly, E.A., 2020. A comparison of eDNA to camera trapping for assessment of terrestrial mammal diversity. Proc. R. Soc. B. 287, 20192353. https://doi.org/10.1098/rspb.2019.2353.

Lesperance, M.L., Allison, M.J., Bergman, L.C., Hocking, M.D., Helbing, C.C., 2021. A statistical model for calibration and computation of detection and quantification limits for low copy number environmental DNA samples. Environ. DNA 3, 970–981.

Levi, T., Allen, J.M., Bell, D., Joyce, J., Russell, J.R., Tallmon, D.A., Vulstek, S.C., Yang, C., Yu, D. W., 2019. Environmental DNA for the enumeration and management of pacific salmon. Mol. Ecol. Resour. 19 (3), 597–608.

Lin, T.-H., Akamatsu, T., Sinniger, F., Harii, S., 2021. Exploring coral reef biodiversity *via* underwater soundscapes. Biol. Conserv. 253, 108901.

Living Lakes Canada, 2021. Nelson, British Columbia, Canada. https://livinglakescanada.ca/. (Accessed 11 October 2021).

Loeza-Quintana, T., Abbott, C.L., Heath, D.D., Bernatchez, L., Hanner, R.H., 2020. Pathway to increase standards and competency of eDNA surveys (PISCES)—advancing collaboration and standardization efforts in the field of eDNA. Environ. DNA 2 (3), 255–260.

Lozano Mojica, J.D., Caballero, S., 2021. Applications of eDNA metabarcoding for vertebrate diversity studies in Northern Colombian water bodies. Front. Ecol. Evol. 8, 16.

Margaryan, A., Noer, C.L., Richter, S.R., Restrup, M.E., Bülow-Hansen, J.L., Leerhøi, F., Langkjær, E.M.R., Gopalakrishnan, S., Carøe, C., Gilbert, M.T.P., et al., 2020. Mitochondrial genomes of Danish vertebrate species generated for the national DNA reference database, DNAmark. Environ. DNA 3 (2), 472–480.

Maruyama, A., Sugatani, K., Watanabe, K., Yamanaka, H., Imamura, A., 2018. Environmental DNA analysis as a non-invasive quantitative tool for reproductive migration of a threatened endemic fish in rivers. Ecol. Evol. 8 (23), 11964–11974.

Matthias, L., Allison, M.J., Maslovat, C.Y., Hobbs, J., Helbing, C.C., 2021. Improving ecological surveys for the detection of cryptic, fossorial snakes using eDNA on and under artificial cover objects. Ecol. Indic. 131, 108187.

McClenaghan, B., Fahner, N., Cote, D., Chawarski, J., McCarthy, A., Rajabi, H., Singer, G., Hajibabaei, M., 2020. Harnessing the power of eDNA metabarcoding for the detection of deep-sea fishes. PLoS One 15 (11), e0236540.

Meschini, M., Machado Toffolo, M., Marchini, C., Caroselli, E., Prada, F., Mancuso, A., Franzellitti, S., Locci, L., Davoli, M., Trittoni, M., et al., 2021. Reliability of data collected by volunteers: a nine-year citizen science study in the Red Sea. Front. Ecol. Evol. 9, 694258.

Minamoto, T., Miya, M., Sado, T., Seino, S., Doi, H., Kondoh, M., Nakamura, K., Takahara, T., Yamamoto, S., Yamanaka, H., et al., 2020. An illustrated manual for environmental DNA research: water sampling guidelines and experimental protocols. Environ. DNA 3, 8–13.

Moushomi, R., Wilgar, G., Carvalho, G., Creer, S., Seymour, M., 2019. Environmental DNA size sorting and degradation experiment indicates the state of *Daphnia magna* mitochondrial and nuclear eDNA is subcellular. Sci. Rep. 9 (1), 12500.

Mulero, S., Boissier, J., Allienne, J.F., Quilichini, Y., Foata, J., Pointier, J.P., Rey, O., 2020. Environmental DNA for detecting *Bulinus truncatus*: a new environmental surveillance tool for schistosomiasis emergence risk assessment. Environ. DNA 2 (2), 161–174.

Murakami, H., Yoon, S., Kasai, A., Minamoto, T., Yamamoto, S., Sakata, M.K., Horiuchi, T., Sawada, H., Kondoh, M., Yamashita, Y., et al., 2019. Dispersion and degradation of environmental DNA from caged fish in a marine environment. Fish. Sci. 85 (2), 327–337.

Nathan, L.M., Simmons, M., Wegleitner, B.J., Jerde, C.L., Mahon, A.R., 2014. Quantifying environmental DNA signals for aquatic invasive species across multiple detection platforms. Environ. Sci. Technol. 48 (21), 12800–12806.

Nature Metrics, 2021. Surrey, United Kingdom. https://www.naturemetrics.co.uk/. (Accessed 11 October 2021).

Nichols, P.K., Timmers, M., Marko, P.B., 2021. Hide 'n seq: direct versus indirect metabarcoding of coral reef cryptic communities. Environ. DNA 00, 1–15.

Nicholson, A., McIsaac, D., MacDonald, C., Gec, P., Mason, B.E., Rein, W., Wrobel, J., Boer, M., Milián-García, Y., Hanner, R.H., 2020. An analysis of metadata reporting in freshwater environmental DNA research calls for the development of best practice guidelines. Environ. DNA 2 (3), 343–349.

Parsons, K.M., Everett, M., Dahlheim, M., Park, L., 2018. Water, water everywhere: environmental DNA can unlock population structure in elusive marine species. R. Soc. Open Sci. 5 (8), 180537.

Phelps, M., 2019. Increasing eDNA capabilities with CRISPR technology for real-time monitoring of ecosystem biodiversity. Mol. Ecol. Resour. 19 (5), 1103–1105.

Randazzo, W., Cuevas-Ferrando, E., Sanjuan, R., Domingo-Calap, P., Sanchez, G., 2020. Metropolitan wastewater analysis for COVID-19 epidemiological surveillance. Int. J. Hyg. Environ. Health 230, 113621.

Robinson, C.V., Porter, T.M., Wright, M.T.G., Hajibabaei, M., 2021. Propylene glycol-based antifreeze is an effective preservative for DNA metabarcoding of benthic arthropods. Freshw. Sci. 40 (1), 77–87.

Rourke, M.L., Fowler, A.M., Hughes, J.M., Broadhurst, M.K., DiBattista, J.D., Fielder, S., Wilkes Walburn, J., Furlan, E.M., 2021. Environmental DNA (eDNA) as a tool for assessing fish biomass: a review of approaches and future considerations for resource surveys. Environ. DNA 00, 1–25.

Saito, T., Doi, H., 2021. Degradation modeling of water environmental DNA: experiments on multiple DNA sources in pond and seawater. Environ. DNA 3, 850–860.

Sigsgaard, E.E., Nielsen, I.B., Bach, S.S., Lorenzen, E.D., Robinson, D.P., Knudsen, S.W., Pedersen, M.W., Jaidah, M.A., Orlando, L., Willerslev, E., et al., 2016. Population characteristics of a large whale shark aggregation inferred from seawater environmental DNA. Nat. Ecol. Evol. 1 (1), 4.

Skinner, M., Murdoch, M., Loeza-Quintana, T., Crookes, S., Hanner, R., 2020. A mesocosm comparison of laboratory-based and on-site eDNA solutions for detection and quantification of striped bass (*Morone saxatilis*) in marine ecosystems. Environ. DNA 2 (3), 298–308.

Spear, M.J., Embke, H.S., Krysan, P.J., Vander Zanden, M.J., 2020. Application of eDNA as a tool for assessing fish population abundance. Environ. DNA 3 (1), 83–91.

Spence, B.C., Rundio, D.E., Demetras, N.J., Sedoryk, M., 2020. Efficacy of environmental DNA sampling to detect the occurrence of endangered coho salmon (*Oncorhynchus kisutch*) in Mediterranean-climate streams of California's central coast. Environ. DNA 3, 727–744.

Stewart, K.A., 2019. Understanding the effects of biotic and abiotic factors on sources of aquatic environmental DNA. Biodivers. Conserv. 28 (5), 983–1001.

Street, R., Malema, S., Mahlangeni, N., Mathee, A., 2020. Wastewater surveillance for COVID-19: an African perspective. Sci. Total Environ. 743, 140719.

Székely, D., Corfixen, N.L., Mørch, L.L., Knudsen, S.W., McCarthy, M.L., Teilmann, J., Heide-Jørgensen, M.P., Olsen, M.T., 2021. Environmental DNA captures the genetic diversity of bowhead whales (Balaena mysticetus) in West Greenland. Environ. DNA 3 (1), 248–260.

Taugbøl, A., Bærum, K.M., Dervo, B.K., Fossøy, F., 2021. The first detection of the fungal pathogen *Batrachochytrium dendrobatidis* in Norway with no evidence of population declines for great crested and smooth newts based on modeling on traditional trapping data. Environ. DNA 3, 760–768.

Thalinger, B., Deiner, K., Harper, L.R., Rees, H.C., Blackman, R.C., Sint, D., Traugott, M., Goldberg, C.S., Bruce, K., 2021. A validation scale to determine the readiness of environmental DNA assays for routine species monitoring. Environ. DNA 3, 823–836.

The Aquatic eDNAtlas Project, 2021. United States Department of Agriculture, USA. https://www.fs.fed.us/rm/boise/AWAE/projects/eDNAtlas/the-edna-atlas-results.html. (Accessed 26 July 2021).

Truelove, N.K., Andruszkiewicz, E.A., Block, B.A., Gilbert, M.T.P., 2019. A rapid environmental DNA method for detecting white sharks in the open ocean. Methods Ecol. Evol. 10 (8), 1128–1135.

Valentin, R.E., Fonseca, D.M., Gable, S., Kyle, K.E., Hamilton, G.C., Nielsen, A.L., Lockwood, J. L., 2020. Moving eDNA surveys onto land: strategies for active eDNA aggregation to detect invasive forest insects. Mol. Ecol. Resour. 20 (3), 746–755.

Vilaça, S.T., Grant, S.A., Beaty, L., Brunetti, C.R., Congram, M., Murray, D.L., Wilson, C.C., Kyle, C.J., 2020. Detection of spatiotemporal variation in ranavirus distribution using eDNA. Environ. DNA 2 (2), 210–220.

Wakelin, S.A., Cave, V.M., Dignam, B.E., D'Ath, C., Tourna, M., Condron, L.M., Zhou, J., Van Nostrand, J.D., O'Callaghan, M., 2016. Analysis of soil eDNA functional genes: potential to increase profitability and sustainability of pastoral agriculture. N. Z. J. Agric. Res. 59 (4), 333–350.

Weltz, K., Lyle, J.M., Ovenden, J., Morgan, J.A.T., Moreno, D.A., Semmens, J.M., 2017. Application of environmental DNA to detect an endangered marine skate species in the wild. PLoS One 12 (6), e0178124.

Williams, M.A., Hernandez, C., O'Sullivan, A.M., April, J., Regan, F., Bernatchez, L., Parle-McDermott, A., 2020. Comparing CRISPR-Cas and qPCR eDNA assays for the detection of Atlantic salmon (*Salmo salar* l.). Environ. DNA 3 (1), 297–304.

Woldt, A., Mcgovern, A., Lewis, T.D., Tuttle-Lau, M., 2020. Quality Assurance Project Plan eDNA Monitoring of Bighead and Silver Carps. U.S. Fish and Wildlife Service, Bloomington, MN, p. 91.

Wood, S.A., Biessy, L., Latchford, J.L., Zaiko, A., von Ammon, U., Audrezet, F., Cristescu, M.E., Pochon, X., 2020. Release and degradation of environmental DNA and RNA in a marine system. Sci. Total Environ. 704, 135314.

Wu, F., Zhang, J., Xiao, A., Gu, X., Lee, W.L., Armas, F., Kauffman, K., Hanage, W., Matus, M., Ghaeli, N., et al., 2020. SARS-COV-2 titers in wastewater are higher than expected from clinically confirmed cases. mSystems 5 (4), 1–9.

Yamahara, K.M., Preston, C.M., Birch, J., Walz, K., Marin, R., Jensen, S., Pargett, D., Roman, B., Ussler, W., Zhang, Y., et al., 2019. *In situ* autonomous acquisition and preservation of marine environmental DNA using an autonomous underwater vehicle. Front. Mar. Sci. 6, 373.

Yan, D., Mills, J.G., Gellie, N.J.C., Bissett, A., Lowe, A.J., Breed, M.F., 2018. High-throughput eDNA monitoring of fungi to track functional recovery in ecological restoration. Biol. Conserv. 217, 113–120.

Yates, M.C., Glaser, D.M., Post, J.R., Cristescu, M.E., Fraser, D.J., Derry, A.M., 2020. The relationship between eDNA particle concentration and organism abundance in nature is strengthened by allometric scaling. Mol. Ecol. 30 (13), 3068–3082.

Zulkefli, N.S., Kim, K.H., Hwang, S.J., 2019. Effects of microbial activity and environmental parameters on the degradation of extracellular environmental DNA from a eutrophic lake. Int. J. Environ. Res. Public Health 16 (18), 3339.

Chapter 6

Informing marine shipping insurance premiums in the Arctic using marine microbial genomics

Mawuli Afenyo[a], Casey R.J. Hubert[b], Srijak Bhatnagar[c], and Changmin Jiang[d]

[a]*Texas A&M University, Galveston, TX, United States,* [b]*Geomicrobiology Group, University of Calgary, Calgary, AB, Canada,* [c]*Faculty of Science and Technology, Athabasca University, Athabasca, AB, Canada,* [d]*Asper Business School, University of Manitoba, Winnipeg, MB, Canada*

Introduction

Oil spills are often described as low-probability, high-consequence events. They do not occur regularly, but when they happen, the consequences can be devastating (Afenyo et al., 2019; Fingas, 2015), resulting in environmental, social, financial, and reputational damage. Oil spills occurring in remote Arctic regions present even greater complications. This is because an oil spill occurring in ice-covered waters makes response efforts and deployment of recovery techniques difficult. The Arctic has recently seen a steady increase in traffic; hence, the potential for oil spills is increasing (Afenyo et al., 2021a,b). Apart from the Siberian oil spill in 2020, there has not been any significant spill in the last 10 years in the Arctic. This therefore begs the question: should we insure vessels going into the Arctic? And should we worry about the potential oil spills from vessels that could happen with accidents?

BP and other resource exploration and production companies opted not to buy insurance for their operations and assets with regard to oil spills, instead diversifying their sources of income and putting money aside to deal with any such occurrence (Abraham, 2011). In a 2015 ruling regarding the 2010 Deepwater Horizon oil spill in the Gulf of Mexico, a court revealed BP had no insurance and could not use the insurance held by its drilling contractor Transocean (Donlon, 2015). In the case of the Mauritius oil spill in 2020, to date, no substantial compensation has been paid to affected communities by

Genomics and the Global Bioeconomy. https://doi.org/10.1016/B978-0-323-91601-1.00006-7

the company responsible. A donation by the government of Japan to its Mauritius counterpart was made (Afenyo et al., 2021a,b), though this may be considered more as a loan and not direct compensation. There is no organization that seems to be responsible when it comes to holding ship owners and ships' operating companies accountable. Having insurance in place would make it easier for appropriate compensation to be realized following oil spills.

The problem, however, may not just be companies refusing to be insured against oil spills. Insurance companies may not have the tools to properly calculate insurance premiums associated with oil spills in places like the Arctic, mainly due to the limited data available. Information on the environments, including genomic information about the marine microbiome and its potential to respond to oil inputs by catalyzing a biodegradation response, would be valuable in this context. Insurance companies would benefit immensely by knowing that some of the oil could be removed through the oil-eating bacteria if those bacteria are present and able to respond in a particular high-risk region, such as shipping lanes or areas where exploratory oil drilling may be under consideration.

Microbiology and genomics have the potential to inform the extent of biodegradation of oil that may be expected in a particular region. Genomic information about biodegradation—pertaining to both the genes involved and the microorganisms harboring those genes—would also enable more constrained estimates of other weathering and transport processes that spilled oil encounters, including evaporation, emulsification, dispersion, photooxidation, spreading, and sinking (Afenyo et al., 2016). These processes depend on each other and in some cases occur simultaneously. For example, if the rate of evaporation is higher, the amount of oil that would be left for microbial populations to biodegrade may be lower, and vice versa. Microbial genomics also has the potential to uncover new biodiversity and potentially new mechanisms for hydrocarbon biodegradation, particularly in the under-explored Arctic marine environment.

Although limited knowledge exists about the Arctic, this is gradually changing. Various Arctic countries are investing heavily in Arctic-specific technology in order to be prepared for any eventuality. For example, Russia has made considerable investments in Arctic resource exploration. As of late 2021, the building of a second liquefied natural gas (LNG) plant located in the Gyda Peninsula in Western Siberia was due to be started, following the success of the Yamal Arctic LNG project, also in Russia. These projects, and future planned LNG projects, are all led by NOVATEK, a Russian oil and gas company (Merkulov, 2020). This industrial activity will mean an increase in tanker ships in this region in addition to the cruise ships and maritime supply ships delivering goods to remote communities (Merkulov, 2020; Afenyo et al., 2021a,b).

The insurance industry is also preparing to determine ways to best position itself to take full advantage of developments in the changing Arctic (IUMI, 2020). According to Helle Hammer, chair of the International Union of Marine

Insurance (IUMI) policy forum, the required risk modeling is extremely difficult. Without the ability to develop actuarial tables and model risks, insuring activities in the Arctic will be challenging. In terms of policy, governments and other regulatory agencies need to put the required laws in place to deal with irregularities. Ship operators as well as resource exploration companies need to liaise with local communities as well as the insurance companies to come up with a formula that will help insure ships and activities in the Arctic.

Such a formula or model will be enhanced by loss component inputs that integrate not only environmental and financial losses but also negative societal implications. A comprehensive model will make the work of the insurance companies easier and help the institutions involved estimate the impacts of potential oil spills.

Marine microbial genomics is typically deployed "after the fact" in post-spill clean-up assessments. Applying these technologies before accidents take place would be prudent (Joye, 2015; Taggart and Clark, 2021). This chapter examines the role of marine microbial genomics in assessing oil biodegradation for determining insurance premiums for shipping operations and natural resource exploration activities. Sequencing DNA from microbiomes inhabiting different marine habitats and environments unlocks a wealth of information about biodiversity and the state of the marine system, including a given microbiome's potential for oil biodegradation. Different kinds of genomic data that may be useful to inform such an activity, including PCR-based amplicon sequencing and shotgun metagenomic sequencing, are presented and examined. This chapter also introduces a socioeconomic model, which serves as an interface for linking the genomics data and insurance premiums. Examples of marine microbial genomics data are presented to show the ways in which genomics can be used to make these kinds of insurance premium assessments possible.

Microbial genomics tools for generating microbiome data

Microbial genomics is a powerful and effective strategy for generating baseline information about natural environments. This information is especially strategic in marine settings that may reasonably be considered to be facing oil spill risks, including in and around oil exploration areas, shipping routes, and maritime ports. The DNA sequencing revolution has positioned microbial genomics perfectly for a fresh approach to baseline data generation using microbial biodiversity metrics. Genomics is more sensitive than other baseline approaches (e.g., manual counting of small animals in marine samples; Paine et al., 2014), because marine microorganisms respond so rapidly to perturbations such as an oil spill or other kinds of pollution. In the case of an Arctic oil spill, cold-adapted oil-degrading bacteria are likely to act as nature's "emergency first responders" by initiating biodegradation long before manual remediation strategies will be initiated.

Baseline information, as the term implies, is often pursued in the context of environmental effects monitoring and enables assessing remediation after a pollution event has taken place. Indeed, researchers conducting microbial genomics in the aftermath of the Deepwater Horizon oil spill have lamented the lack of necessary pre-spill baseline data and strongly recommend determining baselines for the Arctic (Lovejoy, 2013; Joye, 2015).

Baseline information can be a forward-looking tool with predictive value. It is in this context that microbial genomics has relevance for the insurance industry as an input parameter for determining insurance premiums in relation to oil spill clean-up models (e.g., Fig. 6.1), for both the Arctic and other parts of the world. A detailed examination of baseline information can allow the bioremediation potential within a microbiome to be predicted, with a degree of confidence that depends on the extent and sophistication of the DNA sequencing strategy, as explained below. Different approaches and uses of microbial

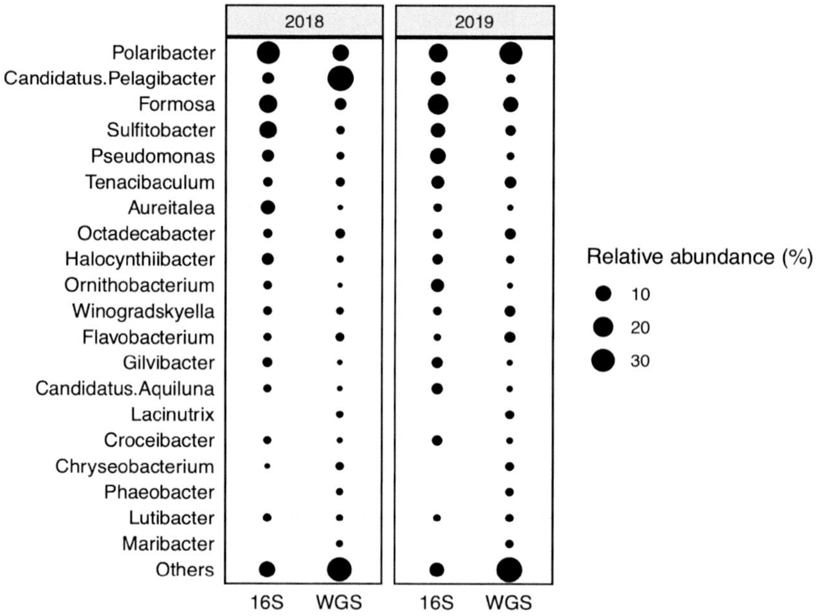

FIG. 6.1 Samples taken from the Labrador Sea, close to an area designated for oil and gas exploration, analyzed with single-gene 16S ribosomal RNA gene amplicon sequencing (16S) and whole-genome shotgun metagenomic sequencing (WGS). Samples were obtained in 2 consecutive years to establish a microbial genomics baseline in this region. The taxonomic composition of the microbial diversity is compared between the two methods. The two 16S libraries contain on average 26,000 DNA sequences, whereas the WGS libraries contain on average 102 million DNA sequences and include much more information on the potential functions of this microbiome. In a related study, samples from the Labrador Sea were used in mock oil spills to enrich oil-degrading bacteria, enabling smaller WGS libraries (30 million sequences) to reveal hydrocarbon biodegradation genes (Murphy et al., 2021).

genomics must be considered in this regard in order to determine necessary advance information for predicting a microbiome's oil spill response potential.

At issue is whether the DNA sequencing assessments are based on single genes or microorganisms' whole genomes in a given sample. It is also important to consider the relative abundance of oil-degrading bacteria in relation to the detection limits of the DNA sequencing strategy that is chosen.

Single-gene amplicon sequencing

Currently, the most common approach for environmental microbiological assessments is to conduct single-gene surveys. Typically, the 16S ribosomal RNA gene is used, as this is the gold-standard taxonomic indicator gene for microorganisms (Amann et al., 1995). This genomic survey involves obtaining environmental DNA, amplifying this specific gene into many copies using polymerase chain reaction (PCR), and then assessing hundreds to thousands (or millions; library size is cost-dependent) of the resulting amplified DNA sequences to determine the proportions of different taxonomic groups (see Fig. 6.1). Current methods usually target a short subregion (e.g., 350 base pairs), which corresponds to about one-quarter of the entire 16S rRNA gene length. Notwithstanding the possibility that the PCR step may not be truly universal in amplifying all microbial groups with equal fidelity (despite this being the objective, nucleotide hybridization chemistry prevents it from being fully realized; Suzuki and Giovannoni, 1996), the result of these PCR "amplicon libraries" is effectively a census of the microorganisms present that reveals proportions of different taxonomic groups.

Interpreting amplicon libraries in the context of oil spills requires the functional capabilities of different organisms to be inferred from their taxonomic affiliation. The result of sequencing about one-quarter (350 base pairs) of the 16S ribosomal RNA gene is effectively providing the names of the organisms and nothing more. Therefore, knowledgeable microbiologists must assess the likelihood that organisms with different taxonomic classifications are capable of oil degradation. This can be done with a degree of confidence, for example, in cases of so-called obligate hydrocarbon-degrading bacteria, i.e., lineages of organisms from which all known examples are able to degrade hydrocarbons. Examples include groups like *Oleiphilus*, *Alcanivorax*, and *Cycloclasticus* (Berry and Gutierrez, 2017). On the other hand, there are very well-known oil-degrading bacteria that belong to groups that are not exclusively oil degraders, such as *Pseudomonas* and *Colwellia*—including the cold-adapted *Colwellia psychrerythraea* (Muriel-Millán et al., 2019; Kleindienst et al., 2015). How should the presence of these organisms be interpreted? While software exists for making inferences based on 16S ribosomal RNA taxonomy, these applications tend to be designed with an emphasis on extensively studied environments, such as the human microbiome (Aßhauer et al., 2015; Langille et al., 2013). A dedicated tool for making similar predictions for oil spills could be a valuable development in this context and help investigators contend with the paucity of information encoded in few hundred base pairs of the 16S rRNA gene.

Whole-genome shotgun metagenomics

The other microbial genomics strategy that is increasingly being employed for understanding natural samples is whole-genome assessments, using "shotgun metagenomics." Instead of honing in on an individual gene as with amplicon sequencing, the shotgun sequencing strategy allows all or many of the genes in any of the genomes present in the sample to be included in a DNA sequencing library. In this method, depending on the budget typically millions to hundreds of millions of sequences are obtained (see Fig. 6.1), and then, software is used to reassemble these genomic fragments into longer stretches (called "contigs") that can be read to predict biological functions encoded by different genes and to assign fragments to different microbial genomes. This enables the direct identification of genes that encode enzymes for catalytic functions, including hydrocarbon biodegradation (Khot et al., 2021), while still providing taxonomic information (Fig. 6.1). Thus, through metagenomics, it is possible to obtain more direct evidence for oil degradation potential within the microbiome.

The average size of the whole genome of a microbe is about 3,500,000 base pairs. Therefore, in terms of the extent of DNA sequencing undertaken (also called "sequencing depth," which is correlated with cost), surveys of 350 base pair single-gene amplicons are a far less expensive way of surveying a large or diverse microbial population. That being said, metagenome libraries do not necessarily need to sequence all of the DNAs present so extensively so as to ensure whole genomes of all organisms are completely pieced back together. In the context of a predictive assessment of biodegradation, it may be adequate to identify and quantify the presence of hydrocarbon biodegradation genes, even if they cannot be arranged and assigned back into corresponding whole genomes. It would be cost-prohibitive to use metagenomics to obtain whole-genome sequences for the same number of organisms surveyed by amplicon sequencing, given the roughly 10,000-fold difference in sequencing depth between the two approaches. Therefore, gene-centric metagenomics with a focus on hydrocarbon biodegradation genes may be attractive. Computational tools for assessing hydrocarbon biodegradation genes in environmental samples are available (Khot et al., 2021).

Overcoming low relative abundance of oil-degrading microorganisms in pristine environments

A challenge associated with these approaches is that oil-degrading bacteria may be numerically rare in the environments that insurance companies are interested in. This in turn means that genomic methods need to have sensitive detection limits. Seawater typically contains 10^5 to 10^6 microorganisms per milliliter (Garneau et al., 2008; Cui et al., 2019), and if there are no naturally occurring hydrocarbons in the vicinity, the relative abundance of oil-degrading bacteria may be below 0.1% or 0.01% of the total population (Murphy et al., 2021). This means 16S ribosomal RNA genes corresponding to oil degradation indicator taxa might only be detected at a frequency of 10^{-3} or 10^{-4}. In the latter scenario, confidently identifying ~10 sequences corresponding to putative oil-degrading bacteria requires a library with 100,000 sequences (as noted above, amplicon libraries are often smaller than this). This can drive up the cost of the DNA sequencing survey. The same principle applies to the already more costly whole-genome shotgun metagenomics strategy.

A way around this issue can be to conduct targeted genomic studies. Genetic probes can be used to modify single-gene studies so that they focus on only certain taxonomic groups at the expense of others (e.g., only survey for members of the genus *Oleiphilus*), or only a specific gene within the genomes of oil-degrading bacteria (e.g., the alkB gene involved in alkane degradation; Wasmund et al., 2009; Jurelevicus et al., 2013). Incubation experiments that recreate oil spill conditions in microcosms (Murphy et al., 2021; Schreiber et al., 2021) or larger mesocosms (Desmond et al., 2021) are another possibility that allows oil-degrading bacteria to proliferate; however, these studies are time- and cost-intensive. Such "mock oil spill" experiments enable detailed studies of microbial populations that become enriched, including their genetic potential and biodegradation phenotype. These important benefits of targeted studies need to be weighed against their drawbacks, including that they can be too selective, i.e., not adequately recreating complex and changing environmental conditions and not necessarily encompassing all of the genes or organisms capable of oil degradation in the environmental samples that are used to establish them. The latter is especially relevant and intriguing in frontier settings like the Arctic where studies are revealing that previously unrecognized lineages of bacteria may be key contributors to the biodegradation of spilled oil (Murphy et al., 2021).

Bioremediation of oil spills

Fig. 6.2 illustrates the relationships among some of the transport and weathering processes, and how genomics information related to biodegradation fits into a bigger picture of the fate and effects of spilled oil. Genomics information can be particularly important for estimating the biodegradation potential. This can be obtained from the genetic information within the marine microbiome in a given area. For example, during expeditions to the Canadian Arctic Ocean aboard the research icebreaker *CCGS Amundsen* and other research vessels, genomics data has been collected using the strategies described above in "Microbial genomics tools for generating microbiome data" section.

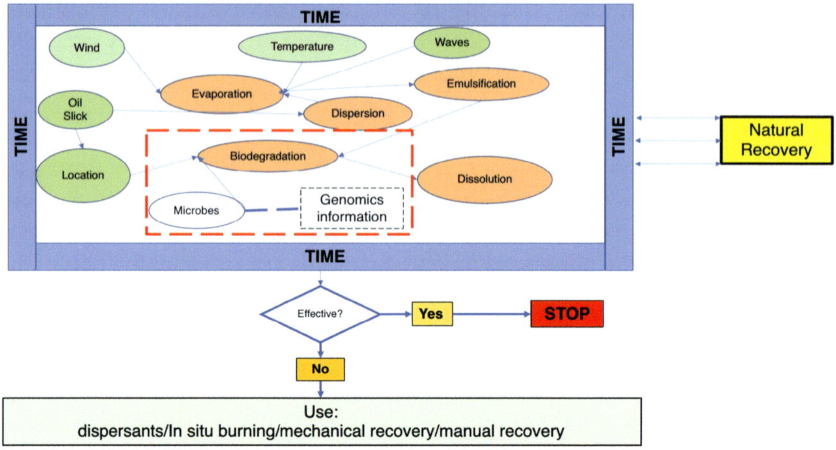

FIG. 6.2 Relationship between weathering and transport processes and genomics information.

Fig. 6.2 shows that when an oil spill occurs, the spilled contamination undergoes weathering and transport processes. One of the key weathering processes is biodegradation. These processes together are responsible for breaking down oil. For example, lighter petroleum components evaporate simultaneously with oil-eating bacteria breaking down compounds that they are genetically predisposed to metabolize. These processes combine to reduce the amount of oil in the environment, with evaporation involving volatilization of compounds into the atmosphere, and biodegradation resulting in the conversion of petroleum compounds to smaller by-products and CO_2.

In reality, evaporation and biodegradation will not completely remove an oil spill in its entirety. If natural processes included in Fig. 6.2 are not able to deal with the oil spill, then human intervention is required. Human intervention methods include the use of dispersants, in situ burning, mechanical recovery (e.g., using booms, skimmers, or pumps), and manual removal (e.g., using shovels or nets). Some of these approaches have serious consequences to the environment. For example, in situ burning and the use of dispersants may introduce even more pollutants into an area (Afenyo et al., 2019, 2021a,b; Fingas, 2015).

In some jurisdictions, governments have enacted laws against the use of some of the oil spill mitigation interventions listed above. Manual removal and mechanical recovery are expensive and risky; they may not always be very effective depending on the type and quantity of oil spilled or the type of environment where the spill occurred. For example, manual removal efforts in the Arctic Ocean are not only dangerous but may be impossible because of the extremely low temperature and inaccessibility. This underscores the importance of natural recovery strategies like evaporation and biodegradation. In this regard, the activation of oil-degrading microbial populations following a spill is referred to as natural attenuation. Applying available genomics information to estimate the rate or extent of biodegradation would be very useful. It should be noted that the natural processes are time-sensitive. While some begin and end in the early days following a spill event, others start later and can continue for many years (Head et al., 2006). Improving these processes, especially biodegradation, increases the chances that oil spill clean-up will effectively continue many years after the pollution event takes place. Ongoing research is being undertaken in this regard including the GEN-ICE project funded by Genome Canada (see https://www.genice.ca/) and the Arctic Response Technology joint industry project led by SINTEF in Norway (see http://www.arcticresponsetechnology.org/).

Genomics and risk assessment

The relationship between genomics and risk assessment is very important. It is sometimes difficult to link the two, such that an improved understanding of risk will make this easier.

Risk is often described as a function of probability, consequences, scenario, and time. Traditionally, the two main components of risk are probability and the consequence. Probability refers to how often a particular event may happen, and the consequence is the outcome of that particular event. In maritime risk assessment, the parameter of vulnerability is added. "Vulnerability" is the probability of exposure to harm and includes the variable of "threat," which may refer to transit by a ship carrying oil. In the context of Arctic shipping, this may be influenced by whether the ship is equipped with single hull, instead of a double hull. It may also refer to rust of some valves, bolts, knobs, and seals, or potentially encountering icebergs or other sea ice during navigation. A third variable of "impact" can be evaluated in terms of potential consequences. These can include the extent of fish kills following oil spills, or the number of families needing to be relocated and disruption to the social makeup of communities.

Information on risk is very important for informing calculations of insurance premiums. For details of the premium calculation, please refer to Afenyo et al. (2021a,b). In insurance, important concepts are the *claim frequency*, which is how often the claim would be demanded, and the *claim severity*, which is the extent of damage of a system or property due to an event. Knowing the claim frequency and claim severity is critical to predicting the premium. There is further complexity in determining the premium, expanded on by Werner et al. (2016) who show the details of the calculation of those parameters. In both Afenyo et al. (2021b) and Werner et al. (2016), there are parameters such as the loss adjustment expense (LAE), the underwriting expense (UWE), and the underwriting profit (UWP). The LAE, UWE, and UWP are determined by other methods, which are not discussed in this chapter. Readers may consult Werner et al. (2016) for details of the calculation of those parameters.

One key parameter in the relationships described above is the loss. This can be determined using risk assessment models developed by Afenyo et al. (2021a). The output of the model gives the losses in dollars and could also be presented in probabilities. The model incorporates the information of biodegradation, which can in turn be assessed by applying microbial genomics. Genomics information about the potential for a favorable microbial response can thus help to determine the loss or the impact due to an accident or oil spill during shipping. Having the right genomics information can enable predictions concerning the biodegradation of crude oil or diesel fuel by microbes inhabiting a particular area. This in turn helps to refine and determine the need for the precision of the other outputs in insurance premium models. It should be noted that while genomics information alone is not enough for establishing insurance premiums, it is important. Information on the presence of different biodegradation genes, determined via microbial genomics, can serve as a key indicator of the potential for biodegradation in an environment, and in this case the Arctic marine environment.

Scientists have published studies regarding modeling and estimating biodegradation in open water (Thrift-Viveros et al., 2015; Vilcáez et al., 2013; Brakstad et al., 2015; Kapellos et al., 2018). McFarlin et al. (2014) also studied biodegradation rates in the Arctic. Reviews by Li and McCay (2016) and Afenyo et al. (2016) present different algorithms and models for biodegradation for oil spill modeling purposes.

Studies demonstrating the incorporation of genomics data into oil spill risk assessment models are more elusive. While the concepts above have been tested to some extent, genomic data types and how to interpret them have not been examined or tested in detail in this context. Having the right genomics data to inform risk assessments in areas of concern will improve the predictive capability of insurance premium modeling. This in turn will inform and enhance the incorporation of other weathering and transport process parameters. In the Arctic, such data is scarce; making efforts to gather such data is extremely valuable with shipping in Arctic seas now on the rise.

Genomics data as input to insurance premium calculations

Recently, Afenyo et al. (2021a,b) published a comprehensive model for assessing the socioeconomic impact of oil spills associated with Arctic shipping. The output is the loss in dollar terms. As shown in Fig. 6.2 and the model in Afenyo et al. (2021a,b), one of the key processes is the biodegradation of oil and the potential of this occurring, which can be informed by assessing the microbiome in a given area using microbial genomics. Fig. 6.3 shows how the models are related to genomics data as an input to the socioeconomic model, and the socioeconomic model in turn gives the loss value required for calculating the insurance premium.

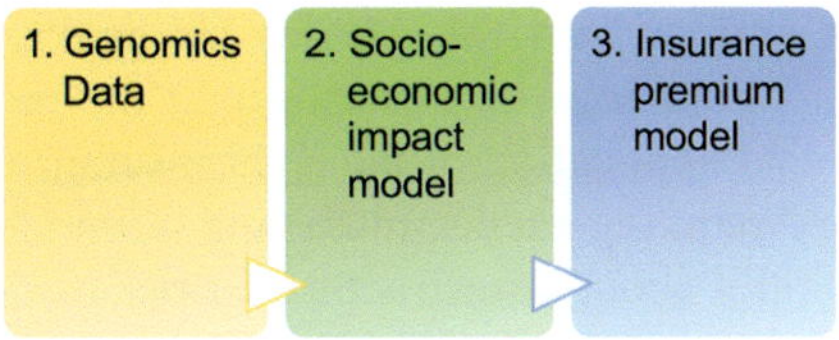

FIG. 6.3 Linkages between genomics data and insurance premium calculation.

As noted above, different kinds of genomic data (Figs. 6.1 and 6.2) can be used as inputs for risk modeling. It may be most likely that the cost-effective single-gene amplicon sequencing surveys are used, resulting in risk coefficients tied to the occurrence and relative abundance of specific taxonomic groups. As explained above, this method is not perfect but still provides an approximation of the potential for oil spill biodegradation based on assessing the presence of microbial groups that are known to normally have this capability. With additional research into the microbial genomics of oil spills, especially in Arctic marine settings, 16S rRNA gene-based models can certainly be refined. It is also possible to invest in more expensive metagenomic studies such that genomics can provide greater confidence to the resulting risk assessment. Alternatively, a hybrid approach whereby large-scale amplicon sequencing surveys are followed by metagenomics on a selected subset of samples of interest can combine the taxonomic depth of the former with functional prediction of the latter.

The cost of both amplicon sequencing and metagenomic sequencing can be determined based on the objectives of the sequencing strategy. The cost difference between the two strategies needs to be considered against potential cost ramifications of having a more precise insurance premium calculation model that metagenomics is more likely to deliver. There may be cases where it will be wise to invest in the shotgun metagenomic sequencing strategy because the risks and liability of clean-up justify the additional expense. Overall, both of these genomics strategies can be used with different confidence levels to inform oil spill risk assessments for insurance premium calculations.

Discussion and conclusions

Genomics data can be used to predict the potential for biodegradation. This kind of information on biodegradation feeds into a larger model, which incorporates interconnected weathering and transport processes (Fig. 6.2). Since some of the processes occur simultaneously and depend on each other, the right genomics information eventually translates into the right model prediction. It is therefore important that governments and the scientific community make an effort to collect genomics information on microbiomes from sensitive areas of interest including Arctic seas. This will allow ongoing refinement of modeling, which will improve the accuracy of insurance premium calculations for ships transiting the Arctic and other regions. Regulating authorities like the insurance bureau, government entities, and the various law-making entities in Arctic nations will play important roles in ensuring that the proper regulations are in place for the efficient application of the right insurance regime. As pointed out by the IUMI, the Arctic represents interesting terrain for insurance companies, and the uncertainties about infrastructure for risk modeling present a challenge. Without the ability to model risk and develop actuarial tables, it is near impossible to come up with insurance premiums. This means that at best, values that would be used as premiums will be extremely either underestimated or overestimated. The strategies suggested here present a realistic way to estimate insurance premiums for a largely unknown place like the Arctic.

Oil spills are inevitable in the Arctic if shipping activities increase and oil and gas exploration commences. There is urgency surrounding the development of comprehensive datasets of genomic information on the Arctic marine microbiome that have predictive value. This information will serve to help protect these regions for both oil spill response purposes and prediction of losses in dollar terms and subsequently establishing insurance premiums. Since microbial communities respond rapidly and measurably to all kinds of perturbations, and not just oil pollution, genomic microbiome surveys will have similar utility in a myriad of contexts.

The approach suggested in this chapter is not without its challenges, including the cost of data collection. Sampling in the Arctic is very expensive and often requires huge investments. While individuals may fund such exercises, Arctic governments have a big role to play. The Canadian government is a good example of investing in scientific expeditions in the Arctic for this kind of data collection. International efforts and databases, such as the *Tara* Oceans Polar Circle initiative (Royo-Llonch et al., 2021), will enable genomics data collected by scientists from different countries working in the Arctic to be shared. These kinds of collaborative efforts will promote better international relations among both Arctic and non-Arctic countries alike.

Acknowledgment

This work was funded by Genome Canada in collaboration with the Government of Manitoba (Genome Prairie) and the Government of Alberta (Genome Alberta). Additional support was provided by Canadian national centers of excellence ArcticNet and MEOPAR.

References

Abraham, K.S., 2011. Catastrophic oil spills and the problem of insurance. Vand. L. Rev. 64 (6), 1769.

Afenyo, M., Ng, A.K.Y., Jiang, C., 2019. Climate change and Arctic shipping: a method for assessing the impacts of oil spills in the Arctic. Transp. Res. D Transp. Environ. 77, 476–490.

Afenyo, M., Ng, A.K.Y., Jiang, C., 2021a. A multi-period model for assessing the socio-economic impact of oil spills in the arctic. Risk Anal. Int. J. 42 (3), 614–633. https://doi.org/10.1111/risa.13773.

Afenyo, M., Khan, F., Veitch, B., 2016. A state-of-the-art review of fate and transport of oil spills in open and ice-covered water. Ocean Eng. 119, 233–248.

Afenyo, M., Li, H., Ng, A.K.Y., Jiang, C., 2021b. An Integrated Insurance Underwriting Tool for Arctic Shipping. International Association of Maritime Economist (IAME), Rotterdam.

Amann, R.I., Ludwig, W., Schleifer, K.H., 1995. Phylogenetic identification and in situ detection of individual microbial cells without cultivation. Microbiol. Rev. 59 (1), 143–169.

Aßhauer, K.P., Wemheuer, B., Daniel, R., Meinicke, P., 2015. Tax4Fun: predicting functional profiles from metagenomic 16S rRNA data. Bioinformatics 31 (17), 2882–2884.

Berry, D., Gutierrez, T., 2017. Evaluating the detection of hydrocarbon-degrading bacteria in 16S rRNA gene sequencing surveys. Front. Microbiol. 8, 896.

Brakstad, O.G., Nordtug, T., Throne-Holst, M., 2015. Biodegradation of dispersed Macondo oil in seawater at low temperature and different oil droplet sizes. Mar. Pollut. Bull. 93, 144–152.

Cui, Y., Chun, S.J., Baek, S.H., Lee, M., Kim, Y., Lee, H.G., Ko, S.R., Hwang, S., Anh, C.Y., Oh, H. M., 2019. The water depth-dependent co-occurrence patterns of marine bacteria in shallow and dynamic southern coast, Korea. Sci. Rep. 9, 9176.

Desmond, D.S., Saltymakova, D., Smith, A., Wolfe, T., Snyder, N., Polcwiartek, K., Bautista, M., Lemes, M., Hubert, C.R.J., Barber, D.G., Isleifson, D., Stern, G.A., 2021. Photooxidation and biodegradation potential of a light crude oil in first year sea ice. Mar. Pollut. Bull. 165, 112154.

Donlon, R.L., 2015. Court: BP has no coverage for oil spill under Transocean insurance. NU Property Casualty. https://www.propertycasualty360.com/2015/02/13/court-bp-has-no-coverage-for-oil-spill-under-trans/?slreturn=20220501212859. (Accessed 1 June 2022).

Fingas, M., 2015. Handbook of Oil Spill Science and Technology. John Wiley & Sons, Somerset.

Garneau, M.E., Roy, S., Lovejoy, C., Gratton, Y., Vincent, W.F., 2008. Seasonal dynamics of bacterial biomass and production in a coastal arctic ecosystem: Franklin Bay, western Canadian Arctic. J. Geophys. Res. 113, C07S91.

Head, I.M., Jones, D.M., Röling, W.F.M., 2006. Marine microorganisms make a meal of oil. Nat. Rev. Microbiol. 4, 173–182.

IUMI, 2020. Things to Ponder for Insurers as Arctic Routes Open Up. International Union of Marine Insurance. https://iumi.com/news/news/things-to-ponder-for-insurers-as-arctic-routes-open-up. (Accessed 8 November 2021).

Joye, S.B., 2015. Deepwater horizon, five years on. Science 349 (6248), 592–593.

Jurelevicus, D., Alvarez, V.M., Peixoto, R., Rosado, A.S., Sedlin, L., 2013. The use of a combination of *alkB* primers to better characterize the distribution of alkane-degrading bacteria. PLoS One 8 (6), e66565.

Kapellos, G.E., Paraskeva, C.A., Kalogerakis, N., Doyle, P.S., 2018. Theoretical insight into the biodegradation of solitary oil microdroplets moving through a water column. Bioengineering 5, 15.

Khot, V., Zorz, J., Gittins, D.A., Chakraborty, A., Bell, E., Bautista, M.A., Paquette, A.J., Hawley, A.K., Novotnik, B., Hubert, C.R.J., Strous, M., Bhatnagar, S., 2021. CANT-HYD: a curated database of phylogeny-derived hidden Markov models for annotation of marker genes involved in hydrocarbon degradation. Front. Microbiol. 12, 764058.

Kleindienst, S., Seidel, M., Ziervogel, K., Grim, S., Loftis, K., Harrison, S., Malkin, S.Y., Perkins, M.J., Field, J., Sogin, M.L., Dittmar, T., Passow, U., Medeiros, P.M., Joye, S.B., 2015. Chemical dispersants can suppress the activity of natural oil-degrading microorganisms. Proc. Natl. Acad. Sci. U. S. A. 112 (48), 14900–14905.

Langille, M.G.I., Zaneveld, J., Caporaso, J.G., McDonald, D., Knights, D., Reyes, J.A., Clemente, J. C., Burkepile, D.E., Vega Thurber, R.L., Knight, R., Beiko, R.G., Huttenhower, C., 2013. Predictive functional profiling of microbial communities using 16S rRNA marker gene sequences. Nat. Biotechnol. 31, 814–821.

Li, Z., McCay, D.F., 2016. Review of Hydrocarbon Biodegradation Rates for Use in Modeling Oil Fate in Seawater. RPS ASA Group, Wakefield, RI.

Lovejoy, C., 2013. Arctic Biodiversity Assessment 2013: Chapter 11, Microorganisms. Conservation of Arctic Flora and Fauna. https://www.arcticbiodiversity.is/index.php/the-report/chapters/microorganisms.

McFarlin, K.M., Prince, R.C., Perkins, R., Leigh, M.B., 2014. Biodegradation of dispersed oil in Arctic seawater at $-1°C$. PLoS One 9, e84297.

Merkulov, V.I., 2020. Analysis of Russian Arctic LNG projects and their development prospects. IOP Conf. Ser.: Mater. Sci. Eng. 940, 012114. https://doi.org/10.1088/1757-899X/940/1/012114.

Muriel-Millán, L.F., Rodríguez-Mejía, J.L., Godoy-Lozano, E.E., Rivera-Gómez, N., Gutierrez-Rios, R.-M., Morales-Guzmán, D., Trejo-Hernández, M.R., Estradas-Romero, A., Pardo-López, L., 2019. Functional and genomic characterization of a *Pseudomonas aeruginosa* strain isolated from the southwestern Gulf of Mexico reveals an enhanced adaptation for long-chain alkane degradation. Front. Microbiol. 6, 572.

Murphy, S.M.C., Bautista, M., Cramm, M.A., Hubert, C.R.J., 2021. Biodegradation of diesel and crude oil by Labrador Sea cold adapted microbial communities. Appl. Environ. Microbiol. 87, e00800–e00821.

Paine, M.D., Debois, E.M., Kilgour, D.W., Tracy, E., Pocklington, P., Crowley, R.D., Williams, U. P., Janes, G.G., 2014. Effects of the Terra Nova offshore oil development on benthic macro-invertebrates over 10 years of development drilling on the grand banks of Newfoundland, Canada. Deep-Sea Res. II 110, 38–64.

Royo-Llonch, M., Sánchez, P., Ruiz-González, C., Salazar, G., Pedrós-Alió, C., Sebastián, M., Labadie, K., Paoli, L., Ibarbalz, F.M., Zinger, L., Churcheward, B., Coordinators, T.O., Chaffron, S., Eveillard, D., Karsenti, E., Sunagawa, S., Wincker, P., Karp-Boss, L., Bowler, C., Acinas, S.G., 2021. Compendium of 530 metagenome-assembled bacterial and archaeal genomes from the polar Arctic Ocean. Nat. Microbiol. 6, 1561–1564.

Schreiber, S., Fortin, N., Tremblay, J., Wasserscheid, J., Sanschagrin, S., Mason, J., Wright, C.A., Spear, D., Johannessen, S.C., Robinson, B., King, T., Lee, K., Greer, C.W., 2021. *In situ* microcosms deployed at the coast of British Columbia (Canada) to study dilbit weathering and associated microbial communities under marine conditions. FEMS Microbiol. Ecol. 97 (7), fiab082.

Suzuki, M.T., Giovannoni, S.J., 1996. Bias caused by template annealing in the amplification of mixtures of 16S rRNA genes by PCR. Appl. Environ. Microbiol. 62 (2), 625–630.

Taggart, D.M., Clark, K., 2021. Lessons learned from 20 years of molecular biological tools in petroleum hydrocarbon remediation. Remediation 31, 83–95.

Thrift-Viveros, D.L., Jones, R., Boufadel, M., 2015. Development of a new oil biodegradation algorithm for NOAA's oil spill modeling suite (GNOME/ADIOS). In: Proceedings of the 38th AMOP Technical Seminar, Vancouver, BC, Canada, 2–4 June 2015. Environment Canada, Ottawa, ON, Canada, pp. 143–152.

Vilcáez, J., Li, L., Hubbard, S.S., 2013. A new model for the biodegradation kinetics of oil droplets: application to the deepwater horizon oil spill in the Gulf of Mexico. Geochem. Trans. 14, 4.

Wasmund, K., Burns, K.A., Kurtböke, D.I., Bourne, D.G., 2009. Novel alkane hydroxylase gene (*alkB*) diversity in sediments associated with hydrocarbon seeps in the Timor Sea, Australia. Appl. Environ. Microbiol. 75 (23), 7391–7398.

Werner, G., Modlin, C., Watson, W.T., 2016. Basic Ratemaking, fifth ed. Casualty Actuarial Society.

Chapter 7

Genomic biosurveillance to protect the world's forest resources

Richard C. Hamelin
Department of Forest and Conservation Sciences, University of British Columbia, Vancouver, British Columbia, Canada

Introduction

Forests cover approximately 30% of the surface of the planet. They are important to achieve the United Nations Sustainable Development Goals by helping to provide ecosystem services such as clean air and clean water. They also provide wildlife habitat and foster high levels of biodiversity. Forests also provide products such as building material, wood products, food, and energy. One extremely important benefit of forests is their ability to capture and store carbon, thereby mitigating the effects of climate change (Pan et al., 2011). Increasingly, it has become clear that forests also keep people healthier, for example, by intercepting and removing particular matters and gaseous pollutants in the air (Nowak et al., 2014). The multitude of ecosystem services provided by forests is increasingly recognized and valued in economic terms so that they can be quantified (Hein et al., 2006; Tallis et al., 2008).

The planet's forests are under threat during the Anthropocene, the era during which humans are changing the planet in irreversible ways (Lewis and Maslin, 2015). We live in the era of megadisturbances, with climate change, exceptional droughts, and invasive pests and diseases creating conditions that could push some forests beyond thresholds of sustainability (Millar and Stephenson, 2015). This could degrade the services provided by forests (Boyd et al., 2013; Donovan et al., 2013; Grebner, 2014). In order to avoid crossing potentially irreversible thresholds of forest degradation, we need to develop better forest health monitoring. Global assessment of forest health is a major scientific challenge that is critical to identify thresholds and hot spots of rapid forest decline (Trumbore et al., 2015). What is needed is a global network that comprises remote sensing as well as intensive monitoring and laboratory and field

Genomics and the Global Bioeconomy. https://doi.org/10.1016/B978-0-323-91601-1.00011-0

experiments to improve modeling and prediction and to focus on limited resources.

Monitoring forest health is a challenging task. One of those challenges is that forest health specialists must gather data and information that allow them to take actions if undesirable pests and pathogens are discovered or if climate change is causing new outbreaks. Rapid action requires accurate and rapid identification of forest enemies. But there is a wide diversity of forest enemies (insects, fungi, oomycetes, viruses, bacteria, and nematodes), each one requiring specific expertise for identification. Pathogens can also be cryptic or they can remain dormant for weeks or months, which can result in asymptomatic transmission of diseases. In addition to being transmitted on the plant itself, pathogens can be transmitted in a variety of different materials such as soil, water, or seeds. They can also jump hosts, adding a layer of complexity to the task of monitoring and surveying their presence.

The development of DNA-based diagnostics and of genome sequencing has changed the way we can conduct forest health surveys. Forest health specialists have largely benefited from advances in medical applications and developed similar approaches for forest health monitoring. Pathogen identification is now regularly conducted by using the polymerase chain reaction (PCR). Increasingly, high-throughput and whole- or targeted genome sequencing is providing a new layer of information that can inform our analyses of sources, pathways of spread, and identify high-risk traded materials or commodities. A regular flowchart for conducting forest health monitoring can now take advantage of supplementing regular microbe isolations and microscopic observations with DNA barcoding (Fig. 1). Increasingly, DNA can be extracted, in the laboratory or directly in the field, from samples collected during surveys or inspections for PCR testing, metabarcoding, or high-throughput sequencing. This chapter will review some of the latest developments in using genomics for forest health biosurveillance.

DNA detection of forest pathogens

The challenge of pest and pathogen identification

Prevention is one of the most important aspects in keeping forests healthy. Once a pest or a pathogen has established and spread from tree to tree, it is often too late to apply mitigation methods that are economically or environmentally achievable. Prevention requires diligent biosurveillance and monitoring, which rely on the identification of pests and pathogens. One of the greatest challenges faced by forest health specialists is to rapidly and accurately identify potential pests and pathogens from a variety of substrates and products. Not only is there a huge variety of insects and pathogens, but they can also have different appearances at different life stages. Some life stages (eggs and larva for insects, spores for pathogens) can have very few distinguishing morphological characters so

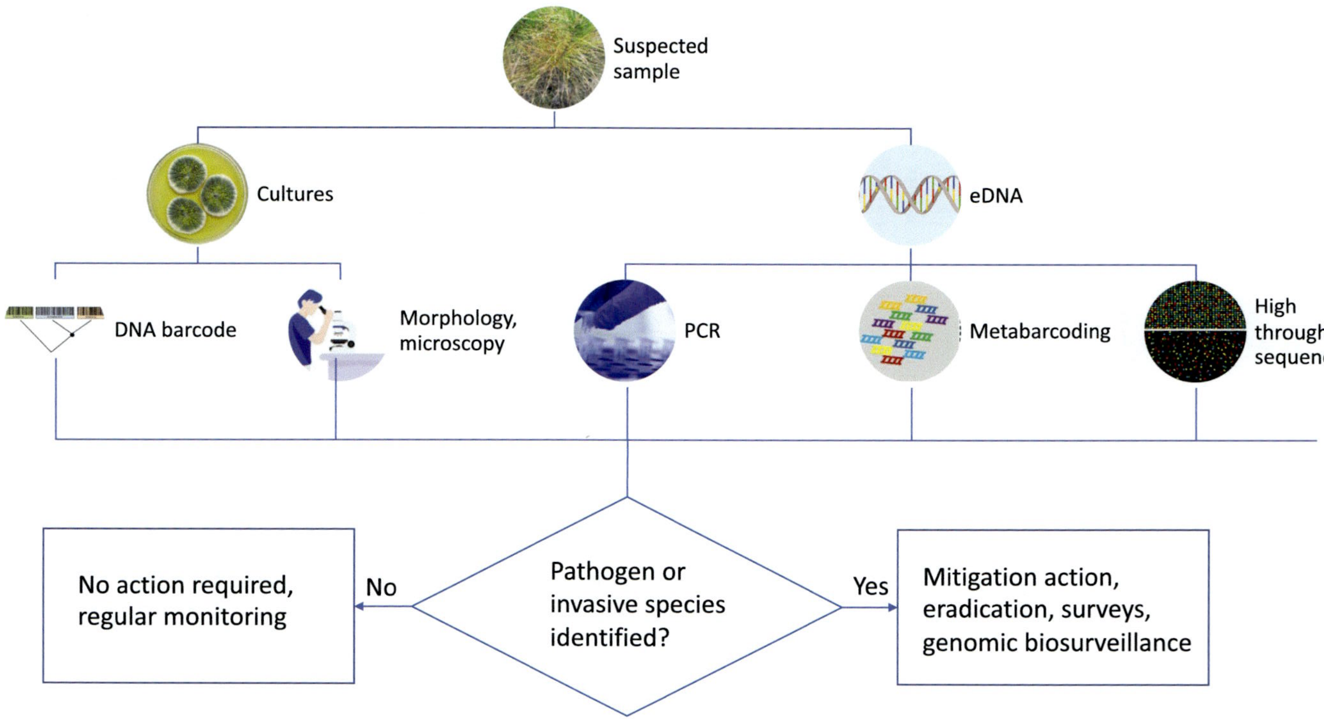

FIG. 1 Flowchart for forest health monitoring and surveillance. Suspected samples harvested during surveys or inspections can be taken to a laboratory for the isolation of fungi and other potential pathogens. Identification can be conducted by morphological observation using microscopes or via DNA barcoding. DNA can also be extracted directly from the suspected samples for PCR if a target pathogen is suspected. When performing diagnostics for pathogen identification, metabarcoding or high-throughput sequencing can be used to capture the entire communities present in a sample. Identification of pathogens or of invasive species triggers mitigation measures, followed by additional surveys. If no pathogen is identified, regular monitoring resumes.

that closely related species cannot be easily differentiated. This could be important because invasive species often have endemic close relatives that can be confounded. In addition, some pathogens can have endophytic stages, meaning that they could be growing within a host tissue without causing symptoms. They can therefore be transmitted asymptomatically and cause disease once conducive conditions trigger an outbreak.

The polymerase chain reaction changed how we detect plant pathogens

Nucleic acid amplification is one of the most useful methods in molecular biology that has applications in the diagnosis of infectious diseases. The polymerase chain reaction (PCR) is the most widely used and reliable method and has improved our ability to detect forest pathogens known to cause severe outbreaks. The application of PCR has greatly improved our ability to detect plant pathogens and modify our mitigation approaches (Martin et al., 2000). A variety of PCR assays have been developed and deployed in diagnostic laboratories for the detection of forest pathogens and pests (Bilodeau et al., 2007; Boutigny et al., 2013; Hamelin et al., 1996, 2000; Lamarche et al., 2015; Schena and Nicosia, 2013; Stewart et al., 2016). PCR is powerful since it allows the detection of very small amounts of DNA from a mixed substrate such as infected leaves, roots, stems, or wood. The method is sensitive, can be highly specific, and can be conducted in high-throughput formats, allowing the user to process large numbers of samples. However, PCR requires obtaining target sequences a priori and can only be used if we know the organism that we are targeting. Unfortunately, potentially invasive species that could threaten forests are often not known prior to their arrival. This was the case of the sudden oak death and the ash dieback, two deadly pathogens that were previously not known to science and have since caused, following their colonization and spread, large outbreaks (Pautasso et al., 2013; Rizzo and Garbelotto, 2003).

DNA barcoding

DNA barcoding has become a very practical tool in diagnostic laboratories to identify potential pathogens or pests during surveys. This method generates a sequence from unknown samples for a predefined DNA region that allows searching databases to identify matches. This powerful approach has revolutionized the way we identify and classify organisms (Hebert et al., 2003). It has been particularly important for the identification of pathogens because of the absence of distinguishing morphological features in some fungi and oomycetes, the causal agents of some of the most important tree diseases (Schoch, 2014; Schoch et al., 2012; Seifert, 2008; Vialle et al., 2009). The establishment of a curated public database at the National Center for Biotechnology Information (NCBI), namely the RefSeq Targeted Loci (RTL) database, contributes to improve the data quality of new sequences and increases confidence in the search outcomes (Schoch, 2014). Examples of DNA barcoding applications include its use in the rust fungi (Pucciniales), a group that comprises important tree and plant pathogens, to investigate host range and cryptic species (Beenken et al., 2017; Feau et al., 2011; Léveillé-Bourret et al., 2021), to correct misidentifications in herbarium samples (Feau et al., 2009), and also to predict potential invasions (Buys et al., 2016).

In spite of the benefits offered by sequencing a single standardized genome region and using common databases, there are limitations of DNA barcoding. The genomes of different taxa evolve at various rates, and the ideal barcoding region for some groups does not provide the desired resolution in other groups (Vialle et al., 2009), thereby complicating standardization. The use of multiple loci was consistently better than using a single standard locus in providing consistent species identification (Roe et al., 2010; Vialle et al., 2009). This is particularly important in mycology, where the DNA barcoding region (the nuclear ribosomal internal transcribed spacer region) does not provide sufficient resolution in speciose genera such as *Penicillium*, *Ophiostoma*, or *Fusarium*, to name a few (Begerow et al., 2010; Roe et al., 2010). Having an accurate species identification can be crucial since close relatives can have completely different characteristics, from pathogenic to endophytic or saprophytic.

To answer this challenge, the mycological community has adopted different barcoding regions for different fungal taxonomic groups and by storing this data in different databases. Ideally, those databases should be curated to ensure reliability in the identification. Having multiple databases is not ideal because such tailored approaches tend to sacrifice quantity for quality (Begerow et al., 2010). A second limitation to DNA barcoding, in particular for pathogens, is the need to have a single organism in culture, or a pure culture, to obtain a DNA sequence. This can be limiting in cases of biotrophic organisms (that cannot grow outside the host), for pathogens that are vectored, or in cases where multiple pathogens cause a disease.

Environmental DNA and metabarcoding for forest pest and pathogen surveillance

High-throughput sequencing of amplicons provides a solution to the challenge of having multiple organisms in samples collected in surveys and inspections (Bik et al., 2012). Amplicon sequencing of environmental DNA (eDNA) is a powerful approach that can be used to generate DNA sequences from complex substrates. This has generated innovative approaches to characterize bacterial, fungal, and oomycete communities from various substrates such as soil, air, and wood using PCR primers targeting the DNA barcoding regions of these microorganisms and conducting high-throughput sequencing (HTS) of those amplicons, so-called metabarcoding (Bálint et al., 2014; Klindworth et al., 2013; Riit et al., 2016). A clear advantage of eDNA metabarcoding is that a list of the microorganisms putatively present in a sample can be obtained without the need to obtain a pure culture.

This is promising and is increasingly applied to plant biosecurity. The data obtained from metabarcoding can provide an early warning system for the detection of invasive pathogens. This could increase survey capacity and detection sensitivity, reduce hands-on time and costs, and assist regulatory agencies (Tremblay et al., 2018). Metagenomics is promising to address critical aspects related to the detection and surveillance of plant pathogens in plant material and foodstuffs (Piombo et al., 2021). Metabarcoding was used to search for the presence of potentially invasive pathogens in pollen pellets collected by honey bees (Tremblay et al., 2019). Metabarcoding has also been useful to identify the cause of some intractable forest declines, a major challenge for ecosystem sustainability. The diversity and composition of fungal and oomycete communities were related to the severity of oak decline, providing clues to the development of biocontrol strategies (Ruiz Gómez et al., 2019). Metabarcoding could also be useful in monitoring fungi associated with insects. Diverse lineages of fungi, including some tree pathogens, were identified by conducting HTS metabarcoding on different tree-attacking insects (Miller et al., 2016).

Yet, there are some challenges with metabarcoding. Metabarcoding still suffers from low precision and a lack of resolution, due to the short length of the sequenced region (Piombo et al., 2021). This matters because species with identical metabarcode sequences can have widely different lifestyles and represent different risks. Improvements in the use of long-read metabarcoding promise to resolve, in part, this issue (Tedersoo et al., 2021; Tedersoo and Anslan, 2019). Yes, careful benchmarking is an important step before adopting any new approach (Kennedy et al., 2018). Similar to the DNA barcoding challenges, the problem of the level of divergence found in the metabarcoding region in different fungal groups, in particular the ITS region, is not likely to be resolved with long reads. Genus-specific amplicon sequencing can be conducted, for example, the Fusarium-specific primers targeting the translation elongation factor 1-α (EF1α) gene. Although this can be useful for profiling Fusarium species composition and dynamics (Cobo-Díaz et al., 2019), it fails to provide information on the entire fungal community and would be limited for pathogen surveillance.

Genomic biosurveillance of forest pathogens

Some pathogens represent urgent threats because of the damage they cause, the hosts they infect, or simply because they have an ability to propagate rapidly. This is the case of the sudden oak and larch death pathogen (*Phytophthora ramorum*; Fig. 2A and B), Dutch elm disease (*Ophiostoma novo-ulmi*; Fig. 2C and D), and ash dieback (*Hymenoscyphus fraxineus*) that are already established in some jurisdictions. These diseases have caused severe damage to trees and are regulated by countries to prevent their introduction or further spread to new regions. These pathogens are efficient invaders, and suspected

FIG. 2 Two important forest diseases. The sudden oak and larch death, caused by *Phytophthora ramorum*, causes tree mortality in oaks and larch in the United Kingdom (A) by inducing cankers (B). The Dutch elm disease has killed billions of elm trees in Europe and North America (C) by causing a wilt that blocks the tree's vascular tissues (D). *(All photographs by R. Hamelin.)*

infections encountered during surveys or inspections often trigger regulatory actions that can include local eradication via the destruction of the infected material, fumigation or fungicide treatment, and trace-back and trace-forward to identify sources and to prevent additional spread.

Whole- or partial genome sequencing can help track the spread of these pathogens, identify pathways of dissemination, and identify sources so that preventative actions can be taken (Hamelin and Roe, 2020; Roe et al., 2018). The first step in this approach is to characterize the genomic variation in the potentially invasive species, including the native range, if it is known, and to map its diversity and characterize its genetic and genomic traits (Hamelin and Roe, 2020; Roe et al., 2018). Invasion routes can be traced with great precision using phylogeographic surveys of native and introduced populations to reconstruct invasion history (Cristescu, 2015). Combining HTS with databases of repository sequences opens the possibility to identify early invasions and to quantify failed ones, which could be important to improve prevention. Global population genomic analysis is in its infancy for tree pathogens, yet data that can help better understand the origin and dynamic of outbreaks has been generated. The genomes hundreds of samples of the pathogens that cause the ash dieback (McMullan et al., 2018), the Dutch elm disease (Hessenauer et al., 2020), the chestnut blight (Demené et al., 2019), and the sudden oak death (Dale et al., 2019; Hamelin et al., 2021) have been sequencing and compared.

Analyses of genomes obtained from populations of samples can inform us about the colonization dynamics, the adaptive potential, and evolutionary events such as recombination or hybridization. Whole-genome sequencing of *Hymenoscyphus fraxineus*, the fungal pathogen that causes ash dieback, revealed that only two divergent individuals founded the European population responsible for the continent-wide outbreak (McMullan et al., 2018). The study also revealed a far greater diversity in pathogenicity-related genes in the native range, prompting the need to increase measures to prevent further introductions. Whole-genome sequences of the Dutch elm disease pathogens revealed that hybridization and introgression among the species and lineages that cause the disease have shaped genomic diversity and impacted fitness-related traits, including pathogenicity (Hessenauer et al., 2020). Genome sequencing of *Phytophthora ramorum*, the causal agent of sudden oak death, revealed that clonal lineages continue to evolve and diverge via a rapidly evolving noncore genome and mitotic recombination (Dale et al., 2019). A genomic biosurveillance approach revealed the presence of a hybrid between two of those lineages that could potentially alter the course of the epidemic (Hamelin et al., 2021).

Portable DNA and genomic testing: The next frontier

All approaches described in the previous sections rely upon extensive laboratory procedures and computing capacity and analyses. Forest health monitoring is often conducted in remote locations where laboratory facilities are not readily available. To collect survey samples and bring them to a laboratory can delay results by days, weeks, and often months. For an efficient response, it is often important to have results rapidly and accurately. Point-of-care or point-of-use methods hold great promises to provide on-site rapid DNA testing that can be used for early detection of plant pathogens. The tools and instruments for portable DNA analyses have been driven mostly by medical or military applications. However, for plant pathogens, the material to be tested is often so different from medical samples that novel approaches are often needed.

There are multiple approaches that can be used to perform DNA detection in the field. Loop-mediated isothermal amplification (LAMP) assays amplify DNA with high specificity, efficiency, and rapidity under isothermal conditions. The assay recognizes the target by six distinct sequences initially and by four distinct sequences afterward, so it affords high selectivity (Notomi et al., 2000). LAMP assays have been developed for quarantine plant pathogen detection as an early warning system (Aglietti et al., 2019) and for several important forest pathogens (Aglietti et al., 2021; Hamilton et al., 2021).

Quantitative PCR, which has traditionally been confined to laboratories, can now be performed in field-portable instruments. There are several advantages of using qPCR compared to other types of assays. Quantitative PCR remains the gold standard for sensitivity and reliability, and there are already a large number of validated assays that can be transferred to design field assays. Therefore, assay design and validation do not have to be repeated, saving time of resources.

Field-ready qPCR has been applied to test for water quality (Billington et al., 2021; Nguyen et al., 2018) and to detect tropical diseases (Rampazzo et al., 2019). Recently, methods for portable detection of forest enemies were developed (Capron et al., 2020). The method uses a crude buffer-based DNA extraction protocol and lyophilized, premade reactions for on-site applications and was demonstrated with pathogens and pests covering a broad taxonomic range, from the Ascomycota and Basidiomycota in the pathogens to the Coleoptera and Lepidoptera for the insects (Capron et al., 2020) (Fig. 3A). DNA detection could be performed from a variety of different tissues, including infected leaves, pathogen spores, or insect legs and antennae (Fig. 3B). The kit necessary to

FIG. 3 Portable PCR detection of forest enemies. The kit necessary to conduct these assays includes a portable qPCR instrument, lyophilized, premade, reactions that only require adding water, extraction reagent, and microtubes (A); the material fits in a backpack and can be carried to remote locations, which is ideal for environmental applications. Crude buffer-based DNA extraction protocol only requires harvesting material from plants, fungi, or insects and heating them in an extraction buffer (B); this does not require equipment and is ideal for on-site applications.

conduct these assays fits in a backpack and can be carried to remote locations, which is ideal for environmental applications.

Outlook, future development, and needs

The genomic revolution is still in its infancy, and the promises of genomic applications in forest protection are only beginning. However, there have already been significant advances and PCR-based applications are now regular tools used by forest health experts. The COVID-19 pandemic has generated public awareness of the power of PCR in detecting infections and of genomics in tracking spread. This new awareness is helping bring similar applications in forestry applications. But similar challenges face medical and forest experts. Data sharing can be sensitive if invasive species are discovered and reported in databases along with the associated metadata. For global biosurveillance to be efficient, having open access databases is a must, yet, publication of data and metadata can be met with resistance by stakeholders because of the potential trade impacts on reporting of potentially invasive species. Yet, the future of forest health monitoring will increasingly demand access to the powerful new genomic tools under development and sharing of the data.

Healthy forests are essential to meet the United Nations Sustainable Development Goals. They provide a source of materials that can be used to make products, including clean energy and bioproducts. They increasingly play a crucial role in our cities by reducing pollution and as natural air-conditioners, reduce the impact of climate change by sequestering carbon, and foster the biodiversity that is essential for life on the planet. Novel approaches such as genomics, combined with modeling and remote sensing, hold great promises to predict future outbreaks and monitor current ones. This will be an important part of the toolbox required to develop real-time global forest health monitoring to protect and preserve our forests.

References

Aglietti, C., Luchi, N., Pepori, A.L., Bartolini, P., Pecori, F., Raio, A., Capretti, P., Santini, A., 2019. Real-time loop-mediated isothermal amplification: an early-warning tool for quarantine plant pathogen detection. AMB Express 9, 1–14.

Aglietti, C., Meinecke, C.D., Ghelardini, L., Barnes, I., van der Nest, A., Villari, C., 2021. Rapid detection of pine pathogens *Lecanosticta acicola*, *Dothistroma pini* and *D. septosporum* on needles by probe-based LAMP assays. Forests 12, 479.

Bálint, M., Schmidt, P.-A., Sharma, R., Thines, M., Schmitt, I., 2014. An Illumina metabarcoding pipeline for fungi. Ecol. Evol. 4, 2642–2653. https://doi.org/10.1002/ece3.1107.

Beenken, L., Lutz, M., Scholler, M., 2017. DNA barcoding and phylogenetic analyses of the genus *Coleosporium* (Pucciniales) reveal that the North American goldenrod rust *C. solidaginis* is a neomycete on introduced and native *Solidago* species in Europe. Mycol. Prog. 16, 1073–1085.

Begerow, D., Nilsson, H., Unterseher, M., Maier, W., 2010. Current state and perspectives of fungal DNA barcoding and rapid identification procedures. Appl. Microbiol. Biotechnol. 87, 99–108.

Bik, H.M., Porazinska, D.L., Creer, S., Caporaso, J.G., Knight, R., Thomas, W.K., 2012. Sequencing our way towards understanding global eukaryotic biodiversity. Trends Ecol. Evol. 27, 233–243.

Billington, C., Abeysekera, G., Scholes, P., Pickering, P., Pang, L., 2021. Utility of a field deployable qPCR instrument for analyzing freshwater quality. Agrosyst. Geosci. Environ. 4, e20223.

Bilodeau, G.J., Lévesque, C.A., de Cock, A., Duchaine, C., Kristjansson, G., Hamelin, R.C., 2007. Molecular detection of *Phytophthora ramorum* by real-time PCR using Taqman, SYBR green and molecular beacons with three genes. Phytopathology 97, 632–642.

Boutigny, A.-L., Guinet, C., Vialle, A., Hamelin, R.C., Andrieux, A., Frey, P., Husson, C., Ioos, R., 2013. Optimization of a real-time PCR assay for the detection of the quarantine pathogen *Melampsora medusae* f. sp. *deltoidae*. Fungal Biol. 117, 389–398.

Boyd, I.L., Freer-Smith, P.H., Gilligan, C.A., Godfray, H.C.J., 2013. The consequence of tree pests and diseases for ecosystem services. Science 342, 1235773. https://doi.org/10.1126/science.1235773.

Buys, M., Flint, H., Miller, E., Yao, H., Caird, A., Ganley, R., 2016. Preparing for the invasion: efficacy of DNA barcoding to discern the host range of myrtle rust (*Puccinia psidii*) among species of Myrtaceae. Forestry 89, 263–270.

Capron, A., Stewart, D., Hrywkiw, K., Allen, K., Feau, N., Bilodeau, G., Tanguay, P., Cusson, M., Hamelin, R.C., 2020. In situ processing and efficient environmental detection (iSPEED) of tree pests and pathogens using point-of-use real-time PCR. PLoS ONE 15, e0226863. https://doi.org/10.1371/journal.pone.0226863.

Cobo-Díaz, J.F., Baroncelli, R., Le Floch, G., Picot, A., 2019. A novel metabarcoding approach to investigate *Fusarium* species composition in soil and plant samples. FEMS Microbiol. Ecol. 95, fiz084.

Cristescu, M.E., 2015. Genetic reconstructions of invasion history. Mol. Ecol. 24, 2212–2225. https://doi.org/10.1111/mec.13117.

Dale, A.L., Feau, N., Everhart, S.E., Dhillon, B., Wong, B., Sheppard, J., Bilodeau, G.J., Brar, A., Tabima, J.F., Shen, D., Brasier, C.M., Tyler, B.M., Grünwald, N.J., Hamelin, R.C., 2019. Mitotic recombination and rapid genome evolution in the invasive forest pathogen *Phytophthora ramorum*. mBio 10. https://doi.org/10.1128/mBio.02452-18. e02452-18.

Demené, A., Legrand, L., Gouzy, J., Debuchy, R., Saint-Jean, G., Fabreguettes, O., Dutech, C., 2019. Whole-genome sequencing reveals recent and frequent genetic recombination between clonal lineages of *Cryphonectria parasitica* in western Europe. Fungal Genet. Biol. 130, 122–133.

Donovan, G.H., Butry, D.T., Michael, Y.L., Prestemon, J.P., Liebhold, A.M., Gatziolis, D., Mao, M. Y., 2013. The relationship between trees and human health. Am. J. Prev. Med. 44, 139–145.

Feau, N., Vialle, A., Allaire, M., Tanguay, P., Joly, D.L., Frey, P., Hamelin, R.C., 2009. Fungal pathogen (mis-) identifications: a case study with DNA barcodes on *Melampsora rusts* of aspen and white poplar. Mycol. Res. 113, 713–724.

Feau, N., Vialle, A., Allaire, M., Maier, W., Hamelin, R.C., 2011. DNA barcoding in the rust genus *Chrysomyxa* and its implications for the phylogeny of the genus. Mycologia 103, 1250–1266.

Grebner, 2014. Forest ecosystem services and the scourge of invasive species. J. For., 112.

Hamelin, R.C., Roe, A.D., 2020. Genomic biosurveillance of forest invasive alien enemies: a story written in code. Evol. Appl. 13, 95–115. https://doi.org/10.1111/eva.12853.

Hamelin, R.C., Berube, P., Gignac, M., Bourassa, M., 1996. Identification of root rot fungi in nursery seedlings by nested multiplex PCR. Appl. Environ. Microbiol. 62, 4026–4031.

Hamelin, R., Bourassa, M., Rail, J., Dusabenyagasani, M., Jacobi, V., Laflamme, G., 2000. PCR detection of *Gremmeniella abietina*, the causal agent of Scleroderris canker of pine. Mycol. Res. 104, 527–532.

Hamelin, R., Bilodeau, G., Heinzelmann, R., Hrywkiw, K., Capron, A., Dort, E., Dale, A., Giroux, E., Carleson, N., Grünwald, N., et al., 2021. Genomic Biosurveillance Detects A Sexual Hybrid in the Sudden Oak Death Pathogen., https://doi.org/10.21203/rs.3.rs-699860/v1.

Hamilton, J., Fraedrich, S., Nairn, C., Mayfield, A., Villari, C., 2021. A field-portable diagnostic approach confirms Laurel Wilt Disease diagnosis in minutes instead of days. Arboricult. Urban For. 47, 98–109.

Hebert, P.D.N., Cywinska, A., Ball, S.L., deWaard, J.R., 2003. Biological identifications through DNA barcodes. Proc. R. Soc. B Biol. Sci. 270, 313–321.

Hein, L., Van Koppen, K., De Groot, R.S., Van Ierland, E.C., 2006. Spatial scales, stakeholders and the valuation of ecosystem services. Ecol. Econ. 57, 209–228.

Hessenauer, P., Fijarczyk, A., Martin, H., Prunier, J., Charron, G., Chapuis, J., Bernier, L., Tanguay, P., Hamelin, R.C., Landry, C.R., 2020. Hybridization and introgression drive genome evolution of Dutch elm disease pathogens. Nat. Ecol. Evol., 1–13.

Kennedy, P.G., Cline, L.C., Song, Z., 2018. Probing promise versus performance in longer read fungal metabarcoding. New Phytol. 217, 973–976.

Klindworth, A., Pruesse, E., Schweer, T., Peplies, J., Quast, C., Horn, M., Glöckner, F.O., 2013. Evaluation of general 16S ribosomal RNA gene PCR primers for classical and next-generation sequencing-based diversity studies. Nucleic Acids Res. 41, e1.

Lamarche, J., Potvin, A., Pelletier, G., Stewart, D., Feau, N., Alayon, D.I.O., Dale, A.L., Coelho, A., Uzunovic, A., Bilodeau, G.J., Brière, S.C., Hamelin, R.C., Tanguay, P., 2015. Molecular detection of 10 of the most unwanted alien forest pathogens in Canada using real-time PCR. PLoS ONE 10, e0134265.

Léveillé-Bourret, É., Eggertson, Q., Hambleton, S., Starr, J.R., 2021. Cryptic diversity and significant cophylogenetic signal detected by DNA barcoding the rust fungi (Pucciniaceae) of Cyperaceae–Juncaceae. J. Syst. Evol. 59, 833–851.

Lewis, S.L., Maslin, M.A., 2015. Defining the Anthropocene. Nature 519, 171–180. https://doi.org/10.1038/nature14258.

Martin, R., James, D., Levesque, C., 2000. Impacts of molecular diagnostic technologies on plant disease management. Annu. Rev. Phytopathol. 38, 207–239.

McMullan, M., Rafiqi, M., Kaithakottil, G., Clavijo, B.J., Bilham, L., Orton, E., Percival-Alwyn, L., Ward, B.J., Edwards, A., Saunders, D.G.O., Garcia Accinelli, G., Wright, J., Verweij, W., Koutsovoulos, G., Yoshida, K., Hosoya, T., Williamson, L., Jennings, P., Ioos, R., Husson, C., Hietala, A.M., Vivian-Smith, A., Solheim, H., MacLean, D., Fosker, C., Hall, N., Brown, J.K.M.,

Swarbreck, D., Blaxter, M., Downie, J.A., Clark, M.D., 2018. The ash dieback invasion of Europe was founded by two genetically divergent individuals. Nat. Ecol. Evol. 2, 1000–1008.
Millar, C.I., Stephenson, N.L., 2015. Temperate forest health in an era of emerging megadisturbance. Science 349, 823–826.
Miller, K.E., Hopkins, K., Inward, D.J., Vogler, A.P., 2016. Metabarcoding of fungal communities associated with bark beetles. Ecol. Evol. 6, 1590–1600.
Nguyen, P.L., Sudheesh, P.S., Thomas, A.C., Sinnesael, M., Haman, K., Cain, K.D., 2018. Rapid detection and monitoring of *Flavobacterium psychrophilum* in water by using a handheld, field-portable quantitative PCR system. J. Aquat. Anim. Health 30, 302–311.
Notomi, T., Okayama, H., Masubuchi, H., Yonekawa, T., Watanabe, K., Amino, N., Hase, T., 2000. Loop-mediated isothermal amplification of DNA. Nucleic Acids Res. 28, e63.
Nowak, D.J., Hirabayashi, S., Bodine, A., Greenfield, E., 2014. Tree and forest effects on air quality and human health in the United States. Environ. Pollut. 193, 119–129.
Pan, Y., Birdsey, R.A., Fang, J., Houghton, R., Kauppi, P.E., Kurz, W.A., Phillips, O.L., Shvidenko, A., Lewis, S.L., Canadell, J.G., et al., 2011. A large and persistent carbon sink in the world's forests. Science 333, 988–993.
Pautasso, M., Aas, G., Queloz, V., Holdenrieder, O., 2013. European ash (*Fraxinus excelsior*) dieback – a conservation biology challenge. Biol. Conserv. 158, 37–49.
Piombo, E., Abdelfattah, A., Droby, S., Wisniewski, M., Spadaro, D., Schena, L., 2021. Metagenomics approaches for the detection and surveillance of emerging and recurrent plant pathogens. Microorganisms 9, 188.
Rampazzo, R.C., Graziani, A.C., Leite, K.K., Surdi, J.A., Biondo, C.A., Costa, M.L., Jacomasso, T., Cereda, M., De Fazio, M., Bianchessi, M.A., et al., 2019. Proof of concept for a portable platform for molecular diagnosis of tropical diseases: on-chip ready-to-use real-time quantitative pcr for detection of *Trypanosoma cruzi* or *Plasmodium* spp. J. Mol. Diagnos. 21, 839–851.
Riit, T., Tedersoo, L., Drenkhan, R., Runno-Paurson, E., Kokko, H., Anslan, S., 2016. Oomycete-specific ITS primers for identification and metabarcoding. MycoKeys 14, 17.
Rizzo, D.M., Garbelotto, M., 2003. Sudden oak death: endangering California and Oregon forest ecosystems. Front. Ecol. Environ. 1, 197–204.
Roe, A.D., Rice, A.V., Bromilow, S.E., Cooke, J.E.K., Sperling, F.A.H., 2010. Multilocus species identification and fungal DNA barcoding: insights from blue stain fungal symbionts of the mountain pine beetle. Mol. Ecol. Resour. 10, 946–959.
Roe, A.D., Torson, A.S., Bilodeau, G., Bilodeau, P., Blackburn, G.S., Cui, M., Cusson, M., Doucet, D., Griess, V.C., Lafond, V., Paradis, G., Porth, I., Prunier, J., Srivastava, V., Tremblay, E., Uzunovic, A., Yemshanov, D., Hamelin, R.C., 2018. Biosurveillance of forest insects: part I—integration and application of genomic tools to the surveillance of non-native forest insects. J. Pest. Sci. https://doi.org/10.1007/s10340-018-1027-4.
Ruiz Gómez, F.J., Navarro-Cerrillo, R.M., Pérez-de-Luque, A., Oβwald, W., Vannini, A., Morales-Rodríguez, C., 2019. Assessment of functional and structural changes of soil fungal and oomycete communities in holm oak declined dehesas through metabarcoding analysis. Sci. Rep. 9, 1–16.
Schena, L., Li Destri Nicosia, M.G., Sanzani, S., Faedda, R., Ippolito, A., Cacciola, S., 2013. Development of quantitative PCR detection methods for phytopathogenic fungi and oomycetes. J. Plant Pathol., 7–24.
Schoch, C., 2014. Finding needles in haystacks: linking scientific names, reference specimens and molecular data for Fungi. Database (Oxford), 1–36.
Schoch, C.L., Seifert, K.A., Huhndorf, S., Robert, V., Spouge, J.L., Levesque, C.A., Chen, W., Fungal Barcoding Consortium, Fungal Barcoding Consortium Author List, 2012. Nuclear

ribosomal internal transcribed spacer (ITS) region as a universal DNA barcode marker for fungi. Proc. Natl. Acad. Sci. U. S. A. 109, 6241–6246.

Seifert, K.A., 2008. Integrating DNA barcoding into the mycological sciences. Persoonia 21, 162–166.

Stewart, D., Zahiri, R., Djoumad, A., Freschi, L., Lamarche, J., Holden, D., Cervantes, S., Ojeda, D. I., Potvin, A., Nisole, A., Béliveau, C., Capron, A., Kimoto, T., Day, B., Yueh, H., Duff, C., Levesque, R.C., Hamelin, R.C., Cusson, M., 2016. A multi-species TaqMan PCR assay for the identification of Asian gypsy moths (*Lymantria* spp.) and other invasive lymantriines of biosecurity concern to North America. PLoS ONE 11, e0160878. https://doi.org/10.1371/journal.pone.0160878.

Tallis, H., Kareiva, P., Marvier, M., Chang, A., 2008. An ecosystem services framework to support both practical conservation and economic development. Proc. Natl. Acad. Sci. 105, 9457–9464.

Tedersoo, L., Anslan, S., 2019. Towards PacBio-based pan-eukaryote metabarcoding using full-length ITS sequences. Environ. Microbiol. Rep. 11, 659–668.

Tedersoo, L., Albertsen, M., Anslan, S., Callahan, B., 2021. Perspectives and benefits of high-throughput long-read sequencing in microbial ecology. Appl. Environ. Microbiol. 87. e00626-21.

Tremblay, É.D., Duceppe, M.-O., Bérubé, J.A., Kimoto, T., Lemieux, C., Bilodeau, G.J., 2018. Screening for exotic forest pathogens to increase survey capacity using metagenomics. Phytopathology 108, 1509–1521.

Tremblay, É.D., Duceppe, M.-O., Thurston, G.B., Gagnon, M.-C., Côté, M.-J., Bilodeau, G.J., 2019. High-resolution biomonitoring of plant pathogens and plant species using metabarcoding of pollen pellet contents collected from a honey bee hive. Environ. DNA 1, 155–175.

Trumbore, S., Brando, P., Hartmann, H., 2015. Forest health and global change. Science 349, 814–818.

Vialle, A., Feau, N., Allaire, M., Didukh, M., Martin, F.N., Moncalvo, J.-M., Hamelin, R.C., 2009. Evaluation of mitochondrial genes as DNA barcode for Basidiomycota. Mol. Ecol. Resour. 9, 99–113.

Chapter 8

Metagenomics: A resilience approach to climate change and conservation of the African Glacier biodiversity

Josiah O. Kuja[a,b], Anne W.T. Muigai[a], and Jun Uetake[c]
[a]*Department of Botany, School of Biological Sciences, Jomo Kenyatta University of Agriculture and Technology, Juja, Kenya,* [b]*Department of Biology, Computational and RNA Biology, University of Copenhagen, Copenhagen, Denmark,* [c]*Field Research Centre for Northern Biosphere, Hokkaido University, Sapporo, Japan*

Introduction

Glaciers are large bodies of dense ice that are formed when ice accumulates faster than it is ablated. They are classified as either temperate or cold depending on their temperature regime (Hodson et al., 2008). Temperate glaciers are characterized with the occurrence of ice at the melting points, while cold glaciers consist of the subglacial temperature variations due to microbial activities and other glacial physiological processes (Anesio and Laybourn-Parry, 2012; Anesio et al., 2017). Glaciers are often inhabited by diverse organisms such as snow algae (Hoham and Remias, 2020), *Cyanobacteria* (Takeuchi et al., 2001), yeast (Branda et al., 2010), bacteria (Segawa et al., 2005), invertebrates, and metazoans (Zawierucha et al., 2021a,b).

Initial studies conducted on the glacier surfaces utilized ice cores to date bacteria, fungi, and algae present in the ice sheets from polar and mountain regions (Christner et al., 2018; Miteva et al., 2015). Most of the glacial studies by then utilized microscopy and spectroscopy for identification and characterization of microorganisms (Hodson et al., 2010; Stibal et al., 2008). Macroscopic filamentous *Cyanobacteria* that dominate the glacier surfaces were therefore the only known glacier prokaryotes (Whitton and Potts, 2007). It was until 2007 that Kaštovská et al. (2007) used fatty acid analyses to identify phylum *Actinobacteria* and 2014 that Edwards et al. (2019) used 16S rRNA amplicon sequences to detect and characterize other bacterial species on the

Genomics and the Global Bioeconomy. https://doi.org/10.1016/B978-0-323-91601-1.00007-9

glacier. Since then, studies have reported *Cyanobacteria* and *Proteobacteria* as the predominant phyla on the glacier surfaces (Edwards et al., 2013; Gokul et al., 2016) and *Actinobacteria, Cyanobacteria, Proteobacteria, Bacteroidetes, Chloroflexi, and Gemmatimonadetes* as both ubiquitous and abundant phyla (Gokul et al., 2016) on the glacier surfaces globally.

The advances and integrations of culture-dependent studies (Hodson et al., 2010; Stibal et al., 2008) into culture-independent studies indicated that genomic tools such as metagenomics, amplicon sequencing, and transcriptomics are indispensable for a full understanding of the glacial flora and fauna through the identification and characterization of cold-adapted organisms. The tools are sensitive, accurate, and efficient for rapid detection of unidentified species. Without the utilization of amplicon sequences, some of the new species would have not been identified, especially from extreme environments. In Arctic glaciers, metagenomics have been used to identify prokaryotes as sources of enzymes and proteins for application in industrial biotechnology due to their innovative physiological and metabolic adaptations to extreme stressors (Saritha et al., 2021). In 2017, the World Enzyme reported a rise in demand for enzymes by 0.5% (Wiltschi et al., 2020), and this is likely to rise given the increased demand for zero carbon and industrialization.

In the recent years, Kenya has utilized genomic tools to study microbial diversity in the low-altitude terrestrial ecosystems (Kambura et al., 2016; Karanja et al., 2020; Kiama et al., 2021), but not in the high-altitude glacial biomes that are rapidly disappearing due to rising temperatures and reduced cloudiness (Prinz et al., 2016). This chapter outlines the significance of amplicon sequencing of the 16S rRNA in the identification and characterization of prokaryotes from the cryoconite of Lewis Glacier and its foreland. Cryoconite is a dark sediment consisting of organic and inorganic substances occurring at the ablation zone inside glacier surface depressions (cryoconite holes) (Cook et al., 2016). We hypothesized that changes in microbial community structure between the glacier (cryoconite holes) and its foreland soil indicate significant environmental changes due to unprecedented climate change. The dark color of the cryoconite granules lowers glacial surface reflectance, hence facilitating localized warming and ice melting. This chapter also outlines the bioprospecting potential of the bacterial species identified on Lewis Glacier in reference to the initial amplicon sequence datasets from the similar glacial environment in the Arctic (Saritha et al., 2021).

Our study on Lewis Glacier and its foreland is the first one in Africa and astride the equator in which amplicon sequences of 16S rRNA gene have been used as an accurate and sensitive approach to detect and characterize metagenomes of the cold-adapted prokaryotic communities. This study also serves as a model for the use of genomics in remote and neglected environments to improve regional bioeconomy, which is in line with the 17 United Nations 2030 agenda for Sustainable Development Goals (SDGs). Genomic analyses of the glacier metagenomes contribute to the achievement of SDG numbers 2, 4, 9, 13, and

15 that advocate for zero hunger, quality education, industry and innovation, climate action, and conserving life on earth, respectively.

Climate change and the Kenyan glaciers

Bioeconomy is the use of renewable biological resources such as land, sea, crops, plants and forests, fish, animals, and microorganisms to produce food, materials, and energy. It provides an alternative to current economic strategies that are predominantly fossil-dependent by synergistically combining natural resource with technologies, markets, people, and policies to address current challenges experienced such as reducing natural resources, climate change, high waste production, biodiversity losses, food and energy insecurities while achieving sustainable growth.

The continued climate change being observed in Africa is a grand challenge to achieving sustainable bioeconomy. The Sixth Assessment Report of the Intergovernmental Panel on Climate Change (IPCC) on August 9, 2021, estimated chances of crossing the global warming level of 1.5°C in the next decades, given the increased emission of greenhouse gases from human activities (Paglia and Parker, 2021). The IPCC report indicated that the rising temperatures may exceed 1.5°C or even 2°C and this will definitely result in sudden rise in land surface temperatures including glacial surfaces, snow caps, and supraglacial layers that contain the cryoconite holes. The situation of rising temperature would, however, be worse for the African continent that is characterized with the tropical and subtropical ecosystems with high mountains holding the last surviving glacial zones. Glacier biodiversity will be exposed to increased heat waves and longer seasons of persistent thermal regimes without snowfalls to facilitate the growth of cold-adapted microorganisms.

An inventory on Mount Kenya done in 2004 revealed that 18 original glaciers found on the mountain had at least reduced by 1.64 km^2 each, at the end of the 19th century approximately 0.27 km^2 in 2004 (Prinz et al., 2018). Eight of the 18 glaciers had, however, completely disappeared by the turn of the 20th century by suffering a substantial loss (Hastenrath, 2006; Prinz et al., 2012), and only a few, the Lewis and Tyndall glaciers, are left. Even the Gregory Glacier that was joined to Lewis Glacier from the leeward side was the last glacier to suffer rapid loss in 2011 (Prinz et al., 2011). The dramatic loss of these glaciers is linked to climate change and increasing destructive human activities, including deforestation along the mountain contours (Agrawala et al., 2005; OECD, 2002).

Lewis Glacier suffers severe heat radiations and persistent climate change as significant threats to the glacial biodiversity. Climate change alters microbial species diversity, community structure, abundance, and several microbial activities that require a conducive moist environment for growth and development (Margesin and Niklinska, 2019). Such dramatic changes interfere with the microbial species distribution and their interaction within the ecosystem.

The complexity of the microbial natural communities means that they can be either beneficial or pathogenic to the environment, other microbial species, or the associated plants. However, all these factors change with increased environmental stress resulting in an altered biodiversity and natural function of terrestrial ecosystems (Nath et al., 2021).

Environmental metagenomics

Metagenomics (derived from the two words "meta"—many and "genomics"—to obtain a genomic sequence) is the study of many microbial communities at the same time, because ordinarily, these communities exist together in their particular environments such as in soils, snow, ice or in human biological systems such as the gut. In this technique, DNA from a community is extracted, pooled together, and then subjected to amplicon sequencing, which is targeted gene sequencing. The major limitation of the microscopic and spectroscopic methods was that sequencing was only possible for the microbes that could be cultured. Due to many factors, only an estimated 1% of microbes found in the natural environment can be cultured. By circumventing the need to first isolate and culture the microbes, difficult-to-culture communities can now be studied through metagenomics that allows for the sampling and analysis of microbial communities collected directly from the environment. The analysis focuses on identifying the microbes that are present, determining their functions in the environment, and understanding the ratios in which they are found. Information resulting from such analyses is key to understand the interactions among microbial communities and the environment.

The amplicon sequence metagenomic studies of the microbial communities on Lewis Glacier, one of the best-studied equatorial glaciers and the largest glacier found in Mount Kenya, were conducted, primarily to determine the ratios of the abundant phyla as well as those that can be used as indicators of environmental changes across the glacier and its foreland. To do this, cryoconite and soil samples obtained from the glacier and glacier foreland were subjected to genomic analysis that encompassed amplicon sequencing of the 16S RNA gene from all the DNAs that were isolated from the samples.

Results of the 16S rRNA amplicon sequences from Lewis Glacier

Genomic analyses of the Lewis Glacier and its foreland samples revealed abundance of 12 prokaryotic phyla. Cyanobacteria were the most prominent phyla in the glacier samples, with a relative abundance of 37% in both the upper and lower glacier sites, while Proteobacteria were the second most prominent phyla in the glacier samples and the most prominent in the foreland samples. The genomic analyses of the samples also showed that the relative abundance of the Cyanobacteria decreased along the glacier foreland with the increase of age from the primary foreland to old soil (Figs. 8.1 and 8.2).

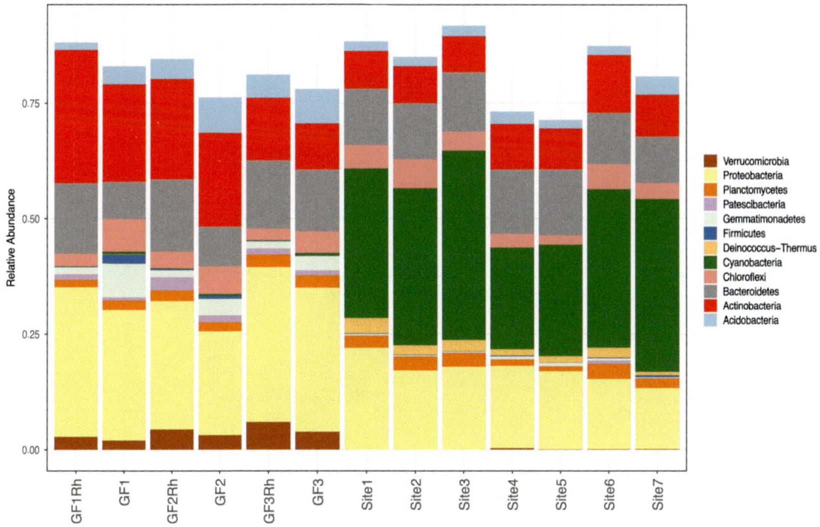

FIG. 8.1 Prokaryotic taxa at phylum level in samples from the Lewis Glacier and its foreland. LG1 = upper Lewis site (*red*), LG2 = lower Lewis site (*blue*), GF = glacier foreland, Rh = rhizosphere (*purple*), soil = black, SF = side foreland.

Further, genomic analyses revealed Bacteroidetes as the third most abundant phyla on the glacier with the relative abundance of 15% on the upper glacier and 14% on the lower glacier, while phyla Actinobacteria were the fourth most abundant with a relative abundance of 12% on the upper glacier and 9% on the lower glacier. The genomic analyses, however, revealed that phyla Verrucomicrobia and Gemmatimonadetes were specific to the foreland samples. Verrucomicrobia were associated with the newly (GF2) formed foreland and the rhizosphere samples (GF2Rh), while Gemmatimonadetes were evenly distributed across the rhizosphere and soil samples (Figs. 8.1 and 8.2). The other six relatively abundant phyla included Chloroflexi, Acidobacteria, Planctomycetes, Deinococcus-Thermus, Firmicutes, and Patescibacteria. They were homogeneously distributed across the sample types with relatively overlapping abundance (Figs. 8.1 and 8.2).

What do these results mean?

Going beyond the identified 12 major phyla across the equatorial glacier and its foreland, results from the genomic analyses revealed changes in the microbial composition and distribution between the glacier and its foreland chronosequence. Such changes are correlated with environmental changes due to rising temperatures and reducing cloudiness. Most of the environmental changes are due to unprecedented climate change, which we have hypothesized to influence

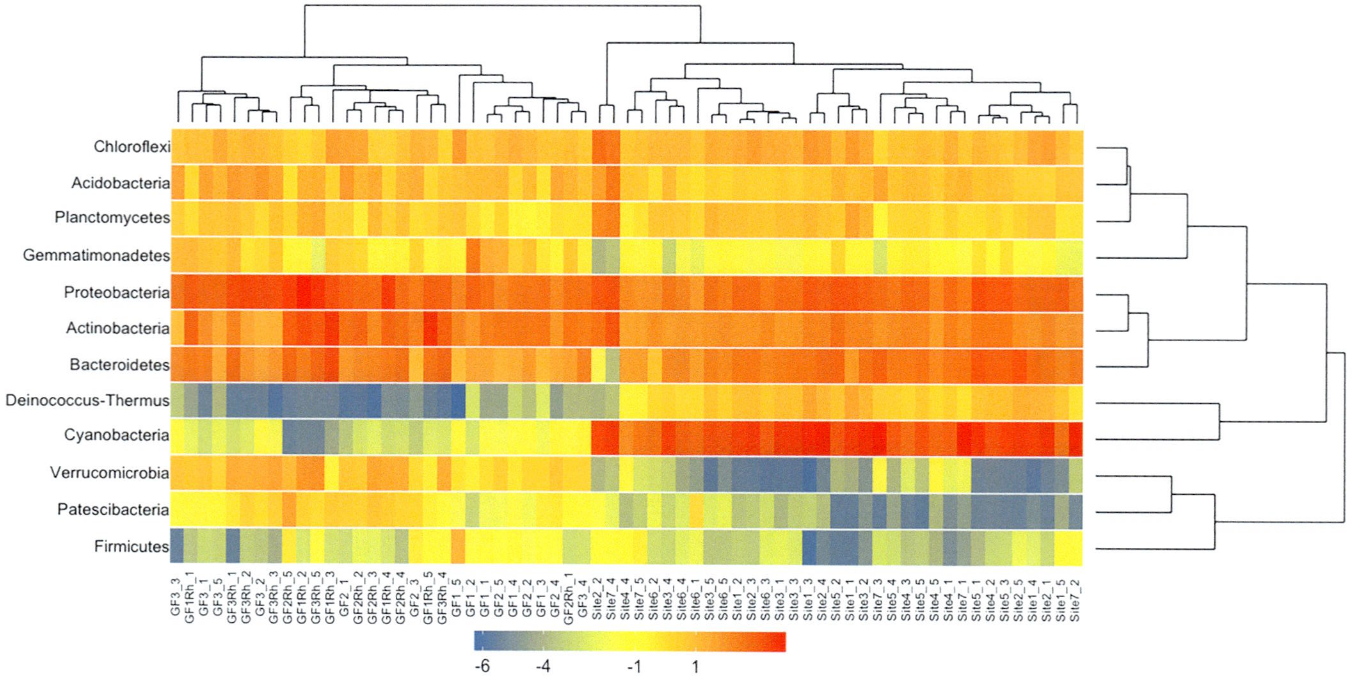

FIG. 8.2 Hierarchical clustering of phyla to assess the relationships between samples and taxa. UL = upper Lewis, LL = lower Lewis, SF = side foreland, GF = glacier foreland.

varying microbial communities between the glacier and its foreland soil. Results from the genomic analyses of the foreland samples were associated with beneficial agricultural practices given their stable association with the plant rhizosphere. In this case, the application of genomic tools in agriculture influences the determination of plant- and crop-beneficial microbes that influence their function, adaptation, survival, and diversity. Using such microbes as alternative biopesticides can increase production with minimal impact on the environment, and human and animal health. Complex endophytic associations and plant-microbiome interactions in the rhizosphere can also determine disease outcomes given the ability of microbes to inhibit colonization of plant roots, shoots, and endospheres by the pathogens.

Cold-adapted microbes (psychrophiles) and food processing

Genomic analyses revealed abundance of stress-resilient prokaryotes that are capable of adapting to unique environment of low liquid water availability, low temperature, low nutrients, and high UV radiation. These prokaryotes are potential sources of cold-active enzymes and freeze-resistant proteins as biocatalyzers and ice-binding molecules, respectively. Cold-active enzymes and freeze-resistant proteins are also thermostable and more productive at low temperatures within food industries. Though enzymes and proteins are obtained from plants and animals, cold-adapted microbes have proven to be resilient sources of cold-active enzymes and freeze-resistant proteins (Mangiagalli et al., 2020; Hamid and Mohiddin, 2018). Ice-binding proteins prevent tissue damage by ice crystals; hence, they are necessary in organ and tissue preservation for biomedical applications. Consequently, ice-binding proteins are important in the preservation of fresh food for sustainable bioeconomy.

The identified prokaryotes inhabit the "seed banks" that act as reservoirs for dormant microorganisms with the potential to revive at favorable environmental conditions (Lennon and Jones, 2011) amidst climate change and rising temperatures. When microorganisms find themselves in such unpredictable conditions, they are unable to grow and reproduce; hence, they enter a state of dormancy (Lennon and Jones, 2011) to reduce energy needs and expenditure. They are therefore able to respond to climate change by affording to wait until favorable changes occur for their growth and reproduction.

Microbes inhabiting the seed banks are also potential indicators of climate change due to their dynamics across the glacier and the foreland samples. In order to determine the organisms that were predominantly controlling the environment, we performed indicator analysis of organisms on environmental samples. Examples of these indicator organisms included the genus *Bradyrhizobium* on the lower Lewis Glacier, the genus *Phormidesmis* on the upper Lewis Glacier, the genus *Nakamurella* on the rhizosphere, the genus *Acidiphilium* on the soil, the genus *Gemmatimonas* on the glacier, and the genus *Blastocatella* on its foreland (Fig. 8.3). Genomic data in our study revealed that the dynamics of

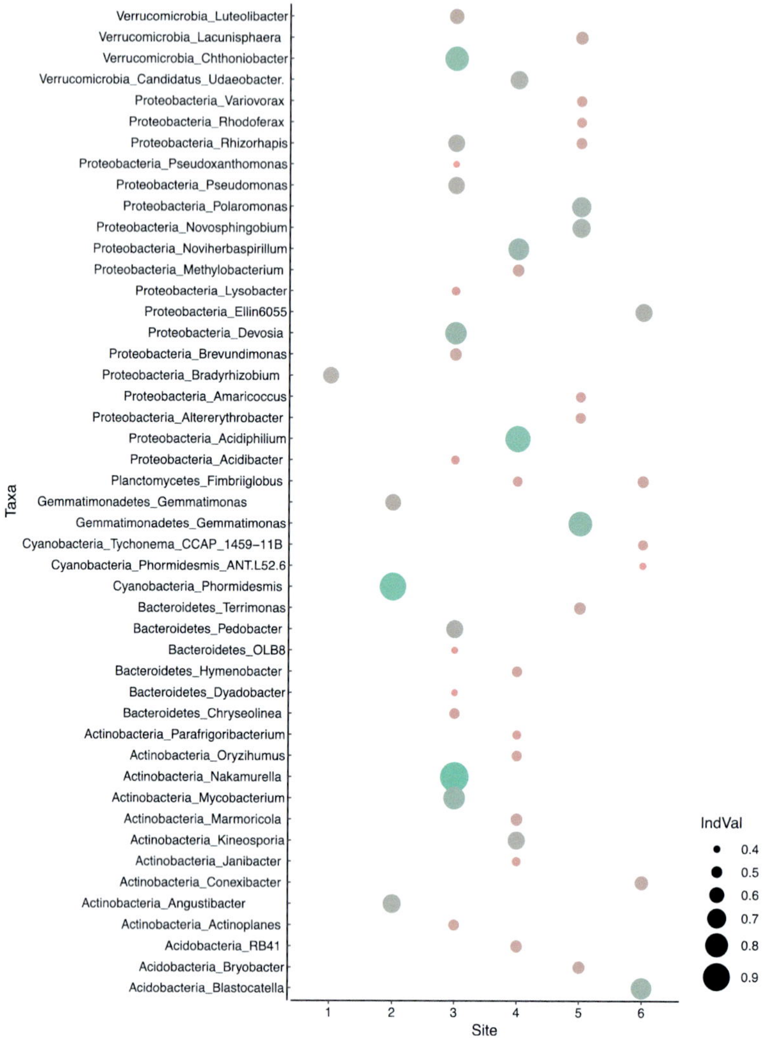

FIG. 8.3 Indicator species on the Lewis Glacier and its foreland at phylum level. Site (1 = lower Lewis, 2 = upper Lewis, 3 = rhizosphere, 4 = soil, 5 = glacier foreland, 6 = side foreland).

the microbial community structure in the equatorial Lewis Glacier and its foreland is species-specific and not generalized to a specific phylum or kingdom of the prokaryote as explained elsewhere (Nemergut et al., 2007). It implies that subsequent studies on the glacier may result in different taxa depending on the rate of recolonization of the foreland and increasing temperatures alongside climate change.

Microbial community structure as indicators of environmental change

Genomic analyses of the 16S rRNA enabled us to identify species associated with particular sample types. For instance, phyla Verrucomicrobia were associated with the rhizosphere and the soil samples, while species that were affiliated to phyla Actinobacteria were associated with glacier samples. Genera like *Tychonema*, *Phormidesmis*, and *Chamaesiphon* that are affiliated to phyla *Cyanobacteria* were associated with the glacier. Their occurrence in high values corresponded to their ability to retain water and produce cold-active pigments that provide sheath protecting fecundity and longevity when carrying out environmental activities. Their functions in the foreland were, however, affected by the fluctuations in water regime, exposure to persistent UV radiations (Rime et al., 2015), and rising temperature due to climate change along the equatorial zones. Studies by Hartmann et al. (2012) and Rime et al. (2015) proved that indicator species analyses are best to analyze the role of organisms in a specific ecosystem. Amplicon sequencing revealed the significant taxonomic shift from the glacier to the foreland soil due to ice melting and recolonization of the foreland by the exogenous species.

Climate change affects the functions and the abundance of microbial community structures due to their differences in sensitivity and tolerance to thermal regimes, physiology, and growth rates (Bardgett et al., 2007; Kumar et al., 2016; Mekala and Polepongu, 2019; Bardgett and Caruso, 2020). Microbial communities respond to increased environmental warming by becoming resistant for a specific period after which they return to their normal composition and functions after the stressful period has passed (Allison and Martiny, 2009; Bissett et al., 2013). They therefore become resilient to the previous environmental stress they have experienced and begin to develop strategies to evolve within their niche. However, persistent stress is likely to cause detrimental changes in the ecosystem functions due to dramatic shifts in microbial community compositions and structure, including situations where entire communities are replaced with new organisms. Rising temperatures alter microbial respiration rates due to their hypersensitivity to sunlight radiations and warmth (Onwuka, 2018; Wang et al., 2020; Zhang et al., 2020).

Microorganisms are poor dispersers, especially in the soil; hence, they respond to climate changes differently from the associated plant colonizers. These shifts in microbial community structures are potential indicators of climate change as the channelized glacier melt deposits serve as starter medium for the establishment of drought-resilient plant communities. The interaction between the plants and the microbial communities suggests a shift in response of soil ecosystem to climate changes. The use of new molecular biology tools such as genomics, sequencing, and bioinformatics offers new strategies for characterizing microbes and has great application in their identification and classification based on the environmental parameters. These technologies when combined with conventional methods such as imagery and other glaciological methods can give strong indicators of how glaciers are changing topologically and in species composition and would provide critical information needed to develop climate mitigation measures through the utilization of assembled genomes for gene function. Spatial-temporal amplicon sequencing and characterization of microbes collected from one environment over time can show changes occurring in the environment in real time. Further, amplicon sequence analyses can also help in the quantification of the effects of climate change.

Microbes and sustainable agricultural ecosystems

Rising temperatures facilitate leafing and flowering earlier in the plant growing season impacting root phenology and rhizosphere development (Allen et al., 2012; Wolkovich et al., 2012; Ettinger et al., 2018; Flynn and Wolkovich, 2018; González-Teuber et al., 2021; van Nuland et al., 2021). Though root phenology varies between different plant species, it is mainly influenced by abiotic factors such as rise in temperature and soil moisture content, which influence the plant-microbe interaction within an ecosystem (Abramoff and Finzi, 2015). The varying abiotic factors are, however, asynchronous under rising temperatures and persistent warming of soil surfaces. Root-shoot phenology therefore influences the rhizosphere-microbial interactions resulting in distinct seasonal dynamics for soil microbial community assemblages, and shift in plant communities affecting productivity and development (Edwards et al., 2018; Khudr et al., 2018; Masenya et al., 2021; Sharma et al., 2019).

Endophytes and plant-associated microbes are essential for achieving sustainable agri-environment for crops and animals. Africa is endowed with a large diversity of endophytes and microbes. The plant-microbe associations are beneficial for their function, adaptation, survival, and species biodiversity. These beneficial microbes can inhibit the colonization of plant roots, shoots, and endospheres by pathogenic microbes, and they have also been reported to influence the growth, development, and adaptation of many plants including African grasses such as *Brachiaria* (Djikeng et al., 2014; Ghimire et al., 2015), which are native to the continent and resilient to drought and climate change. Brachiaria grass is also resistant to diseases that affect other forage crops.

Similarly, plant-microbe association has been reported to the efficacy of herbal medicine. Phytochemical constituents of medicinal plants correlate with the activity of endophytic microbes. Important therapeutic compounds have been identified from microbe-plant interactions. Using such microbes as alternative biopesticides can increase production with minimal impact on the environment, and human and animal health. Complex endophytic associations and plant-microbiome interactions in the rhizosphere can determine disease outcomes and phenological developments. The identification of *Brachiaria* beneficial microbes for drought-resistant and improved plant nutrition would promote the livelihood and productivity of livestock across Africa.

The association of plants and microbiomes is inevitable even in extreme environments, e.g., high-altitude tropical glacier ecosystems and low-altitude terrestrial ecosystems. As more and more genomes are sequenced worldwide, the amount of sequence data, both raw and analyzed, continues to grow. These datasets, that are often very large, need to be stored and managed effectively in order to act as sources of information. In comparison with the rest of the world, the amount of genomic data being generated from Africa remains low. This is caused by lack of relevant sequencing infrastructure and technological expertise. The result of this is that only a few strains of such important microbes have been isolated and sequenced for inclusion in the genome databases, which are mostly located outside the continent. Challenges posed by climate change mean that these unique biomes and plant species are disappearing before they are fully identified, characterized, collected, and preserved.

Conclusion

The new and emerging molecular tools such as the analysis of 16S rRNA amplicon sequences have revealed the existence of new species belonging to major phyla that would not have otherwise been detected, for example, the Cyanobacteria, Proteobacteria, Bacteroidetes, and Actinobacteria that are potential sources of cold-active enzymes and freeze-resistant proteins as resilient biocatalyst in food processing and cryo-preservation in biomedical sectors, respectively. Amplicon sequencing is important in the identification of species diversity and their community structure. It is an important metagenomic approach to identify species that would otherwise not have been detected by the culture-dependent methods. Amplicon sequencing is essential for the emerging economies that require the ability to harness natural resources for improved bioeconomy and sustainable development.

African countries need to embrace genomic applications to leverage agricultural production and utilization of bioresources like the cold-active enzymes and freeze-resistant proteins from remote and neglected ecosystems such as the glacier. Through the implementation and adoption of amplicon sequence and long-read sequence analyses in Africa, drought-resilient plant-associated microbes can be identified and utilized for industrial applications. Currently, the distribution of sequencing devices is skewed to South Africa for diverse genomic and sequencing approaches. Other countries, including Kenya, are working toward meeting the economic demands that rely on genomic tools and robust sequencing techniques to utilize microbial communities and other natural resources as sources of bioprospecting products.

Acknowledgments

This study was supported by the Ministry of Education, Culture, Sports, Science and Technology, Japan Society for the Promotion of Science Grant-in-Aid for Scientific Research

(C) 15K07234 and by the International Foundation for Science Grant-in-Aid for Scientific Research No. 1-2-A-6101-1. We also thank Kenya Wildlife Service (KWS) for the access permit to the sampling area in Mt. Kenya.

References

Abramoff, R.Z., Finzi, A.C., 2015. Are above- and below-ground phenology in sync? New Phytol. https://doi.org/10.1111/nph.13111.

Agrawala, S., Gigli, S., Raksakulthai, V., Hemp, A., Moehner, A., Conway, D., el Raey, M., Ahmed, A.U., Risbey, J., Baethgen, W., Martino, D., 2005. Climate change and natural resource management: key themes from case studies. In: Bridge Over Troubled Waters: Linking Climate Change and Development. OECD iLibrary.

Allen, J.M., Ault, T.R., Betancourt, J.L., Bolmgren, K., Cleland, E.E., Cook, B.I., Crimmins, T.M., Davies, T.J., Kraft, N.J.B., Mazer, S.J., McCabe, G.J., McGill, B.J., Parmesan, C., Pau, S., Regetz, J., Salamin, N., Schwartz, M.D., Travers, S.E., Wolkovich, E.M., 2012. Warming experiments underpredict plant phenological responses to climate change. Nature 485.

Allison, S.D., Martiny, J.B.H., 2009. Resistance, resilience, and redundancy in microbial communities. In: In the Light of Evolution. The National Academies Press, Washington, DC.

Anesio, A.M., Laybourn-Parry, J., 2012. Glaciers and ice sheets as a biome. Trends Ecol. Evol. https://doi.org/10.1016/j.tree.2011.09.012.

Anesio, A.M., Lutz, S., Chrismas, N.A.M., Benning, L.G., 2017. The microbiome of glaciers and ice sheets. npj Biofilms Microbiomes, 0–1. https://doi.org/10.1038/s41522-017-0019-0.

Bardgett, R.D., Caruso, T., 2020. Soil microbial community responses to climate extremes: resistance, resilience and transitions to alternative states. Philos. Trans. R. Soc. Lond., B, Biol. Sci. https://doi.org/10.1098/rstb.2019.0112.

Bardgett, R.D., Richter, A., Bol, R., Garnett, M.H., Bäumler, R., Xu, X., Lopez-Capel, E., Manning, D.A. C., Hobbs, P.J., Hartley, I.R., Wanek, W., 2007. Heterotrophic microbial communities use ancient carbon following glacial retreat. Biol. Lett. 3, 487–490. https://doi.org/10.1098/rsbl.2007.0242.

Bissett, A., Brown, M.v., Siciliano, S.D., Thrall, P.H., 2013. Microbial community responses to anthropogenically induced environmental change: towards a systems approach. Ecol. Lett. 16. https://doi.org/10.1111/ele.12109.

Branda, E., Turchetti, B., Diolaiuti, G., Pecci, M., Smiraglia, C., Buzzini, P., 2010. Yeast and yeast-like diversity in the southernmost glacier of Europe (Calderone Glacier, Apennines, Italy). FEMS Microbiol. Ecol. https://doi.org/10.1111/j.1574-6941.2010.00864.x.

Christner, B.C., Lavender, H.F., Davis, C.L., Oliver, E.E., Neuhaus, S.U., Myers, K.F., Hagedorn, B., Tulaczyk, S.M., Doran, P.T., Stone, W.C., 2018. Microbial processes in the weathering crust aquifer of a temperate glacier. Cryosphere 12. https://doi.org/10.5194/tc-12-3653-2018.

Cook, J.M., Edwards, A., Bulling, M., Mur, L.A.J., Cook, S., Gokul, J.K., Cameron, K.A., Sweet, M., Irvine-Fynn, T.D.L., 2016. Metabolome-mediated biocryomorphic evolution promotes carbon fixation in Greenlandic cryoconite holes. Environ. Microbiol. 18. https://doi.org/10.1111/1462-2920.13349.

Djikeng, A., Rao, I.M., Njarui, D., Mutimura, M., Caradus, J., Ghimire, S.R., Johnson, L., Cardoso, J.A., Ahonsi, M., Kelemu, S., 2014. Climate-smart Brachiaria grasses for improving livestock production in East Africa. Trop. Grassl.-Forrajes Trop. 2. https://doi.org/10.17138/tgft(2)38-39.

Edwards, A., Debbonaire, A.R., Nicholls, S.M., Rassner, S.M.E., Sattler, B., Cook, J.M., Davy, T., Soares, A., Mur, L.A.J., Hodson, A.J., 2019. In-field metagenome and 16S rRNA gene amplicon nanopore sequencing robustly characterize glacier microbiota. biorxivs.org. https://doi.org/10.1101/073965.

Edwards, A., Pachebat, J.A., Swain, M., Hegarty, M., Hodson, A.J., Irvine-Fynn, T.D.L., Rassner, S. M.E., Sattler, B., 2013. A metagenomic snapshot of taxonomic and functional diversity in an alpine glacier cryoconite ecosystem. Environ. Res. Lett. 8 (3), 035003. https://doi.org/10.1088/1748-9326/8/3/035003.

Edwards, J.A., Santos-Medellín, C.M., Liechty, Z.S., Nguyen, B., Lurie, E., Eason, S., Phillips, G., Sundaresan, V., 2018. Compositional shifts in root-associated bacterial and archaeal microbiota track the plant life cycle in field-grown rice. PLoS Biol. 16. https://doi.org/10.1371/journal.pbio.2003862.

Ettinger, A.K., Gee, S., Wolkovich, E.M., 2018. Phenological sequences: how early-season events define those that follow. Am. J. Bot. 105. https://doi.org/10.1002/ajb2.1174.

Flynn, D.F.B., Wolkovich, E.M., 2018. Temperature and photoperiod drive spring phenology across all species in a temperate forest community. New Phytol. 219. https://doi.org/10.1111/nph.15232.

Ghimire, S., Njarui, D., Mutimura, M., Cardoso, J., Johnson, L., Gichangi, E., Teasdale, S., Odokonyero, K., Caradus, J., Rao, I., Djikeng, A., 2015. Climate-smart Brachiaria for improving livestock production in East Africa : emerging opportunities. In: Proceedings of 23rd International Grassland Congress 2015-Keynote Lectures.

Gokul, J.K., Hodson, A.J., Saetnan, E.R., Irvine-Fynn, T.D.L., Westall, P.J., Detheridge, A.P., Takeuchi, N., Bussell, J., Mur, L.A.J., Edwards, A., 2016. Taxon interactions control the distributions of cryoconite bacteria colonizing a High Arctic ice cap. Mol. Ecol. 25. https://doi.org/10.1111/mec.13715.

González-Teuber, M., Palma-Onetto, V., Aguilera-Sammaritano, J., Mithöfer, A., 2021. Roles of leaf functional traits in fungal endophyte colonization: potential implications for host–pathogen interactions. J. Ecol. https://doi.org/10.1111/1365-2745.13678.

Hamid, B., Mohiddin, F.A., 2018. Cold-active enzymes in food processing. In: Enzymes in Food Technology. Springer, Singapore, pp. 383–400, https://doi.org/10.1007/978-981-13-1933-4_19.

Hartmann, M., Howes, C.G., Vaninsberghe, D., Yu, H., Bachar, D., Christen, R., Henrik Nilsson, R., Hallam, S.J., Mohn, W.W., 2012. Significant and persistent impact of timber harvesting on soil microbial communities in northern coniferous forests. ISME J. 6. https://doi.org/10.1038/ismej.2012.84.

Hastenrath, S., 2006. Diagnosing the decaying glaciers of equatorial East Africa. Meteorol. Z. 15, 265–271. https://doi.org/10.1127/0941-2948/2006/0106.

Hodson, A., Anesio, A.M., Tranter, M., Fountain, A.G., Osborn, M., Hodson, A., Fountain, A., Priscu, J., Laybourn-Parry, J., Sattler, B., 2008. Glacial ecosystems recommended citation. Glacial ecosystems. Ecol. Monogr. 78, 41–67.

Hodson, A., Roberts, T.J., Engvall, A.-C., Holmén, K., Mumford, P., 2010. Glacier ecosystem response to episodic nitrogen enrichment in Svalbard, European High Arctic. Biogeochemistry. https://doi.org/10.1007/s10533-009-9384-y.

Hoham, R.W., Remias, D., 2020. Snow and glacial algae: a Review1. J. Phycol. https://doi.org/10.1111/jpy.12952.

Kambura, A.K., Mwirichia, R.K., Kasili, R.W., Karanja, E.N., Makonde, H.M., Boga, H.I., 2016. Bacteria and Archaea diversity within the hot springs of Lake Magadi and Little Magadi in Kenya. BMC Microbiol. 16, 136.

Karanja, E.N., Fliessbach, A., Adamtey, N., Kambura, A.K., Musyoka, M., Fiaboe, K., Mwirichia, R., 2020. Diversity and structure of prokaryotic communities within organic and conventional farming systems in central highlands of Kenya. PLoS One 15, e0236574.

Kaštovská, K., Stibal, M., Šabacká, M., Černá, B., Šantrůčková, H., Elster, J., 2007. Microbial community structure and ecology of subglacial sediments in two polythermal Svalbard glaciers characterized by epifluorescence microscopy and PLFA. Polar Biol. 30, 277–287. https://doi.org/10.1007/s00300-006-0181-y.

Khudr, M.S., Purkiss, S.A., Hager, R., 2018. Indirect ecological effects interact with community genetic effects in a host–parasite system and dramatically reduce parasite burden. Proc. R. Soc. B Biol. Sci. 285. https://doi.org/10.1098/rspb.2018.0891.

Kiama, C.W., Njire, M.M., Kambura, A.K., Mugweru, J.N., Matiru, V.N., Wafula, E.N., Kagali, R. N., Kuja, J.O., 2021. Prokaryotic diversity and composition within equatorial lakes Olbolosat and Oloiden in Kenya (Africa). Curr. Res. Microb. Sci. 2, 100066.

Kumar, V., Rawat, A.K., Amule, F.C., 2016. Climate change impact on beneficial soil microbial community: a review. Ecol. Environ. Conserv. 22.

Lennon, J.T., Jones, S.E., 2011. Microbial seed banks: the ecological and evolutionary implications of dormancy. Nat. Rev. Microbiol. https://doi.org/10.1038/nrmicro2504.

Mangiagalli, M., Brocca, S., Orlando, M., Lotti, M., 2020. The "cold revolution". Present and future applications of cold-active enzymes and ice-binding proteins. New Biotechnol. 55, 5–11. https://doi.org/10.1016/j.nbt.2019.09.003.

Margesin, R., Niklinska, M.A., 2019. Editorial: elevation gradients: microbial indicators of climate change? Front. Microbiol. 10, 1–2. https://doi.org/10.3389/fmicb.2019.02405.

Masenya, K., Thompson, G.D., Tekere, M., Makhalanyane, T.P., Pierneef, R.E., Rees, D.J.G., 2021. Pathogen infection influences a distinct microbial community composition in sorghum RILs. Plant Soil 463. https://doi.org/10.1007/s11104-021-04875-3.

Mekala, S., Polepongu, S., 2019. Impact of climate change on soil microbial community. In: Plant Biotic Interactions: State of the Art., https://doi.org/10.1007/978-3-030-26657-8_3.

Miteva, V., Rinehold, K., Sowers, T., Sebastian, A., Brenchley, J., 2015. Abundance, viability and diversity of the indigenous microbial populations at different depths of the NEEM Greenland ice core. Polar Res. https://doi.org/10.3402/polar.v34.25057.

Nath, S., Shyanti, R.K., Nath, Y., 2021. Influence of anthropocene climate change on biodiversity loss in different ecosystems. In: Global Climate Change. Elsevier.

Nemergut, D.R., Anderson, S.P., Cleveland, C.C., Martin, A.P., Miller, A.E., Seimon, A., Schmidt, S.K., 2007. Microbial community succession in an unvegetated, recently deglaciated soil. Microb. Ecol. 53, 110–122. https://doi.org/10.1007/s00248-006-9144-7.

OECD, 2002. Special Issue on Climate Change Linking Climate Change Responses with Development Planning: Some Case Studies. OECD Papers 4.

Onwuka, B., 2018. Effects of soil temperature on some soil properties and plant growth. Adv. Plants Agric. Res. 8. https://doi.org/10.15406/apar.2018.08.00288.

Paglia, E., Parker, C., 2021. The intergovernmental panel on climate change: guardian of climate science. In: Guardians of Public Value., https://doi.org/10.1007/978-3-030-51701-4_12.

Prinz, R., Fischer, A., Nicholson, L., Kaser, G., 2011. Seventy-six years of mean mass balance rates derived from recent and re-evaluated ice volume measurements on tropical Lewis Glacier, Mount Kenya. Geophys. Res. Lett. https://doi.org/10.1029/2011GL049208.

Prinz, R., Nicholson, L., Kaser, G., 2012. Variations of the Lewis glacier, Mount Kenya, 2004-2012. Erdkunde. https://doi.org/10.3112/erdkunde.2012.03.05.

Prinz, R., Nicholson, L.I., Mölg, T., Gurgiser, W., Kaser, G., 2016. Climatic controls and climate proxy potential of Lewis glacier, Mt. Kenya. Cryosphere 10. https://doi.org/10.5194/tc-10-133-2016.

Prinz, R., Ladner, M., Nicholson, L.I., Kaser, G., 2018. Mapping the loss of Mt. Kenya' s glaciers : an example of the challenges of satellite monitoring of very small glaciers. Geosciences. https://doi.org/10.3390/geosciences8050174.

Rime, T., Hartmann, M., Brunner, I., Widmer, F., 2015. Vertical distribution of the soil microbiota along a successional gradient in a glacier forefield. Mol. Ecol., 1091–1108. https://doi.org/10.1111/mec.13051.

Saritha, M., Kumar, P., Panwar, N.R., Burman, U., 2021. Physiological and metabolic basis of microbial adaptations under extreme environments. Extreme Environ. https://doi.org/10.1201/9780429343452-11.

Segawa, T., Miyamoto, K., Ushida, K., Agata, K., Okada, N., Kohshima, S., 2005. Seasonal change in bacterial flora and biomass in mountain snow from the Tateyama Mountains, Japan, analyzed by 16S rRNA gene sequencing and real-time PCR. Appl. Environ. Microbiol. 71. https://doi.org/10.1128/AEM.71.1.123-130.2005.

Sharma, S., Magotra, S., Ganjoo, S., Andrabi, T., Gupta, R., Sharma, S., Vakhlu, J., 2019. Dynamics of plant microbiome and its effect on the plant traits. In: Microbial Diversity in Ecosystem Sustainability and Biotechnological Applications: Volume 2. Soil & Agroecosystems., https://doi.org/10.1007/978-981-13-8487-5_12.

Stibal, M., Tranter, M., Benning, L.G., Rehák, J., 2008. Microbial primary production on an Arctic glacier is insignificant in comparison with allochthonous organic carbon input. Environ. Microbiol. 10, 2172–2178.

Takeuchi, N., Kohshima, S., Seko, K., 2001. Structure, formation, and darkening process of albedo-reducing material (Cryoconite) on a Himalayan glacier: a granular algal mat growing on the glacier. Arct. Antarct. Alp. Res. 33, 115–122. https://doi.org/10.2307/1552211.

van Nuland, M.E., Ware, I.M., Schadt, C.W., Yang, Z., Bailey, J.K., Schweitzer, J.A., 2021. Natural soil microbiome variation affects spring foliar phenology with consequences for plant productivity and climate-driven range shifts. New Phytol. https://doi.org/10.1111/nph.17599.

Wang, Y., Ma, A., Liu, G., Ma, J., Wei, J., Zhou, H., Brandt, K.K., Zhuang, G., 2020. Potential feedback mediated by soil microbiome response to warming in a glacier forefield. Glob. Chang. Biol. 26. https://doi.org/10.1111/gcb.14936.

Whitton, B.A., Potts, M., 2007. The Ecology of Cyanobacteria: Their Diversity in Time and Space. Springer Science & Business Media.

Wiltschi, B., Cernava, T., Dennig, A., Casas, M.G., Geier, M., Gruber, S., et al., 2020. Enzymes revolutionize the bioproduction of value-added compounds: from enzyme discovery to special applications. Biotechnol. Adv. 40. https://doi.org/10.1016/j.biotechadv.2020.107520.

Wolkovich, E.M., Cook, B.I., Allen, J.M., Crimmins, T.M., Betancourt, J.L., Travers, S.E., Pau, S., Regetz, J., Davies, T.J., Kraft, N.J.B., Ault, T.R., Bolmgren, K., Mazer, S.J., McCabe, G.J., McGill, B.J., Parmesan, C., Salamin, N., Schwartz, M.D., Cleland, E.E., 2012. Warming experiments underpredict plant phenological responses to climate change. Nature 485. https://doi.org/10.1038/nature11014.

Zawierucha, K., Porazinska, D.L., Ficetola, G.F., Ambrosini, R., Baccolo, G., Buda, J., Ceballos, J. L., Devetter, M., Dial, R., Franzetti, A., Fuglewicz, U., Gielly, L., Łokas, E., Janko, K., Novotna Jaromerska, T., Kościński, A., Kozłowska, A., Ono, M., Parnikoza, I., Pittino, F., Poniecka, E., Sommers, P., Schmidt, S.K., Shain, D., Sikorska, S., Uetake, J., Takeuchi, N., 2021a. A hole in the nematosphere: tardigrades and rotifers dominate the cryoconite hole environment, whereas nematodes are missing. J. Zool. 313, 18–36. https://doi.org/10.1111/jzo.12832.

Zawierucha, K., Uetake, J., Shain, D., 2021b. A Compendium of Tropical Ice Masses, Reference Module in Earth Systems and Environmental Sciences. Elsevier Inc, https://doi.org/10.1016/b978-0-12-821139-7.00033-7.

Zhang, X., Kuzyakov, Y., Zang, H., Dippold, M.A., Shi, L., Spielvogel, S., Razavi, B.S., 2020. Rhizosphere hotspots: root hairs and warming control microbial efficiency, carbon utilization and energy production. Soil Biol. Biochem. 148. https://doi.org/10.1016/j.soilbio.2020.107872.

Further reading

Abed, R.M.M., al Kharusi, S., Schramm, A., Robinson, M.D., 2010. Bacterial diversity, pigments and nitrogen fixation of biological desert crusts from the Sultanate of Oman. FEMS Microbiol. Ecol. 72. https://doi.org/10.1111/j.1574-6941.2010.00854.x.

Bajerski, F., Wagner, D., 2013. Bacterial succession in Antarctic soils of two glacier forefields on Larsemann Hills, East Antarctica. FEMS Microbiol. Ecol. 85. https://doi.org/10.1111/1574-6941.12105.

Barry, R.G., Gan, T.Y., 2011. The Global Cryosphere: Past, Present, and Future., https://doi.org/10.1017/CBO9780511977947.

Bera, T., Inglett, K.S., Liu, G.D., 2020. Effects of solid oxygen fertilizers and biochars on nitrous oxide production from agricultural soils in Florida. Sci. Rep. 10. https://doi.org/10.1038/s41598-020-78198-1.

Bolker, B.M., Brooks, M.E., Clark, C.J., Geange, S.W., Poulsen, J.R., Stevens, M.H.H., White, J.-S. S., 2009. Generalized linear mixed models: a practical guide for ecology and evolution. Trends Ecol. Evol. 24, 127–135.

Botnen, S.S., Davey, M.L., Aas, A.B., Carlsen, T., Thoen, E., Heegaard, E., Vik, U., Dresch, P., Mundra, S., Peintner, U., Taylor, A.F.S., Kauserud, H., 2019. Biogeography of plant root-associated fungal communities in the North Atlantic region mirrors climatic variability. J. Biogeogr. 46. https://doi.org/10.1111/jbi.13613.

Brook, B.W., 2008. Synergies between climate change, extinctions and invasive vertebrates. Wildl. Res. https://doi.org/10.1071/WR07116.

Campbell, A., Kapos, V., Scharlemann, J.P.W., Bubb, P., Chenery, A., Coad, L., Dickson, B., Doswald, N., Khan, M.S.I., Kershaw, F., Rashid, M., 2009. Review of the Literature on the Links Between Biodiversity and Climate Changes: Impacts, Adaptation and Mitigation. CBD Technical Series 42.

Cavicchioli, R., Ripple, W.J., Timmis, K.N., Azam, F., Bakken, L.R., Baylis, M., Behrenfeld, M.J., Boetius, A., Boyd, P.W., Classen, A.T., Crowther, T.W., Danovaro, R., Foreman, C.M., Huisman, J., Hutchins, D.A., Jansson, J.K., Karl, D.M., Koskella, B., Mark Welch, D.B., Martiny, J. B.H., Moran, M.A., Orphan, V.J., Reay, D.S., Remais, J.v., Rich, V.I., Singh, B.K., Stein, L.Y., Stewart, F.J., Sullivan, M.B., van Oppen, M.J.H., Weaver, S.C., Webb, E.A., Webster, N.S., 2019. Scientists' warning to humanity: microorganisms and climate change. Nat. Rev. Microbiol. 17, 569–586. https://doi.org/10.1038/s41579-019-0222-5.

Chaudhry, V., Rehman, A., Mishra, A., Chauhan, P.S., Nautiyal, C.S., 2012. Changes in bacterial community structure of agricultural land due to long-term organic and chemical amendments. Microb. Ecol. 64. https://doi.org/10.1007/s00248-012-0025-y.

Chiri, E., Nauer, P.A., Henneberger, R., Zeyer, J., Schroth, M.H., 2015. Soil-methane sink increases with soil age in forefields of alpine glaciers. Soil Biol. Biochem. 84. https://doi.org/10.1016/j.soilbio.2015.02.003.

Cook, J., 2016. Cryoconite : The Dark Biological Secret of the Cryosphere., https://doi.org/10.1177/0309133315616574.

Crowther, T.W., Thomas, S.M., Maynard, D.S., Baldrian, P., Covey, K., Frey, S.D., van Diepen, L.T.A., Bradford, M.A., 2015. Biotic interactions mediate soil microbial feedbacks to climate change. Proc. Natl. Acad. Sci. U. S. A. 112. https://doi.org/10.1073/pnas.1502956112.

Cuffey, K.M., Paterson, W.S.B., 2010. The Physics of Glaciers, fourth ed. Elsevier.

Cullen, N.J., Mölg, T., Kaser, G., Hussein, K., Steffen, K., Hardy, D.R., 2006. Kilimanjaro Glaciers: recent areal extent from satellite data and new interpretation of observed 20th century retreat rates. Geophys. Res. Lett. 33. https://doi.org/10.1029/2006GL027084.

Cullen, N.J., Sirguey, P., Mölg, T., Kaser, G., Winkler, M., Fitzsimons, S.J., 2013. A century of ice retreat on Kilimanjaro: the mapping reloaded. Cryosphere 7. https://doi.org/10.5194/tc-7-419-2013.

DeBruyn, J.M., Nixon, L.T., Fawaz, M.N., Johnson, A.M., Radosevich, M., 2011. Global biogeography and quantitative seasonal dynamics of Gemmatimonadetes in soil. Appl. Environ. Microbiol. 77, 6295–6300. https://doi.org/10.1128/AEM.05005-11.

Downie, C., 1964. Glaciations of Mount Kilimanjaro, northeast Tanganyika. Bull. Geol. Soc. Am. 75. https://doi.org/10.1130/0016-7606(1964)75[1:GOMKNT]2.0.CO;2.

Duane, W.J., Pepin, N.C., Losleben, M.L., Hardy, D.R., 2008. General characteristics of temperature and humidity variability on Kilimanjaro, Tanzania. Arct. Antarct. Alp. Res. 40. https://doi.org/10.1657/1523-0430(06-127)[DUANE]2.0.CO;2.

Dubey, R.K., Tripathi, V., Abhilash, P.C., 2015. Book review: principles of plant-microbe interactions: microbes for sustainable agriculture. Front. Plant Sci. 6. https://doi.org/10.3389/fpls.2015.00986.

Dulić, T., Meriluoto, J., Palanački Malešević, T., Gajić, V., Važić, T., Tokodi, N., Obreht, I., Kostić, B., Kosijer, P., Khormali, F., Svirčev, Z., 2017. Cyanobacterial diversity and toxicity of biocrusts from the Caspian lowland loess deposits, North Iran. Quat. Int. 429. https://doi.org/10.1016/j.quaint.2016.02.046.

Dutta, H., Dutta, A., 2016. The microbial aspect of climate change. Energy Ecol. Environ. 1, 209–232. https://doi.org/10.1007/s40974-016-0034-7.

Edwards, K., 2004. Primary succession and ecosystem rehabilitation. Restor. Ecol. 12. https://doi.org/10.1111/j.1061-2971.2004.12301.x.

Garcia-Pichel, F., Belnap, J., 2001. Small-Scale Environments and Distribution of Biological Soil Crusts., https://doi.org/10.1007/978-3-642-56475-8_16.

Ghilamicael, A.M., Budambula, N.L.M., Anami, S.E., Mehari, T., Boga, H.I., 2017. Evaluation of prokaryotic diversity of five hot springs in Eritrea. BMC Microbiol. 17, 203. https://doi.org/10.1186/s12866-017-1113-4.

Hastenrath, S., 1983. Net balance, surface lowering, and ice-flow pattern in the interior of Lewis Glacier, Mount Kenya, Kenya. J. Glaciol. 29, 392–402.

Hastenrath, S., 2009. Past glaciation in the tropics. Quat. Sci. Rev. 28. https://doi.org/10.1016/j.quascirev.2008.12.004.

Hock, R., 2005. Glacier melt: a review of processes and their modelling. Prog. Phys. Geogr. https://doi.org/10.1191/0309133305pp453ra.

Hodson, A., Brock, B., Pearce, D., Laybourn-Parry, J., Tranter, M., 2015. Cryospheric ecosystems: a synthesis of snowpack and glacial research. Environ. Res. Lett. 10. https://doi.org/10.1088/1748-9326/10/11/110201.

Irvine-Fynn, T.D.L., Hodson, A.J., Moorman, B.J., Vatne, G., Hubbard, A.L., 2011. Polythermal glacier hydrology: a review. Rev. Geophys. 49, 1–37. https://doi.org/10.1029/2010RG000350.1.INTRODUCTION.

Jansson, J.K., Hofmockel, K.S., 2020. Soil microbiomes and climate change. Nat. Rev. Microbiol. https://doi.org/10.1038/s41579-019-0265-7.

Keeler, A.M., Rose-Person, A., Rafferty, N.E., 2021. From the ground up: building predictions for how climate change will affect belowground mutualisms, floral traits, and bee behavior. Clim. Change Ecol. 1. https://doi.org/10.1016/j.ecochg.2021.100013.

Kühnel, R., Roberts, T.J., Björkman, M.P., Isaksson, E., Aas, W., Holmén, K., Ström, J., 2011. 20-year climatology of NO_3^- and NH_4^+ wet deposition at Ny-Ålesund, Svalbard. Adv. Meteorol. 2011, 1–10. https://doi.org/10.1155/2011/406508.

Kuja, J., 2019. Metagenomics and Composition of Prokaryotes on Lewis Glacier and its Foreland in Mount Kenya, Nyeri County (Microbiology). Jkuat.ac.ke.

Kuja, J.O., Makonde, H.M., Boga, H.I., Muigai, A.T.W., Uetake, J., 2018a. Phylogenetic diversity of prokaryotes on the snow-cover of Lewis glacier in Mount Kenya. Afr. J. Microbiol. Res. 12, 574–579. https://doi.org/10.5897/AJMR2017.8750.

Kuja, J.O., Makonde, H.M., Muigai, A.T., Omire, A., Boga, H.I., Uetake, J., 2018b. The status of Lewis glacier of Mount Kenya and the threat to novel microbial communities. Int. J. Micro. Myco. 7, 6–13.

Langford, H., Hodson, A., Banwart, S., Bøggild, C., 2010. The microstructure and biogeochemistry of Arctic cryoconite granules. Ann. Glaciol. 51, 87–94. https://doi.org/10.3189/172756411795932083.

Lipson, D.A., Schadt, C.W., Schmidt, S.K., 2002. Changes in soil microbial community structure and function in an alpine dry meadow following spring snow melt. Microb. Ecol. 43, 307–314. https://doi.org/10.1007/s00248-001-1057-x.

Margesin, R., Gander, S., Zacke, G., Gounot, A.M., Schinner, F., 2003. Hydrocarbon degradation and enzyme activities of cold-adapted bacteria and yeasts. Extremophiles 7, 451–458. https://doi.org/10.1007/s00792-003-0347-2.

Marleau, J.N., Jin, Y., Bishop, J.G., Fagan, W.F., Lewis, M.A., 2011. A stoichiometric model of early plant primary succession. Am. Nat. 177. https://doi.org/10.1086/658066.

Meola, M., Lazzaro, A., Zeyer, J., 2014. Diversity, resistance and resilience of the bacterial communities at two alpine glacier forefields after a reciprocal soil transplantation. Environ. Microbiol. 16. https://doi.org/10.1111/1462-2920.12435.

Mizuno, K., Fujita, T., 2014. Vegetation succession on Mt. Kenya in relation to glacial fluctuation and global warming. J. Veg. Sci. 25, 559–570. https://doi.org/10.1111/jvs.12081.

Mölg, T., Georges, C., Kaser, G., 2003. The contribution of increased incoming shortwave radiation to the retreat of the Rwenzori glaciers, East Africa, during the 20th century. Int. J. Climatol. 23. https://doi.org/10.1002/joc.877.

Mölg, T., Cullen, N.J., Hardy, D.R., Kaser, G., Klok, L., 2008. Mass balance of a slope glacier on Kilimanjaro and its sensitivity to climate. Int. J. Climatol. 28. https://doi.org/10.1002/joc.1589.

Mölg, T., Cullen, N.J., Hardy, D.R., Winkler, M., Kaser, G., 2009. Quantifying climate change in the tropical midtroposphere over East Africa from glacier shrinkage on Kilimanjaro. J. Clim. 22. https://doi.org/10.1175/2009JCLI2954.1.

Mote, P.W., Kaser, G., 2007. The shrinking glaciers of Kilimanjaro: can global warming be blamed? Am. Sci. 95. https://doi.org/10.1511/2007.66.318.

Mur, L.R., Skulberg, O.M., Utkilen, H., 1999. Cyanobacteria in the environment. In: Toxic Cyanobacteria in Water: A Guide to Their Public Health Consequences, Monitoring and Management. WHO (Chapter 2).

Naylor, D., Sadler, N., Bhattacharjee, A., Graham, E.B., Anderton, C.R., McClure, R., Lipton, M., Hofmockel, K.S., Jansson, J.K., 2020. Soil microbiomes under climate change and implications for carbon cycling. Annu. Rev. Environ. Resour. https://doi.org/10.1146/annurev-environ-012320-082720.

Nelson, F.E., 2012. The global cryosphere: past, present and future. Polar Geogr. 35. https://doi.org/10.1080/1088937x.2012.665398.

Nemergut, D.R., Townsend, A.R., Sattin, S.R., Freeman, K.R., Fierer, N., Neff, J.C., Bowman, W.D., Schadt, C.W., Weintraub, M.N., Schmidt, S.K., 2008. The effects of chronic nitrogen fertilization on alpine tundra soil microbial communities: implications for carbon and nitrogen cycling. Environ. Microbiol. 10, 3093–3105. https://doi.org/10.1111/j.1462-2920.2008.01735.x.

Parizek, B.R., Alley, R.B., Anandakrishnan, S., Conway, H., 2002. Sub-catchment melt and long-term stability of ice stream D, West Antarctica. Geophys. Res. Lett. 29. https://doi.org/10.1029/2001gl014326.

Powell, J.T., Chatziefthimiou, A.D., Banack, S.A., Cox, P.A., Metcalf, J.S., 2015. Desert crust microorganisms, their environment, and human health. J. Arid Environ. 112. https://doi.org/10.1016/j.jaridenv.2013.11.004.

Pugnaire, F.I., Morillo, J.A., Peñuelas, J., Reich, P.B., Bardgett, R.D., Gaxiola, A., Wardle, D.A., van der Putten, W.H., 2019. Climate change effects on plant-soil feedbacks and consequences for biodiversity and functioning of terrestrial ecosystems. Sci. Adv. https://doi.org/10.1126/sciadv.aaz1834.

Rabatel, A., Francou, B., Soruco, A., Gomez, J., Cáceres, B., Ceballos, J.L., Basantes, R., Vuille, M., Sicart, J.E., Huggel, C., Scheel, M., Lejeune, Y., Arnaud, Y., Collet, M., Condom, T., Consoli, G., Favier, V., Jomelli, V., Galarraga, R., Ginot, P., Maisincho, L., Mendoza, J., Ménégoz, M., Ramirez, E., Ribstein, P., Suarez, W., Villacis, M., Wagnon, P., 2013. Current state of glaciers in the tropical Andes: a multi-century perspective on glacier evolution and climate change. Cryosphere 7. https://doi.org/10.5194/tc-7-81-2013.

Ravolainen, V., Soininen, E.M., Jónsdóttir, I.S., Eischeid, I., Forchhammer, M., van der Wal, R., Pedersen, Å., 2020. High Arctic ecosystem states: conceptual models of vegetation change to guide long-term monitoring and research. Ambio 49. https://doi.org/10.1007/s13280-019-01310-x.

Schütte, U.M.E., Abdo, Z., Foster, J., Ravel, J., Bunge, J., Solheim, B., Forney, L.J., 2010. Bacterial diversity in a glacier foreland of the high Arctic. Mol. Ecol. 19, 54–66. https://doi.org/10.1111/j.1365-294X.2009.04479.x.

Sigler, W.v., Crivii, S., Zeyer, J., 2002. Bacterial succession in glacial forefield soils characterized by community structure, activity and opportunistic growth dynamics. Microb. Ecol. 44, 306–316. https://doi.org/10.1007/s00248-002-2025-9.

Stibal, M., Lawson, E.C., Lis, G.P., Mak, K.M., Wadham, J.L., Anesio, A.M., 2010. Organic matter content and quality in supraglacial debris across the ablation zone of the Greenland ice sheet. Ann. Glaciol. 51, 1–8. https://doi.org/10.3189/172756411795931958.

Takeuchi, N., Nishiyama, H., Li, Z., 2010. Structure and formation process of cryoconite granules on Ürümqi glacier No. 1, Tien Shan, China. Ann. Glaciol. 51, 9–14.

Takeuchi, N., Sakaki, R., Uetake, J., Nagatsuka, N., Shimada, R., Niwano, M., Aoki, T., 2018. Temporal variations of cryoconite holes and cryoconite coverage on the ablation ice surface of Qaanaaq glacier in Northwest Greenland. Ann. Glaciol. https://doi.org/10.1017/aog.2018.19.

Telling, J., Boyd, E.S., Bone, N., Jones, E.L., Tranter, M., Macfarlane, J.W., Martin, P.G., Wadham, J.L., Lamarche-Gagnon, G., Skidmore, M.L., Hamilton, T.L., Hill, E., Jackson, M., Hodgson, D.A., 2015. Rock comminution as a source of hydrogen for subglacial ecosystems. Nat. Geosci. 8. https://doi.org/10.1038/ngeo2533.

ter Braak, C.J.F., Schaffers, A.P., 2004. Co-correspondence analysis: a new ordination method to relate two community compositions. Ecology 85. https://doi.org/10.1890/03-0021.

Thompson, L.G., Brechera, H.H., Mosley-Thompson, E., Hardy, D.R., Mark, B.G., 2009. Glacier loss on Kilimanjaro continues unabated. Proc. Natl. Acad. Sci. U. S. A. 106. https://doi.org/10.1073/pnas.0906029106.

Thompson, L.G., Davis, M.E., Mosley-Thompson, E., Porter, S.E., Corrales, G.V., Shuman, C.A., Tucker, C.J., 2021. The impacts of warming on rapidly retreating high-altitude, low-latitude glaciers and ice core-derived climate records. Glob. Planet. Chang. 203. https://doi.org/10.1016/j.gloplacha.2021.103538.

Tran, P., Ramachandran, A., Khawasik, O., Beisner, B.E., Rautio, M., Huot, Y., Walsh, D.A., 2018. Microbial life under ice: metagenome diversity and in situ activity of Verrucomicrobia in seasonally ice-covered Lakes. Environ. Microbiol. 20. https://doi.org/10.1111/1462-2920.14283.

Tscherko, D., Rustemeier, J., Richter, A., Wanek, W., Kandeler, E., 2003. Functional diversity of the soil microflora in primary succession across two glacier forelands in the Central Alps. Eur. J. Soil Sci. 54, 685–696. https://doi.org/10.1046/j.1351-0754.2003.0570.x.

Uetake, J., Tanaka, S., Hara, K., Tanabe, Y., Samyn, D., Motoyama, H., Imura, S., Kohshima, S., 2014. Novel biogenic aggregation of moss gemmae on a disappearing African glacier. PLoS One 9, 112510. https://doi.org/10.1371/journal.pone.0112510.

van der Putten, W.H., Bardgett, R.D., Bever, J.D., Bezemer, T.M., Casper, B.B., Fukami, T., Kardol, P., Klironomos, J.N., Kulmatiski, A., Schweitzer, J.A., Suding, K.N., van de Voorde, T.F.J., Wardle, D.A., 2013. Plant – soil feedbacks in a changing world plant – soil feedbacks: the past, the present and future challenges. J. Ecol. 101.

Veettil, B.K., Kamp, U., 2019. Global disappearance of tropical mountain glaciers: observations, causes, and challenges. Geosciences (Switzerland). https://doi.org/10.3390/geosciences9050196.

Velásquez, A.C., Castroverde, C.D.M., He, S.Y., 2018. Plant–pathogen warfare under changing climate conditions. Curr. Biol. https://doi.org/10.1016/j.cub.2018.03.054.

Vimercati, L., Darcy, J.L., Schmidt, S.K., 2019. The disappearing periglacial ecosystem atop Mt. Kilimanjaro supports both cosmopolitan and endemic microbial communities. Sci. Rep. 9. https://doi.org/10.1038/s41598-019-46521-0.

Vuille, M., Francou, B., Wagnon, P., Juen, I., Kaser, G., Mark, B.G., Bradley, R.S., 2008. Climate change and tropical Andean glaciers: past, present and future. Earth Sci. Rev. 89. https://doi.org/10.1016/j.earscirev.2008.04.002.

Walker, L.R., del Moral, R., 2003. Primary Succession and Ecosystem Rehabilitation., https://doi.org/10.1017/cbo9780511615078.

Yang, G.L., Hou, S.G., le Baoge, R., Li, Z.G., Xu, H., Liu, Y.P., Du, W.T., Liu, Y.Q., 2016. Differences in bacterial diversity and communities between glacial snow and glacial soil on the Chongce ice cap, West Kunlun Mountains. Sci. Rep. 6. https://doi.org/10.1038/srep36548.

Yoshitake, S., Uchida, M., Koizumi, H., Kanda, H., Nakatsubo, T., 2010. Production of biological soil crusts in the early stage of primary succession on a high Arctic glacier foreland. New Phytol. 186, 451–460. https://doi.org/10.1111/j.1469-8137.2010.03180.x.

Young, J.A.T., Hastenrath, S., 1991. Glaciers of the Middle East and Africa—glaciers of Africa. In: US Geological Survey Professional Paper 1386 G.

Zawierucha, K., Shain, D.H., 2019. Disappearing Kilimanjaro snow—are we the last generation to explore equatorial glacier biodiversity? Ecol. Evol. 9, 8911–8918. https://doi.org/10.1002/ece3.5327.

Zawierucha, K., Ostrowska, M., Vonnahme, T.R., Devetter, M., Nawrot, A.P., Lehmann, S., Kolicka, M., 2016. Diversity and distribution of tardigrada in arctic cryoconite holes. J. Limnol. 75. https://doi.org/10.4081/jlimnol.2016.1453.

Zawierucha, K., Gąsiorek, P., Buda, J., Uetake, J., Janko, K., Fontaneto, D., 2018. Tardigrada and rotifera from moss microhabitats on a disappearing Ugandan glacier, with the description of a new species of water bear. Zootaxa 4392, 311–328. https://doi.org/10.11646/zootaxa.4392.2.5.

Zawierucha, K., Buda, J., Fontaneto, D., Ambrosini, R., Franzetti, A., Wierzgoń, M., Bogdziewicz, M., 2019. Fine-scale spatial heterogeneity of invertebrates within cryoconite holes. Aquat. Ecol. https://doi.org/10.1007/s10452-019-09681-9.

Zeng, Y., Koblížek, M., 2017. Phototrophic gemmatimonadetes: a new "Purple" branch on the bacterial tree of life. In: Modern Topics in the Phototrophic Prokaryotes: Environmental and Applied Aspects., https://doi.org/10.1007/978-3-319-46261-5_5.

Zhang, S., Hou, S., Baoge, R., Xu, H., Liu, Y., Li, Z., 2016. Difference of community structure among culturable bacteria in different glacial samples on Chongce ice cap. Wei sheng wu xue bao = Acta Microbiol. Sin. 56.

Zumsteg, A., Luster, J., Göransson, H., Smittenberg, R.H., Brunner, I., Bernasconi, S.M., Zeyer, J., Frey, B., 2012. Bacterial, archaeal and fungal succession in the forefield of a receding glacier. Microb. Ecol. 63. https://doi.org/10.1007/s00248-011-9991-8.

Part III

Genomics as a driver of the bioeconomy in agriculture

Chapter 9

The impact of biotechnology and genomics on an ancient crop: *Cannabis sativa*

Erin J. Gilchrist[a], Shumin Wang[b], and Teagen D. Quilichini[c]
[a]*Molecular Diagnostics, Anandia Laboratories, Vancouver, BC, Canada,* [b]*Strain Engineering, Willow Biosciences Inc., Vancouver, BC, Canada,* [c]*Aquatic and Crop Resource Development Research Centre, National Research Council Canada, Saskatoon, SK, Canada*

The convoluted evolutionary history of cannabis

Cannabis-human interactions

The utilization of cannabis by humans has spanned millennia, shaping the evolutionary trajectory of the plant and its pharmacological offerings. Pollen fossil records indicate the putative origin of cannabis in Central Asia approximately 28 million years ago (McPartland et al., 2019), with archeological evidence supporting its medicinal and ritualistic use by ancient peoples, including herbal medicine texts and artworks from 16th century BCE and psychoactive cannabis remains in 2500- to 2700-year-old tombs (Jiang et al., 2016; Ren et al., 2019; Russo et al., 2008; Zias et al., 1993). Physical evidence of cannabis' use as a food and spun textile date back 10,000 and 5600 years, respectively (Kobayashi et al., 2008; Zhang and Gao, 1999). The wide use of cannabis for fiber, food, and medicine continued from ancient times through to the 1800s in Europe, Asia, Africa, and the Americas. In North America, Indigenous peoples have grown cannabis for cultural, material, and medicinal uses for the last 1200 years, a practice that continued after the arrival of Europeans because of the latter's interest in the plant's extraordinary fibers (Lev, 2018). Today, cannabis has become a global commodity ranked as the most widely used drug in the world, with an estimated 200 million cannabis users in 2019 between the ages of 15 and 64 (United Nations Office on Drugs and Crime, 2021a,b). The global market share is valued at an estimated \$20–\$50 million (Ajay, 2021; Haligonia.ca, 2021), including an estimated \$4.7 billion for the industrial hemp market that includes an array of applications in bioenergy, fiber, health food, and value-added products (ResearchAndMarkets.com, 2021).

Genomics and the Global Bioeconomy. https://doi.org/10.1016/B978-0-323-91601-1.00009-2

From its broad traditional uses, cannabis entered a period of strict prohibition in the early 1900s as alcohol restrictions gained support among North American settlers and prompted a movement to identify and control other drugs of concern. In 1922, a Canadian prohibitionist, Emily Murphy, published a racist and antidrug book purporting that "marihuana" incited violence and other manias. Following Murphy's claims, Canada became the first Western country to ban cannabis, listing it in the Opium and Narcotic Control Act alongside heroin and cocaine (Lev, 2018). In the 1930s, Harry Anslinger, the first commissioner of the U.S. Treasury Department's Federal Bureau of Narcotics, cautioned that cannabis was a dangerous and unfamiliar plant that, among other evils, would make people of non-European heritage "think they were as good as" European-Americans. Thus, in 1937, the U.S. government proclaimed cannabis an illegal drug, and by the 1940s, the medicinal purposes for cannabis use were removed from the U.S. Pharmacopeia (Holland, 2010; Kinder, 1981). Cannabis prohibition did not receive support from the American Medical Association, with doctors and pharmacists noting the correlative nature of the data presented by Anslinger, which was guided by racist ideology and opinion. Despite rising cannabis prohibition in North America and Europe, cannabis remained legal in much of the world until 1961 when the United Nations developed the "Single Convention on Narcotic Drugs" prohibiting signatories from growing or supplying drugs (including cannabis) for nonmedical usage. One hundred and eighty-six countries are currently signatories of this convention (United Nations Office on Drugs and Crime, 2021a).

Despite the legislated prohibition and negative publicity of cannabis in the early-mid 1900s, illicit efforts to breed, cultivate, and sell cannabis persisted "underground" for several decades. Ongoing research documenting the effectiveness of cannabis as a medicine, including the discovery of the endocannabinoid system and expanded understanding of human physiological responses to cannabinoids (Mechoulam et al., 2014; Pisanti and Bifulco, 2017), facilitated the reduction of cannabis restrictions in many countries, and supported shifts in public opinion away from the dangers of the plant, to its potential value. By the 1990s, several countries and states had authorized legal cannabis for medical use. In 2013, Uruguay legalized cannabis for general use as part of an effort to reduce drug-related crimes, although implementation of this action took several years due to financial and political opposition from the United States (Maybin, 2019).

Continued relaxation of restrictions occurred in June of 2018, when the World Health Organization's Expert Committee on Drug Dependence (ECDD) voted, based on scientific evidence, to remove the nonpsychoactive component of cannabis, known as cannabidiol (CBD), from its international drug control mandate. Whole cannabis extracts and tetrahydrocannabinol (THC)-based products remain in the 1961 Single Convention on Narcotic Drugs (World Health Organization, 2019). Canada has produced medical cannabis legally since 2001, growing from a single manufacturer to 116 approved licensed producers by September 2018. On October 17, 2018, cannabis was legalized for adult recreational use in Canada and is now regulated in a manner similar to alcohol by most provinces. Cannabis product use and sales are legal for recreational or medical use in 49 different countries and decriminalized in several others (https://en.wikipedia.org/wiki/Legality_of_cannabis). The relaxation of restrictive legislation supports the progress of scientific research in cannabis plant biology and its potential as a medicine.

The impact of human interventions on cannabis phytochemistry

Human cultivation, selection, and dispersal of cannabis have intimately shaped modern-day cannabis plants, facilitating the development of varieties optimized for the production of seeds for food and oil (Fig. 9.1C), bast fibers from its stalks for rope and textiles (Fig. 9.1D), and phytochemical-bearing flowers for their psychoactive and medicinal properties. Although populations of wild cannabis persist across Eurasia and North America, their minute production of the therapeutically valued compounds is in stark contrast to cannabis that has been cultivated for its mind-altering effects (McPartland and Small, 2020; Zhang et al., 2018). The herbal potency of the psychoactive chemical known as THC (described below) has increased fourfold since 1995 (Freeman et al., 2021; United Nations Office on Drugs and Crime, 2021a,b) in spite of long-standing cannabis prohibition. As a result of prohibition, selection efforts centered largely on a limited set of readily observable traits, including plant size, architecture, timing of flowering, and specific psychoactive and organoleptic properties (including flavor, aroma, and texture). Breeding of cannabis for the illicit market has influenced many aspects of the modern cultivars' chemistry, development, and genetics, producing an array of cannabis varieties with elevated psychoactive compound production, as described below.

The distinctive aroma of cannabis is largely conveyed through its production of terpenes, volatile molecules found widely in the plant kingdom. Terpenes also impart medicinal effects and may act synergistically with cannabinoids, as suggested by the varied medicinal effects that varieties of cannabis with different levels or combinations of cannabinoids, terpenes, and flavonoids exhibit. This effect has been coined the "entourage" effect (Russo, 2011), but remains poorly understood. Together with cannabinoids, terpenes form a viscous resin that is made and stored on the outer surface of the plant in fragile glands called trichomes (Fig. 9.1E). Trichomes are found on most above-ground surfaces of the cannabis plant, but are concentrated on female flowers. The cannabinoid and terpene-rich resin synthesized by the plant fills the delicate balloon-shaped gland head of the trichomes, where it is thought to have evolved as a defensive mechanism for the plant, releasing a sticky concoction on invaders when disturbed (Brousseau et al., 2021; Hu et al., 2021). Although evidence supports the evolution of resin-filled trichomes as a defensive mechanism for other trichome gland-bearing plants such as tobacco, tomato, basil, and mint

FIG. 9.1 Anatomic features of dioecious *Cannabis* sp. (A) The apex of a dioecious female cannabis plant, with central, early-stage flowers. (B) The apex of a dioecious male cannabis plant, bearing pollen-producing flowers. (C) Mature, dry cannabis seeds. (D) Cannabis plant stem, with two leaf petioles and female flowers indicated by *arrows*. The stem is the source of bast fibers in hemp-type plants. (E) Live-cell image of a cannabis trichome. The gland head *(bracketed)* contains a cavity filling with cannabinoid and terpene compounds, subtended by cells that synthesize and secrete these chemical mixtures. The image is produced by a two-photon microscope to allow internal features of the trichome to be visualized. Coloring reflects the wavelength of light emitted by the trichome, as described by Livingston et al. (2020).

(Amme et al., 2005; Glas et al., 2012), centuries of cannabis breeding selection suggests humans have influenced the evolutionary trajectory of cannabis' trichomes to favor increased resin production enriched in cannabinoids and terpenes (Livingston et al., 2020).

For the medical and recreational use of cannabis, the composition and potency of cannabinoids, and arguably terpenes, is of primary importance. Phytocannabinoids, or cannabinoids made by the plant, are a diverse group of fatty molecules characterized as terpenophenolics (Andre et al., 2016). These include the two major cannabinoids, tetrahydrocannabinolic acid (THCA) and cannabidiolic acid (CBDA), which can be converted by heat to their active forms, THC and CBD, in a process called decarboxylation (Fig. 9.2). It is only the

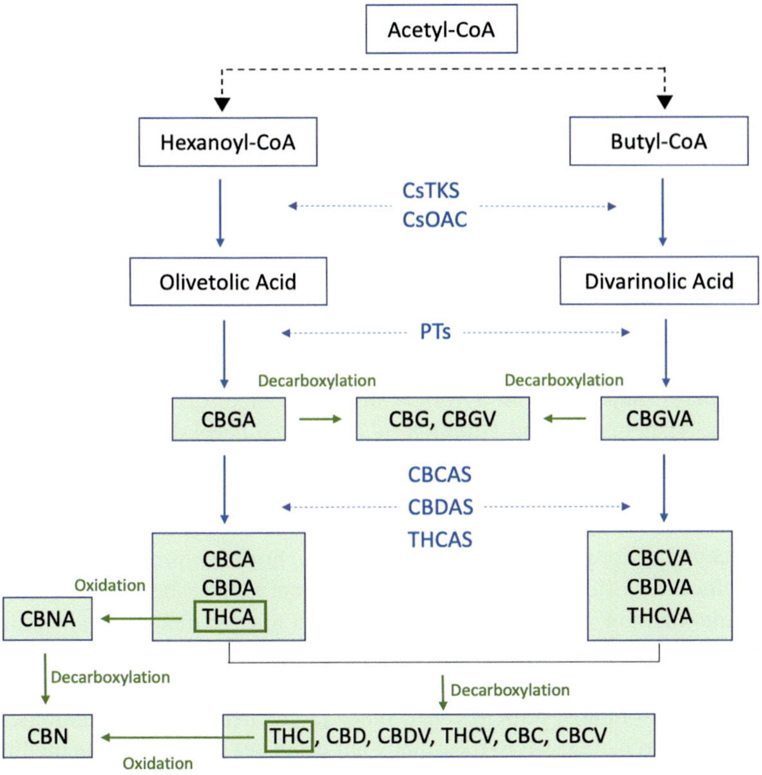

FIG. 9.2 Cannabinoid biosynthetic pathway. A schematic diagram of the cannabinoid biosynthetic pathway showing enzymes (denoted in *blue*), nonenzymatic reactions (denoted in *green*), and cannabinoid products (depicted in shaded *green* boxes). In the first step, acetyl-CoA is converted to hexanoyl-CoA or butyl-CoA by acyl-activating enzymes that are common to all living organisms. The next enzymatic step uses two enzymes, tetraketide synthase (CsTKS) and olivetolic acid synthase (CsOAC), to produce olivetolic acid (OA) or divarinolic acid (DA), depending on which substrate the enzyme uses from the previous step. OA and DA are then converted by aromatic prenyltransferase enzymes (PTs) to CBGA and CBGVA, respectively. Cannabinoid-specific enzymes then transform these precursor molecules into specific cannabinoids. Three examples are shown here: CBCA synthase (CBCAS), CBDA synthase (CBDAS), and THCA synthase (THCAS), which convert the precursor into CBCA/CBDA/THCA or CBCVA/CBDVA/THCVA. Subsequently, these compounds undergo decarboxylation to produce CBD, CBDV, THC, THCV, CBC, or CBCV if they are exposed to heat or radiation, or spontaneously break down. Oxidation, or natural degradation, turns THCA into CBNA or THC into CBN. CBN can also be produced by the decarboxylation of CBNA. Abbreviations for cannabinoids listed: cannabichromene (CBC); cannabichromenic acid (CBCA); cannabidiol (CBD); cannabidiolic acid (CBDA); cannabidivarin (CBDV); cannabidivarinic acid (CBDVA); cannabigerol (CBG); cannabigerolic acid (CBGA); cannabigerovarin (CBGV); cannabigerovarinic acid (CBGVA); cannabinol (CBN); cannabinolic acid (CBNA); tetrahydrocannabinol (THC); tetrahydrocannabinolic acid (THCA); tetrahydrocannabivarin (THCV); tetrahydrocannabivarinic acid (THCVA).

decarboxylated version of cannabinoids that have been shown to have therapeutic effects in humans, through their binding to the cannabinoid receptors, CB1 and CB2, in the endocannabinoid system (Devane et al., 1988; Matsuda et al., 1990; Munro et al., 1993).

To date, more than 100 phytocannabinoids have been identified (Hill et al., 2012; Montone et al., 2020). In addition to THC and CBD, minor cannabinoids, including cannabigerol (CBG), cannabichromene (CBC), cannabinol (CBN), cannabidivarin (CBDV), and tetrahydrocannabivarin (THCV) have been studied for their pharmacological properties (Bielawiec et al., 2020; Pertwee, 2006) (Fig. 9.2). The therapeutic applications for cannabinoids are broad and varied, including the relief of pain, inflammation, nausea, suppressed appetite, spasticity, mood, memory, and insomnia (Cohen et al., 2019). The endocannabinoid system plays major roles in a variety of neurological disorders, such as Parkinson's, Alzheimer's, and Huntington's diseases, necessitating research into the therapeutic potential of cannabinoids for these conditions (De Petrocellis et al., 2011; Andre et al., 2016; Bielawiec et al., 2020; Cristino et al., 2020). THC dominates the chemical landscape of most modern-day cannabis varieties because of long-term selection for its psychoactive properties. CBD has also gained significant interest for its medicinal effects because it does not cause intoxication, and is therefore considered a safer treatment for children or those who do not wish to experience the psychotropic effects of THC. Evidence for the medicinal effects of other cannabinoids, as well as the importance of cannabinoids, terpenes, and other phytochemical combinations, is a growing research area still in its infancy but promises to broaden the medicinal uses of cannabis (Badolia, 2021; Rodriguez et al., 2021; Thomas and Kayser, 2019).

Breeding during prohibition generally favored selection of cultivars or strains with noteworthy impacts on the human experience, including potency and psychoactivity, and this continues today as the legal cannabis industry strives to increase the strength or specificity of cannabis products to appeal to consumers or meet demands of the medical market (Naim-Feil et al., 2021). Many of these observable traits impact trichome biology and are likely to favor the selection of plants bearing large and abundant trichomes capable of housing ample psychoactive and organoleptic phytochemicals. A recent study profiling trichome-specific metabolites has identified substantial increases in the accumulation of fragrant monoterpenes, alongside cannabinoids, over the course of trichome maturation (Livingston et al., 2020). Beyond directly impacting the chemical profiles of cannabis cultivars, selective pressure favoring cannabinoid and terpene production is likely to have impacted a number of plant traits, modifying the genetic programming of the plant to support abundant and large flower production, maximized trichome counts and size, and optimized onset and duration of flowering. Further research in this area will help decipher the impact of these human selective pressures on the genetic regulation of development in cannabis.

Selection for a finite set of traits, including elevated cannabinoid production, has resulted in decreased genetic diversity of modern-day cultivated cannabis (Welling et al., 2016). Reductions in cannabis diversity have significant implications for successful breeding programs, including potential negative impacts on plant fitness. Diversified chemical profiles and resilience traits in cannabis are therefore likely targets for future breeding programs and may include the introgression of traits from wild cultivars or the bioengineering of novel traits. With much of past selection efforts occurring "underground" and in the absence of scientific guidance, there is some uncertainty as to what defines the boundaries between cannabis "cultivars," "varieties," and "strains," as evidenced by genetic variance between cannabis products on the market bearing the same name (Henry et al., 2018; Reimann-Philipp et al., 2020; Sawler et al., 2015; Schwabe and McGlaughlin, 2019; Smith et al., 2021). With the relatively recent growth of the legal medical and adult-use recreational cannabis markets, high-throughput genetic and chemical testing for cannabis has been developed primarily by U.S. instrument manufacturers and testing laboratories, but faces challenges in regulatory consistency, method validation, and the quality of testing (Cassiday, 2016; Wayne Labs, 2019; ; Zimmer, 2020). Nonetheless, the development of these methods has been instrumental in establishing regulatory and quality control measures across the globe, including precise genetics, cannabinoid content, and pesticide testing, to ensure consumers are informed and receive products that are safe for human consumption. These scientific approaches, including metabolic and genetic profiling, also hold tremendous potential as tools to support and inform cannabis cultivation and breeding research.

Advancing cannabis research with genetics

Redefining cannabis classifications

Despite prohibition, genomic research on cannabis has advanced over the last two decades, including the development of several molecular markers for this species (Barcaccia et al., 2020). The first cannabis genome sequence was published in 2011 (van Bakel et al., 2011), followed by several additions to the publicly available sequence databases, which have improved annotation and provided high-quality genome assemblies (available through the National Center for Biotechnology Information, https://www.ncbi.nlm.nih.gov/assembly/?term=cannabis). In addition, several private companies have generated cannabis genome sequences that are partially available to the public, but are often protected by copyrights, trademarks, or patents (https://phylos.bio/galaxy, https://www.medicinalgenomics.com/cannabis-genome, https://www.nrgene.com/cannagene). A number of genotyping-by-sequencing (GBS) studies (Henry et al., 2018; Lynch et al., 2016; Sawler et al., 2015) and genome-wide association studies (GWAS) (Campbell et al., 2019; Petit et al., 2020;

Welling et al., 2020) have used these data to identify single-nucleotide polymorphisms (SNPs) associated with genomic regions or genetic traits in cannabis. In addition, sequencing of expressed genes from cannabis trichomes has supported efforts to characterize the specialized biochemical activities in these structures (Huchelmann et al., 2017; Lange, 2015; Livingston et al., 2020; Zager et al., 2019). These genomic resources offer valuable tools for ongoing research aimed at addressing key issues facing cannabis cultivation, including the control of metabolite production, plant pathogen resistance, and the consistency of cultivars propagated from seed. While there is some evidence that genetic modification of the plant through bioengineering is being attempted (Ahmed et al., 2021; Beard et al., 2021; Hesami et al., 2021; Roscow, 2017), public opposition to genetic modification technology remains strong, especially among consumers of natural medicines such as cannabis. Thus, most cannabis breeding and research being reported uses conventional breeding technology where cultivars with specific phenotypes are crossed and offspring with appealing traits are selected for production or further breeding. It should be noted that different jurisdictions regulate the intellectual property rights of novel plants differently, with the United States and some countries in Europe focusing patent protection on specific methods used to create the novel plant, whereas in Canada the regulatory system focuses on the characteristics of the final plant product (McKenzie, 2020).

An interesting discussion among cannabis consumers and patients centers on the use of the designations "indica" or "sativa" as a descriptive addition to the strain name. Cannabis suppliers often use the "indica" and "sativa" nomenclature to describe the medicinal or therapeutic effects of a specific cultivar. However, genomic data have shown that both indica and sativa varieties belong to the species *C. sativa* (McPartland, 2018) and that these colloquial names cannot be assigned to conserved genetic differences between cultivars (Alimohammadi and Bagherieh-Najjar, 2009; Hazekamp et al., 2016; Jin et al., 2021b; Sawler et al., 2015; Schwabe and McGlaughlin, 2019). Thus, although the terms "indica" and "sativa" are well accepted in the cannabis consumer market and serve a useful role in marketing and in medical prescription, the terms were popularized during cannabis prohibition, and genomic research has shown that it is not clear how valid these assignments are or on what they are based.

Another challenge for the cannabis industry is the use of accurate, meaningful names for different cannabis strains or varietal cultivars. Because the development and breeding of most cultivars took place when secrecy was required, names were often assigned haphazardly and inconsistently. As a result, genotype testing has shown that different cultivars with the same name may have very different genetics and chemistry (Gilchrist et al., 2021b; Henry et al., 2018; Reimann-Philipp et al., 2020; Schwabe and McGlaughlin, 2019; Smith et al., 2021). This is a serious issue, particularly for medical cannabis prescribers and consumers who require product consistency. Genetic testing of cannabis using genotyping offers a dependable system to substantiate strain or cultivar authenticity. However, given the consumer focus on the chemistry of the plant (potency, smell, and taste), commercial cannabis operations have been reluctant to identify cultivars based on genotype, preferring to use chemical testing for potency and terpene content instead. This is, in part, based on the fact that the chemistry methods used for analysis in cannabis are the same as those used for other plants, whereas the genetic methods are specific to the genome sequences of each species and are therefore just starting to become routine in cannabis.

Chemical profiling of cannabis, although a valuable indicator of product potency and therapeutic effects, is subject to variation due to factors such as environmental conditions of plant growth or processing, the method of chemical profiling, and the handling and preparation of materials for analysis. In contrast, the genetic makeup of a plant, or its genotype, is consistent and independent of sampling methods or environmental conditions. One method of genotyping that is applicable to cannabis involves the use of simple sequence repeats (SSRs), also called microsatellites, to generate a DNA fingerprint for the plant. SSR genotyping involves the use of polymerase chain reaction (PCR) to amplify specific regions of DNA that are variable from cultivar to cultivar (Alghanim and Almirall, 2003; Gao et al., 2014; Gilchrist et al., 2021b; Houston et al., 2016; Howard et al., 2008). It is a well-established method that has successfully been used on fungi, plants, invertebrates, and vertebrates, for crop and livestock varietal authentication and for human forensic DNA testing (Bernardi et al., 2016; Norrgard, 2008). In contrast to pan-genomic methods like next-generation sequencing (NGS), or single nucleotide polymorphism (SNP) methods, SSR genotyping is fast and inexpensive, providing results that are easily accessible and do not require complex algorithms or software for data interpretation. While SNP methods are more useful for determining genomic diversity if several thousand randomly distributed markers are used, SSR genotyping has been shown to be as accurate as SNP methods for the purposes of confirming cultivar identity, ascertaining parentage, or proving the provenance of genetic material (García et al., 2018). This genotyping method provides a rapid and inexpensive form of genetic authentication, which could be used to provide proof of cultivar identification and improve transparency for patients and consumers, given the confounding nomenclature currently used in the cannabis industry.

Sex determination

Cannabis is predominantly dioecious, meaning it produces separate male and female plants. As in humans, sex in cannabis is determined by X and Y sex chromosomes, with dioecious female plants carrying two X chromosomes and males carrying one X and one Y chromosome. Sex determination in cannabis likely evolved in the ancestral lineage of the Cannabaceae family from which hops and cannabis descended (Prentout et al., 2021). Dioecious male and female cannabis plants are indistinguishable in the vegetative stage of growth (before flowering), but can be readily distinguished by their flowers upon entering the reproductive phase (Fig. 9.1A and B). Monoecious cannabis cultivars also exist, producing both male and female flowers on the same plant and carrying an XX sex chromosome set. As in several other plants, female cannabis plants may be induced to produce male flowers through the exogenous application of plant hormone inhibitors, or silver ions (Galoch 1978; Mohan Ram and Sett 1982; Lubell and Brand 2018). Interestingly, the ability of dioecious or monoecious XX plants to produce both male and female flowers indicates that the gene(s) for male flower production cannot be located on the Y chromosome.

Monoecious and dioecious cannabis are typically grown for different applications. Monoecious cultivars are useful in hemp cultivation, especially when grown for seeds, because of the ability for every plant to cross-pollinate and set seed. However, cannabis cultivated for medical or recreational purposes typically involves dioecious female plants because the production of valued cannabinoids is highest on the unfertilized female flowers (Livingston et al. 2020; Jin et al. 2020; Richins et al. 2018). Male plants do not produce high levels of cannabinoids, and their release of copious amounts of fine pollen can reduce the value of the harvested product due to seed production. Thus, to ensure female-only cannabis cultivation, there is a need to eliminate male plants early in development. Propagating all-female cannabis from clones (rather than from seed) is the method of choice for most indoor cannabis operations. However, the issue of early sex identification becomes important in the management of breeding programs and growth operations using seed propagation methods. To mitigate the fact that identifying males is visually challenging in the vegetative stage, several molecular methods have been developed to identify young male plants prior to flowering or the release of pollen. Cannabis sex identification methods based on SNP differences between XX and XY plants have demonstrated efficacy (Gilchrist et al., 2021a; Toth et al., 2020) along with methods based on "Male-Associated DNA in Cannabis" (MADC) regions that generate different PCR amplicon sizes in XX and XY plants (Mandolino et al., 1999; Shao et al., 2003). These methods have been commercialized by several companies (https://anandia.ca/, https://www.medicinalgenomics.com/gender-detection/, https://leafworks.com/products/cannabis-hemp-dna-sex-test, https://phylos.bio/products/plant-sex-test).

Development of seed-based cultivars

Most medical and recreational cannabis cultivars are currently clonally propagated, producing populations of genetically identical plants. Clonal propagation overcomes the genetic volatility of metabolite production and minimizes variability in cannabinoid and terpene contents, but also fosters reduced genetic diversity. The development of seed-based cultivars for cannabis enables growers to move away from labor-intensive clonal propagation methods and is an approach of growing interest in the cannabis industry (Barcaccia et al.,

2020; CanBreed, 2021; Casanoa, 2019; Crawford et al., 2021; Royal Queen Seeds, 2021). The outcrossing of wind-pollinated hemp varieties supports healthy genetic heterogeneity, but is avoided in medical-grade cannabis where such heterogeneity is likely to create chemical variation and inconsistent products. Hence, there is interest in developing stable, F1 hybrid seed lines as used in other crop plants (Barcaccia et al., 2020; Casanoa, 2019; Crawford et al., 2021). Generating F1 hybrid seeds requires highly inbred, genetically homozygous lines to serve as the parents (Riggs, 1988). The F1 hybrid seeds produced from the cross between the inbred parents then carry the same copy of each gene inherited from a specific parent, leading to a level of genetic homogeneity that some hope will supplant the need for clonal propagation in cannabis.

The difficulty with this approach is several-fold. First, cannabis has a fairly high level of inbreeding depression, which is the reduction in viability that occurs when breeding two closely related individuals (Kurtz et al., 2020; Punja and Holmes, 2020). This will need to be overcome in order to successfully generate inbred parent lines. In another medicinal plant, Artemisia, development of hybrid seeds has been problematic due to high levels of self-incompatibility (Wetzstein et al., 2018) and has resulted in the continued use of cloning as the propagation method of choice. In addition, hybrid seeds that have been used in other annual crops like peppers, corn, and tomato require a long and dedicated period of research, which, in the largely hedonistic and rapidly evolving cannabis industry, is unlikely to be practical. Further, the influence of epigenetic factors on gene expression affects the phenotype of seed-grown cultivars in ways that are not yet well understood. Epigenetics is the influence of nongenetic factors on the physical appearance or phenotype of an organism, and the epigenetic control of gene expression in cannabis has not been well studied. Because of epigenetic factors, the generation of hybrid seeds will produce progeny that are not identical to each other in phenotype, in spite of their similar genetic backgrounds (Khanday and Sundaresan, 2021). The successes seen in corn, tomato, or peppers with the development of F1 hybrid seeds may not translate to nonhemp cannabis, where the main objective is consistent metabolite production. Small differences in genetics affecting the production of cannabinoids and terpenes may therefore be amplified by environmental and epigenetic inconsistencies, yielding exaggerated differences in metabolic phenotypes in cannabis grown from seed. Given these challenges, a practical consideration might be the production of clonal seed lines through apomixis, a process of generating seeds without going through fertilization (Albertini et al., 2010; Fiaz et al., 2021; Khanday and Sundaresan, 2021). Apomixis occurs naturally in some plants, and all of the seeds produced by apomixis are clones of the parent plant. This process does not occur naturally in cannabis and therefore has challenges that will need to be overcome but are worthy of consideration. One final concern about seed propagation in cannabis is that growers generally find that seeds do not remain viable for long periods of time after harvesting and there is limited data available to support cannabis seed viability (Small and Brookes, 2012).

The boom of cannabis biotechnology and testing

Using biotechnology to select target traits and support bioengineering

Legalization and commercialization of cannabis has broadened the market for consumers, requiring the diversification of cannabis traits such as flavor and aroma, and the level and mixture of psychoactive and nonpsychoactive cannabinoids. In contrast to the reliance on consumer preference data which informs the breeding efforts of most food crops, data on consumer preferences are limited for cannabis and require consideration of medical properties in addition to recreational predilections (BDSA, 2021; Peckenpaugh, 2020; Pehota, 2020). Work on synthetic cannabinoids and the production of other cannabis plant metabolites has already begun (https://demetrix.com/,https://sanobiotec.com/synthetic-vs-natural-cannabinoids/, https://hyasynthbio.com/, https://www.willowbio.com/#about-us), and is likely to affect breeding goals since some metabolites may be easier to grow or purify using microorganisms or other sources. Data on consumer preferences for the forms of consumption (e.g., edibles, vapes, oils, etc.) may also influence cannabis breeding goals since the extraction steps used affect the composition of the final product. These considerations may result in tailored cannabis breeding efforts that accommodate different preparative, processing, and market uses.

Target trait: Mold and mildew resistance

Disease caused by viral, fungal, and bacterial pathogens in cannabis remains poorly understood (Punja, 2018), and pathogen-resistance breeding has been slow to develop as a result. To address the severe losses caused by potential pathogen infections, cannabis growers' efforts have largely focused on eliminating sources of infection rather than on the costly and time-consuming research of resistance breeding. Mold and mildew resistance in plants can be conferred by resistance genes and/or by the architectural morphology of the plant (Ando et al., 2007; Zhou and Zhang, 2020). *Golovinomyces* and *Botrytis* species are fungal pathogens that cause powdery mildew and gray mold, respectively, on the leaves and flowers of cannabis plants and are endemic in many growth operations. While exposure to these fungi does not cause infection in humans, their spores may be harmful when inhaled by immunocompromised individuals (Jerushalmi et al., 2020; Thompson III et al., 2017). Unfortunately, few fungicides have been approved for use on cannabis, and even fewer are very effective against these pathogens. However, mold and mildew infections are fairly easy to detect, and resistance is a trait that was a focus of cannabis breeding efforts well before legalization, as evidenced by the presence of mold and mildew resistant cultivars (https://moldresistantstrains.com/). Biotechnology tools hold promise for continuing this research in a variety of ways. The adaptation of exogenous, RNA or peptide-based pathogen control methods used in other crops to cannabis should be applicable and will not cause harm to humans when consumed (Duanis-Assaf et al., 2021; Höfle et al., 2020; Ormancey et al., 2021; Wang et al., 2017; Wang and Jin, 2017). These methods use RNA molecules or peptides (small protein particles), which are synthesized in vitro and applied to the surface of plants via spraying or drenching. The RNA or peptides are then taken up either by the plant, supporting improved plant immunity, or by the pathogenic organism, causing diminished viability or infectiousness (Qiao et al., 2021; Wang et al., 2017; Werner et al., 2020). In addition to external control of pathogen infection, cannabis genes that are known to affect powdery mildew resistance in other plant species have been identified (Mihalyov and Garfinkel, 2021; Pépin et al., 2021), along with the characterization of genetically controlled morphologies that support mold resistance (Jin et al., 2021a). Conventional breeding efforts may benefit from the inclusion of genetic engineering to fine-tune the cannabis plant's immunity to general or specific fungi (Liu et al., 2021). The identification and continued study of these traits will support breeding programs that ensure robust plant growth and consistent yields, and minimize plant losses and the potential risks to human health caused by these diseases.

Target trait: Virus resistance

Another area of growing interest in the cannabis industry is viruses, some of which have been reported to cause crop damage or yield reduction (Simmons, 2020). Research in this area is compounded by the common correlation of "sick-looking" plants with the laboratory detection of a virus, but no causation data. One of the few studies on this topic is summarized in an abstract by Hartowicz et al. from a scientific conference in 1971, in which presented data identified many viruses capable of infecting cannabis, yet only four were specifically named as causing severe symptoms: tobacco ringspot virus, tomato ringspot virus, tobacco streak virus, and cucumber mosaic virus. Interestingly, one of the most commonly detected viruses in cannabis, tobacco mosaic virus, did not cause any symptoms (Hartowicz et al., 1971). Another virus for which there is considerable historic lore, cannabis cryptic virus, has similarly been shown not to be correlated with the physical symptoms for which it has repeatedly been associated (Righetti et al. 2018). There are some viruses that have been demonstrated to cause disease and damage in cannabis, however. Damage to cannabis crops by Hop Latent Viroid (Warren et al., 2019), lettuce chlorosis virus (Hadad et al., 2019), and beet curly top virus (Giladi et al., 2020) has recently been proven, for example. Because viruses require the host plant's natural metabolism to reproduce, effective innate resistance is rare and usually virus-specific (Kang et al., 2005; Truniger and Aranda, 2009). For this reason, conventional breeding methods have been replaced by genetic engineering in many crops (Bastet et al., 2018; Kalinina et al., 2019; Zhao et al., 2020) and may be something that is explored in cannabis for mitigating damages caused by these pathogens, in addition to the exogenous application of RNA-based treatments as described in the section above (Taliansky et al., 2021). Armed with new tools for discovering the underlying causes of susceptibility in cannabis, and growing knowledge of the fungal and viral pathogens of concern to indoor and outdoor cannabis cultivation, the future looks bright for developing resilient cannabis cultivars.

Trait target: Metabolite production

Dissecting the genetic underpinnings of cannabis' metabolite production presents a major challenge, as the plant is capable of producing tremendous chemical diversity and the production pathways for cannabinoids and terpenes are intertwined (Booth et al., 2020; Zager et al., 2019). Genetic markers for THC and CBD dominance were identified in the early 2000s (De Meijer et al., 2003) but the redundant and repetitive nature of the genes responsible for the synthesis of these and other cannabinoids has been challenging to unravel, and control over the level of cannabinoid and terpene production, in general, remains elusive (Grassa et al., 2018; Laverty et al., 2019; Weiblen et al., 2015). Until recently, few studies presented comprehensive cataloguing of cannabis plant metabolites (Aliferis and Bernard-Perron, 2020; Hazekamp et al., 2016; Mudge et al., 2019), but continued metabolomic studies to characterize the chemical profiles present in various cannabis cultivars should support breeding efforts to target the genes responsible for the coveted biochemical components.

In addition to THC and CBD, interest has been expressed in boosting the production of minor cannabinoids, including CBN, CBG, CBC, THCV, and up to 100 others (Jin et al., 2020; Richins et al., 2018; Tomko et al., 2020) (Fig. 9.2). Many minor cannabinoids have been shown to have physiological effects in humans that differ from THC or CBD, although conclusive data is sorely lacking (De Petrocellis et al., 2011; Rodriguez et al., 2021; Thomas and Kayser, 2019; Tomko et al., 2020). Given the profitability that CBD-based products have generated in recent years, elevating minor cannabinoid production is seen as a desirable target for increasing market breadth, even though the evidence for their overall effectiveness remains limited and the genes responsible for their synthesis have not all been characterized (Black, 2018; Darling, 2021). Although most of the biosynthetic genes in these pathways are known, control of their expression levels is poorly understood, prompting the exploration of synthetic production of these metabolites in the laboratory or in noncannabis species. Identifying the genes required for the production and levels of each of the many cannabinoid compounds will require complex efforts to characterize their endogenous functions in the plant, potential impacts on pathogen resistance or other agronomic factors, and mechanisms that fine-tune their expression in various cultivars and growth conditions.

Bioengineering solutions for eco-conscious cannabinoid production

The legal cannabis industries have developed rapidly to support medical and adult-use recreational markets in Canada, Georgia, Mexico, South Africa, Uruguay, several U.S. states, and the Australian Capital Territory (https://en.wikipedia.org/wiki/Legality_of_cannabis). The high demand for compliant and consistent cannabinoid production and increasing public support for commercial practices that mitigate the effects of climate change will need to be addressed as these markets continue to grow. Thus, the high energy and material inputs associated with cannabis plant cultivation are likely to face competition from new bioengineering solutions that offer efficient and precise production of individual cannabinoids. Synthetic biology using other organisms to produce these products via novel enzymes and fermentation technologies may offer the ability to produce pure cannabinoids at industrial scales, especially for minor cannabinoids that are typically produced in minute quantities by cannabis plants (Carvalho et al., 2017).

Growing legal cannabis plants has a number of challenges as it is typically highly regulated and requires special growth conditions to produce good-quality, trichome-covered flowers with significant levels of cannabinoids. Synthetic biology offers an alternate approach to the production of these pharmaceutically relevant compounds that relies on heterologous systems rather than cannabis plants. Despite the many challenges faced in moving cannabinoid production out of the cannabis plant, the potential for higher purity, lower environmental costs, and fewer regulatory constraints makes heterologous systems an attractive alternative (Luo et al., 2019; Zirpel et al., 2018). Cannabinoid biosynthesis has been well studied (Blatt-Janmaat and Qu, 2021; Gülck and Møller, 2020) and involves three critical steps to form these specialized metabolites (Fig. 9.2). The first is the formation of olivetolic acid (OA) in two steps, using first tetraketide synthase (CsTKS), then olivetolic acid cyclase (CsOAC) (Gagne et al., 2012). Cannabigerolic acid (CBGA) is then synthesized by aromatic prenyltransferases (PTs) to form the backbone of the cannabinoid structure (Blatt-Janmaat and Qu, 2021; Tahir et al., 2021). The last step uses enzymes that are specific to individual cannabinoids, for example, THCA synthase converts CBGA into THCA, and CBDA synthase converts CBGA into CBDA. These genes have been introduced into various organisms using molecular and genetic tools, including bacteria (*Escherichia coli*), tobacco (*Nicotiana*), and yeast (*Saccharomyces cerevisiae*) (Ayakar et al., 2020; Geissler et al., 2018; Gülck et al., 2020; Luo et al., 2019). *E. coli* has proven to be a challenging system for cannabinoid production due to improper protein folding for terminal cannabinoid cyclases, but co-culturing with other organisms such as yeast may resolve this issue (Wang et al., 2020). Efforts to produce cannabinoids in tobacco, an established crop plant, have not seen much success to date (Gülck et al., 2020). Currently, yeast is the most well-established platform for heterologous cannabinoid production. In this system, yeast are engineered to carry cannabis biosynthetic genes and fed different fatty acid substrates to support cannabinoid molecule biosynthesis (Luo et al., 2019; Zirpel et al., 2015, 2017, 2018). The recent discovery of nonplant polyketide synthase genes that enable the production of OA, the precursor to all cannabinoids, in high titers in fungal cell cultures (Okorafor et al., 2021) has supported the emergence of new strategies for synthetic cannabinoid production, including the search for genes in other organisms that may allow increased cannabinoid production beyond what is possible with endogenous cannabis genes. There is still considerable room for improvement in the quest to produce high titers and industrial-scale fermentation of cannabinoids in yeast, however, enzyme activities need to be optimized and methods for better intracellular fatty acid availability improved.

Sustainability and environmental impacts

Cannabis, like other crops, requires pest controls, the application of fertilizers, and optimized environmental conditions to ensure high-yielding, healthy plant production. But as a controlled substance, commercial cannabis cultivation faces unique challenges to ensure that high-quality, safe, and consistent product is obtained. A high percentage of cannabis growth in temperate climates occurs in secure indoor growth facilities that rely on controlled environmental conditions with artificial lighting, carbon dioxide (CO_2) supplementation, automated watering schemes, and energy-intensive HVAC systems to control odors and maintain optimal temperatures and humidity (Lu, 2021). The carbon footprint for cannabis production has been estimated to be 10,000 times greater than that of crops like corn, emitting CO_2 in the range of 2000–5000 kg for every kg of dried flower produced by indoor facilities across the United States (Riskin, 2021; Summers et al., 2021). Legal indoor cannabis cultivation further requires integrated pest management practices as the use of fungicides or biologic controls must comply with regulatory policies. Although legal outdoor cannabis operations offer dramatic reductions in the energy inputs required, outdoor cultivation comes with its own set of challenges. Outdoor growth often requires irrigation to support cannabis' weak drought tolerance, as well as significant fertilizer and pesticide applications (Warren, 2015). The flowering of most cannabis cultivars is dependent on the precise timing of day-light cycles, requiring 12 h of light and 12 h of dark to initiate flowering, which has further challenged outdoor growth in temperate climates. Some auto-flowering cultivars exist, in which the timing of flowering is not dependent on the light cycle under which the plants are grown (Dowling et al., 2021). Successful outdoor cultivation in temperate climates currently requires auto-flowering cultivars, limiting the varieties that are available for this type of production.

In addition to the resources required to produce high-yielding plants, waste generated in the processing and packaging of cannabis is of concern. This consists of large amounts of plant waste material that is discarded after the floral buds have been harvested, including leaves, stalks, and roots. Cannabinoid production is only concentrated in the flowers of the plant and therefore is typically the only part of the plant that is used. For legal reasons, this waste typically cannot be composted because of trace levels of cannabinoids. Requiring mixing of waste with nonplant matrix (e.g., kitty litter and dish soap) to render it unusable for consumption. Further, legislative requirements strictly regulate the labeling and packaging of cannabis products for distribution, creating large amounts of single-use plastic waste.

The environmental impact of outdoor cannabis growth remains challenging to calculate, due to the high number of illegal operations that persist, and the noncompliant use of pest management and waste disposal often associated with these operations. Changes in environmental, social, and corporate governance (ESG) practices and their effect on attracting consumers offer hope for pushing the cannabis industry toward more environmentally friendly practices (Drotleff, 2021; Mitchell, 2020). Government initiatives to establish regulations that address environmental concerns from both outdoor and indoor grows are also anticipated to promote more sustainable cannabis cultivation practices, and are one reason for advocating for broader legalization of this crop.

Testing and regulating cannabis: the challenge of persisting illegal markets

Testing is required for all cannabis commodities in most legal markets, to ensure safe products for human consumption. Cannabis materials are tested for the presence of pesticides, microbial contamination, and potency. Regulations around testing are variable and contaminant testing limits are often determined by instrument sensitivities rather than by medical research on the danger of the contaminants for humans. Of course, "better safe than sorry" is a good adage, but the expense of testing and the move toward more compliance would be better justified if the research into what was "safe" was forefront and funded by the bodies setting testing limits. Some of the disadvantages of legally grown cannabis are higher costs, more limited selection of genetics, and growing conditions optimized for total yield rather than customized for the production of aromatic or psychoactive metabolites specific to individual cultivars. The major advantage of legal cannabis over illegally grown product is the fact that it is legal. That definition, on its own, will attract a higher percentage of customers than illegally grown and sold products. In the end, it is clear that the legal and illegal production of cannabis will continue to be intertwined for the foreseeable future.

Illegal breeding in cannabis has been ongoing for decades and has produced many of the popular cultivars in the legal market today. In Canada, as incentive to bring more of the legacy cultivars into the legal cannabis market, licensed producers are allowed to bring unlicensed cannabis genetics in from the illegal market without penalty, but only at the time when they obtain their first license and claim these cultivars as varieties they intend to grow. There is no legitimate way of incorporating unlicensed, novel cultivars into breeding or cultivation programs after this initial licensing period. As a result, very limited seed varieties are available for legal purchase in Canada, while unregulated seed exchanges continue to be popular among growers (Cain, 2019). Illegal cannabis is cheaper to grow and sell because it avoids licensing and required testing of its products and the expensive formalities that go along with ensuring growth facilities and products are clean and safe. Illegally sold cannabis has been shown to have increased levels of pesticide and microbial contamination as well as unreliable levels of THC (potency) as compared to legally sold cannabis (Botelho et al., 2021; Pusiak et al., 2021). Altogether, although the illegal market benefits from lower costs, it provides greater risks to human health, environmental damage, and the potential for organized criminal groups to profit and control the industry.

Conclusions

Cannabis is a fascinating plant about which we still have much to learn. It has been cultivated by humans for millennia, but modern science is only beginning to uncover its unique and beneficial properties. Differences of opinion abound about the direction that cannabis genomics research should take, but some common themes appear. Research into how the plant makes the hundreds of molecules that give it its medicinal properties must continue, including into the therapeutic properties of minor cannabinoids and of specific mixtures of cannabinoids, terpenes, and other metabolites. Better understanding of the mechanisms that produce and regulate the array of cannabinoids and terpenes in cannabis will support targeted improvement of the cannabis crop and accelerate efforts to move cannabinoid synthesis into alternate systems. Genetics and biotechnology hold tremendous potential to inform breeding efforts for improved metabolite production, pathogen resistance, and propagation methods beyond cloning. Better understanding of epigenetic, environmental, and nongenetic factors on gene expression in cannabis will provide insight into the interaction between the genomics of the plant and its phenotype, or physical characteristics. Testing of legal cannabis products is performed using high-throughput methods that are only possible because of advances in biotechnology and instrumentation, and continue to be improved. While illegal cannabis operations can avoid the costs and regulatory burdens of thorough testing of cannabis products, the risks to health and safety, for many consumers, far outweigh reduced costs. Finally, the legal and illegal markets for cannabis will continue to evolve and shift as recognition for the value of safe, sustainable, ethically grown cannabis plant products increases globally.

Acknowledgments

The trichome image presented was prepared by T.D.Q. in the University of British Columbia's BioImaging Facility, under the direction of Professor A. Lacey Samuels.

Statement of interest

E.J.G. is an employee of Anandia Laboratories, which offers molecular diagnostic tests for cannabis, including the tests for sex identification and genotyping mentioned in this chapter. T.D.Q. and S.W. declare no conflicts of interest.

References

Ahmed, S., Gao, X., Jahan, M.A., Adams, M., Wu, N., Kovinich, N., 2021. Nanoparticle-based genetic transformation of *Cannabis sativa*. J. Biotechnol. 326, 48–51. https://doi.org/10.1016/J.JBIOTEC.2020.12.014.

Ajay, D., 2021. Global Cannabis Market is Estimated to Grow at a CAGR of 27.0% from 2021 to 2030. MarketWatch. https://www.marketwatch.com/press-release/global-cannabis-market-is-estimated-to-grow-at-a-cagr-of-270-from-2021-to-2030-2021-07-12.

Albertini, E., Barcaccia, G., Mazzucato, A., Sharbel, T., Falcinelli, M., 2010. Apomixis in the era of biotechnology. In: Plant Developmental Biology—Biotechnological Perspectives. vol. 1, pp. 405–436, https://doi.org/10.1007/978-3-642-02301-9_20.

Alghanim, H.J., Almirall, J.R., 2003. Development of microsatellite markers in *Cannabis sativa* for DNA typing and genetic relatedness analyses. Anal. Bioanal. Chem. 376 (8), 1225–1233. https://doi.org/10.1007/s00216-003-1984-0.

Aliferis, K.A., Bernard-Perron, D., 2020. Cannabinomics: application of metabolomics in cannabis (*Cannabis sativa* L.) research and development. Front. Plant Sci. 11, 554. https://doi.org/10.3389/FPLS.2020.00554.

Alimohammadi, M., Bagherieh-Najjar, M.B., 2009. Agrobacterium-mediated transformation of plants: basic principles and influencing factors. Afr. J. Biotechnol. 8 (20), 5142–5148.

Amme, S., Rutten, T., Melzer, M., Sonsmann, G., Vissers, J.P.C., Schlesier, B., Mock, H.-P., 2005. A proteome approach defines protective functions of tobacco leaf trichomes. Proteomics 5 (10), 2508–2518. https://doi.org/10.1002/pmic.200401274.

Ando, K., Grumet, R., Terpstra, K., Kelly, J.D., 2007. Manipulation of plant architecture to enhance crop disease control. CAB Rev. 2 (026), 1–8. https://doi.org/10.1079/PAVSNNR20072026. Perspectives in Agriculture, Veterinary Science, Nutrition and Natural Resources.

Andre, C.M., Hausman, J.F., Guerriero, G., 2016. Cannabis sativa: the plant of the thousand and one molecules. Front. Plant Sci. 7, 19.

Ayakar, S.R., Hallam, S.J., Hossain, S., Pawar, S., Roy, P., Srivastava, S.K., Yadav, V., 2020. Metabolic Engineering of *E. coli* for the Biosynthesis of Cannabinoid Products (Patent No. 16/644484). USPTO.

Badolia, I., 2021. Cannabinoids future research. In: Cannabinoids and Pain, pp. 325–332, https://doi.org/10.1007/978-3-030-69186-8_40.

Barcaccia, G., Palumbo, F., Scariolo, F., Vannozzi, A., Borin, M., Bona, S., 2020. Potentials and challenges of genomics for breeding *Cannabis cultivars*. In: Frontiers in Plant Science. vol. 11. Frontiers Media S.A, p. 1472, https://doi.org/10.3389/fpls.2020.573299.

Bastet, A., Lederer, B., Giovinazzo, N., Arnoux, X., German-Retana, S., Reinbold, C., Brault, V., Garcia, D., Djennane, S., Gersch, S., Lemaire, O., Robaglia, C., Gallois, J.-L., 2018.

Trans-species synthetic gene design allows resistance pyramiding and broad-spectrum engineering of virus resistance in plants. Plant Biotechnol. J. 16 (9), 1569–1581. https://doi.org/10.1111/PBI.12896.

BDSA, 2021. BDSA Talks Seasonal Sales Strategies, Consumer Behavior, and How to Win at cannabis Retail—MJ Brand Insights. MJ Brand Insights. https://mjbrandinsights.com/bdsa-talks-seasonal-sales-strategies-consumer-behavior-and-how-to-win-at-cannabis-retail/.

Beard, K.M., Boling, A.W.H., Bargmann, B.O.R., 2021. Protoplast isolation, transient transformation, and flow-cytometric analysis of reporter-gene activation in *Cannabis sativa* L. Ind. Crop. Prod. 164, 113360. https://doi.org/10.1016/j.indcrop.2021.113360.

Bernardi, J., Colli, L., Ughini, V., Busconi, M., 2016. Use of microsatellites to study agricultural biodiversity and food traceability. In: Microsatellite Markers. IntechOpen, pp. 20–45, https://doi.org/10.5772/64863.

Bielawiec, P., Harasim-Symbor, E., Chabowski, A., 2020. Phytocannabinoids: useful drugs for the treatment of obesity? Special focus on cannabidiol. Front. Endocrinol. 11, 114. https://doi.org/10.3389/fendo.2020.00114.

Black, L., 2018. Hunting for Rare Cannabinoids—Green Guide Spring 2018—The Stranger. The Stranger. https://www.thestranger.com/green-guide-spring-2018/2018/04/11/26024638/hunting-for-rare-cannabinoids.

Blatt-Janmaat, K., Qu, Y., 2021. The biochemistry of phytocannabinoids and metabolic engineering of their production in heterologous systems. Int. J. Mol. Sci. 22 (5), 1–19. https://doi.org/10.3390/ijms22052454.

Booth, J.K., Yuen, M.M.S., Jancsik, S., Madilao, L.L., Page, J.E., Bohlmann, J., 2020. Terpene synthases and terpene variation in *Cannabis sativa*. Plant Physiol. 184 (1), 130–147. https://doi.org/10.1104/PP.20.00593.

Botelho, D., Boudreau, A., Rackov, A., Rehman, A., Phillips, B., Hay, C., Broad, K., Crowhurst, P., Cockburn, S., Smith, T., Balanova, B., 2021. Analysis of illicit and legal Cannabis products for a suite of chemical and microbial contaminants. In: RPC Science and Engineering Engineering, pp. 1–12.

Brousseau, V.D., Wu, B.-S., MacPherson, S., Morello, V., Lefsrud, M., 2021. Cannabinoids and terpenes: how production of photo-protectants can be manipulated to enhance *Cannabis sativa* L. phytochemistry. Front. Plant Sci. 12, 620021. https://doi.org/10.3389/FPLS.2021.620021.

Cain, P., 2019. Almost a Year After Legalization, Why Do Legal Sellers Offer Home Growers So Little?—National | Globalnews.ca. Global News. https://globalnews.ca/news/5866615/marijuana-cannabis-home-grow/.

Campbell, B.J., Berrada, A.F., Hudalla, C., Amaducci, S., McKay, J.K., 2019. Genotype × environment interactions of industrial hemp cultivars highlight diverse responses to environmental factors. Agrosyst. Geosci. Environ. 2 (1), 1–11. https://doi.org/10.2134/AGE2018.11.0057.

CanBreed, 2021. F1 Hybrid Seeds. http://www.can-breed.com/page7.html.

Carvalho, Â., Hansen, E.H., Kayser, O., Carlsen, S., Stehle, F., 2017. Designing microorganisms for heterologous biosynthesis of cannabinoids. FEMS Yeast Res. 17 (4), 37. https://doi.org/10.1093/FEMSYR/FOX037.

Casanoa, S., 2019. Development of ornamental *Cannabis sativa* L. cultivars: phytochemical, morphological, genetic characterization and propagation aspects. Acta Horticult. 1263, 283–290. https://doi.org/10.17660/ACTAHORTIC.2019.1263.37.

Cassiday, L., 2016, October. The Highs and Lows of Cannabis Testing. The American Oil Chemists' Society.

Cohen, K., Weizman, A., Weinstein, A., 2019. Positive and negative effects of Cannabis and cannabinoids on health. Clin. Pharmacol. Ther. 105 (5), 1139–1147. https://doi.org/10.1002/CPT.1381.

Crawford, S., Rojas, B.M., Crawford, E., Otten, M., Schoenenberger, T.A., Garfinkel, A.R., Chen, H., 2021. Characteristics of the diploid, triploid, and tetraploid versions of a cannabigerol-dominant F1 hybrid industrial hemp cultivar, *Cannabis sativa* 'stem cell CBG.'. Genes 12 (6), 923. https://doi.org/10.3390/GENES12060923.

Cristino, L., Bisogno, T., Di Marzo, V., 2020. Cannabinoids and the expanded endocannabinoid system in neurological disorders. Nat. Rev. Neurol. 16 (1), 9–29.

Darling, C., 2021. Cannabis Cannabinoids Explained: Types, Effects, Legality. Leafly. https://theweedblog.com/marijuana-science/how-many-different-cannabinoids-are-there-in-marijuana.

De Meijer, E.P.M., Bagatta, M., Carboni, A., Crucitti, P., Moliterni, V.M.C., Ranalli, P., Mandolino, G., 2003. The inheritance of chemical phenotype in *Cannabis sativa* L. Genetics 163, 335–346. https://doi.org/10.1007/s10681-009-9894-7.

De Petrocellis, L., Ligresti, A., Moriello, A.S., Allarà, M., Bisogno, T., Petrosino, S., Stott, C.G., Di Marzo, V., 2011. Effects of cannabinoids and cannabinoid-enriched Cannabis extracts on TRP channels and endocannabinoid metabolic enzymes. Br. J. Pharmacol. 163 (7), 1479–1494. https://doi.org/10.1111/j.1476-5381.2010.01166.x.

Devane, W.A., Dysarz 3rd, F.A., Johnson, M.R., Melvin, L.S., Howlett, A.C., 1988. Determination and characterization of a cannabinoid receptor in rat brain. Mol. Pharmacol. 34 (5), 605–613.

Dowling, C.A., Melzer, R., Schilling, S., 2021. Timing is everything: the genetics of flowering time in *Cannabis sativa*. Biochemist 43 (3), 34–38. https://doi.org/10.1042/BIO_2021_138.

Drotleff, L., 2021. Why ESG is Important to Cannabis Businesses and How to Get Started. Hemp Industry Daily.

Duanis-Assaf, D., Galsurker, O., Davydov, O., Maurer, D., Feygenberg, O., Sagi, M., Poverenov, E., Fluhr, R., Alkan, N., 2021. Double-stranded RNA targeting fungal ergosterol biosynthesis pathway controls Botrytis cinerea and postharvest gray mold. Plant Biotechnol. J. https://doi.org/10.1111/PBI.13708. Accepted Author Manuscript.

Fiaz, S., Wang, X., Younas, A., Alharthi, B., Riaz, A., Ali, H., 2021. Apomixis and strategies to induce apomixis to preserve hybrid vigor for multiple generations. GM Crops Food 12 (1), 57–70. https://doi.org/10.1080/21645698.2020.1808423.

Freeman, T.P., Craft, S., Wilson, J., Stylianou, S., ElSohly, M., Di Forti, M., Lynskey, M.T., 2021. Changes in delta-9-tetrahydrocannabinol (THC) and cannabidiol (CBD) concentrations in cannabis over time: systematic review and meta-analysis. Addiction 116 (5), 1000–1010. https://doi.org/10.1111/ADD.15253.

Gagne, S.J., Stout, J.M., Liu, E., Boubakir, Z., Clark, S.M., Page, J.E., 2012. Identification of olivetolic acid cyclase from *Cannabis sativa* reveals a unique catalytic route to plant polyketides. Proc. Natl. Acad. Sci. 109 (31), 12811–12816. https://doi.org/10.1073/pnas.1200330109.

Galoch, E., 1978. The hormonal control of sex differentiation in dioecious plants of hemp (*Cannabis sativa*): the influence of plant growth regulators on sex expression in male and female plants. Acta Soc. Bot. Pol. 47 (1–2), 153–162. https://doi.org/10.5586/asbp.1978.013.

Gao, C., Xin, P., Cheng, C., Tang, Q., Chen, P., Wang, C., Zang, G., Zhao, L., 2014. Diversity analysis in *Cannabis sativa* based on large-scale development of expressed sequence tag-derived simple sequence repeat markers. PLoS One 9 (10), e110638. https://doi.org/10.1371/journal.pone.0110638.

García, C., Guichoux, E., Hampe, A., 2018. A comparative analysis between SNPs and SSRs to investigate genetic variation in a juniper species (*Juniperus phoenicea* ssp. turbinata). Tree Genet. Genomes 14 (6), 1–9. https://doi.org/10.1007/S11295-018-1301-X.

Geissler, M., Volk, J., Stehle, F., Kayser, O., Warzecha, H., 2018. Subcellular localization defines modification and production of Δ9-tetrahydrocannabinolic acid synthase in transiently transformed Nicotiana benthamiana. Biotechnol. Lett. 40 (6), 981–987. https://doi.org/10.1007/s10529-018-2545-0.

Giladi, Y., Hadad, L., Luria, N., Cranshaw, W., Lachman, O., Dombrovsky, A., 2020. First report of beet curly top virus infecting *Cannabis sativa* in Western Colorado. In: The American Phytopathological Society, Plant Disease Notes. vol. 104., https://doi.org/10.1094/PDIS-08-19-1656-PDN (3).

Gilchrist, E.J., Hegebarth, D., Wang, S., Quilichini, T.D., Sawler, J., Toh, S.Y., Foley, C., Page, J.E., 2021a. A rapid method for sex identification in *Cannabis sativa* using high resolution melt (HRM) analysis. Botany. accepted, October, 2021.

Gilchrist, E.J., Toh, S.Y., Foley, C., Ssebuliba, E., Zhu, S., 2021b. Genetic authentication of cannabis cultivars using SSR genotyping: Cannabis DNA fingerprinting. Unpublished.

Glas, J.J., Schimmel, B.C.J., Alba, J.M., Escobar-Bravo, R., Schuurink, R.C., Kant, M.R., 2012. Plant glandular trichomes as targets for breeding or engineering of resistance to herbivores. Int. J. Mol. Sci. 13 (12), 17077–17103. https://doi.org/10.3390/IJMS131217077.

Grassa, C.J., Wenger, J.P., Dabney, C., Poplawski, S.G., Motley, S.T., Michael, T.P., Schwartz, C.J., Weiblen, G.D., 2018. A complete Cannabis chromosome assembly and adaptive admixture for elevated cannabidiol (CBD) content. BioRxiv, 458083. https://doi.org/10.1101/458083.

Gülck, T., Møller, B.L., 2020. Phytocannabinoids: origins and biosynthesis. Trends Plant Sci. 25 (10), 985–1004. https://doi.org/10.1016/j.tplants.2020.05.005.

Gülck, T., Booth, J.K., Carvalho, Â., Khakimov, B., Crocoll, C., Motawia, M.S., Møller, B.L., Bohlmann, J., Gallage, N.J., 2020. Synthetic biology of cannabinoids and cannabinoid glucosides in Nicotiana benthamiana and *Saccharomyces cerevisiae*. J. Nat. Prod. 83 (10), 2877–2893. https://doi.org/10.1021/ACS.JNATPROD.0C00241.

Hadad, L., Luria, N., Smith, E., Sela, N., Lachman, O., Dombrovsky, A., 2019. Lettuce chlorosis virus disease: a new threat to Cannabis production. Viruses 11 (9), 802. https://doi.org/10.3390/V11090802.

Haligonia.ca., 2021. How Big Is the Marijuana Market in 2021. Haligonia.Ca.

Hartowicz, L.E., Knutson, H., Paulsen, A., Eaton, B.J., Eshbaugh, E., 1971. Possible biological control of wild hemp. In: Proceedings North Central Weed Control Conference. 26, p. 69.

Hazekamp, A., Tejkalová, K., Papadimitriou, S., 2016. Cannabis: from cultivar to Chemovar II—a metabolomics approach to Cannabis classification. Cannabis Cannabinoid Res. 1 (1), 202–215. https://doi.org/10.1089/can.2016.0017.

Henry, P., Hilyard, A., Johnson, S., Orser, C., 2018. Predicting chemovar cluster and variety verification in vegetative cannabis accessions using targeted single nucleotide polymorphisms. PeerJ 6, e27442v1. Preprints https://doi.org/10.7287/PEERJ.PREPRINTS.27442V1.

Hesami, M., Najafabadi, M.Y., Adamek, K., Torkamaneh, D., Jones, A.M.P., 2021. Synergizing off-target predictions for in silico insights of CENH3 knockout in Cannabis through CRISPR/Cas. Molecules 26 (7), 2053. https://doi.org/10.3390/MOLECULES26072053.

Hill, A.J., Williams, C.M., Whalley, B.J., Stephens, G.J., 2012. Phytocannabinoids as novel therapeutic agents in CNS disorders. Pharmacol. Ther. 133 (1), 79–97. https://doi.org/10.1016/j.pharmthera.2011.09.002.

Höfle, L., Biedenkopf, D., Werner, B.T., Shrestha, A., Jelonek, L., Koch, A., 2020. Study on the efficiency of dsRNAs with increasing length in RNA-based silencing of the Fusarium CYP51 genes. RNA Biol. 17 (4), 463–473. https://doi.org/10.1080/15476286.2019.1700033.

Holland, J., 2010. In: Holland, J. (Ed.), The Pot Book: A Complete Guide to Cannabis. Simon and Shuster.

Houston, R., Birck, M., Hughes-Stamm, S., Gangitano, D., 2016. Evaluation of a 13-loci STR multiplex system for *Cannabis sativa* genetic identification. Int. J. Legal Med. 130 (3), 635–647. https://doi.org/10.1007/s00414-015-1296-x.

Howard, C., Gilmore, S., Robertson, J., Peakall, R., 2008. *Application of New DNA Markers for Forensic Examination of Cannabis sativa Seizures—Developmental Validation of Protocols and a Genetic Database* (Monograph, Issue 29). National Drug law Enforcement Research Fund, An Initiative of the National Drug Strategy. Australian Institute of Criminology, Canberra.

Hu, L., Zhang, K., Wu, Z., Xu, J., Erb, M., 2021. Plant volatiles as regulators of plant defense and herbivore immunity: molecular mechanisms and unanswered questions. Curr. Opin. Insect Sci. 44, 82–88. https://doi.org/10.1016/j.cois.2021.03.010.

Huchelmann, A., Boutry, M., Hachez, C., 2017. Plant glandular trichomes: natural cell factories of high biotechnological interest. Plant Physiol. 175 (1), 6–22. https://doi.org/10.1104/pp.17.00727.

Jerushalmi, S., Maymon, M., Dombrovsky, A., Freeman, S., 2020. Fungal pathogens affecting the production and quality of medical Cannabis in Israel. Plan. Theory 9 (7), 882. https://doi.org/10.3390/plants9070882.

Jiang, H., Wang, L., Merlin, M.D., Clarke, R.C., Pan, Y., Zhang, Y., Xiao, G., Ding, X., 2016. Ancient Cannabis burial shroud in a central Eurasian cemetery. Econ. Bot. 70 (3), 213–221. https://doi.org/10.1007/S12231-016-9351-1.

Jin, D., Dai, K., Xie, Z., Chen, J., 2020. Secondary metabolites profiled in Cannabis inflorescences, leaves, stem barks, and roots for medicinal purposes. Sci. Rep. 10 (1), 1–14. https://doi.org/10.1038/s41598-020-60172-6.

Jin, D., Henry, P., Shan, J., Chen, J., 2021a. Identification of phenotypic characteristics in three chemotype categories in the genus Cannabis. HortScience 56 (4), 481–490. https://doi.org/10.21273/HORTSCI15607-20.

Jin, D., Henry, P., Shan, J., Chen, J., 2021b. Classification of cannabis strains in the Canadian market with discriminant analysis of principal components using genome-wide single nucleotide polymorphisms. PLoS One 16 (6), e0253387. https://doi.org/10.1371/JOURNAL.PONE.0253387.

Kalinina, N.O., Khromov, A., Love, A.J., Taliansky, M.E., 2019. CRISPR applications in plant virology: virus resistance and beyond. Phytopathology 110 (1), 18–28. https://doi.org/10.1094/PHYTO-07-19-0267-IA.

Kang, B.-C., Yeam, I., Jahn, M.M., 2005. Genetics of plant virus resistance. Annu. Rev. Phytopathol. 43 (1), 581–621. https://doi.org/10.1146/annurev.phyto.43.011205.141140.

Khanday, I., Sundaresan, V., 2021. Plant zygote development: recent insights and applications to clonal seeds. Curr. Opin. Plant Biol. 59, 101993. https://doi.org/10.1016/J.PBI.2020.101993.

Kinder, D.C., 1981. Bureaucratic cold warrior: Harry J. Anslinger and illicit narcotics traffic. Pac. Hist. Rev. 50 (2), 169–191. https://doi.org/10.2307/3638725.

Kobayashi, M., Momohara, A., Okitsu, S., et al., 2008. Fossil hemp fruits in the earliest Jomon period from the Okinoshima site, Chiba Prefecture. Shokuseishi kenkyū 16, 11–18.

Kurtz, L.E., Mahoney, J.D., Brand, M.H., Lubell-Brand, J.D., 2020. Comparing genotypic and phenotypic variation of selfed and outcrossed progeny of hemp. HortScience 55 (8), 1206–1209. https://doi.org/10.21273/HORTSCI15061-20.

Lange, B.M., 2015. The evolution of plant secretory structures and emergence of terpenoid chemical diversity. Annu. Rev. Plant Biol. 66 (1), 139–159. https://doi.org/10.1146/annurev-arplant-043014-114639.

Laverty, K.U., Stout, J.M., Sullivan, M.J., Shah, H., Gill, N., Holbrook, L., Deikus, G., Sebra, R., Hughes, T.R., Page, J.E., Van Bakel, H., 2019. A physical and genetic map of *Cannabis sativa* identifies extensive rearrangements at the THC/CBD acid synthase loci. Genome Res. 29 (1), 146–156. https://doi.org/10.1101/gr.242594.118.

Lev, E., 2018. Canada's Path from Prohibition: A Cannabis Timeline. Lift & Co. Magazine. https://lift.co/magazine/canadas-path-from-prohibition-a-cannabis-timeline.

Liu, X., Ao, K., Yao, J., Zhang, Y., Li, X., 2021. Engineering plant disease resistance against biotrophic pathogens. Curr. Opin. Plant Biol. 60, 101987. https://doi.org/10.1016/J.PBI.2020.101987.

Livingston, S.J., Quilichini, T.D., Booth, J.K., Wong, D.C.J., Rensing, K.H., Laflamme-Yonkman, J., Castellarin, S.D., Bohlmann, J., Page, J.E., Samuels, A.L., 2020. Cannabis glandular trichomes alter morphology and metabolite content during flower maturation. Plant J. 101 (1), 37–56. https://doi.org/10.1111/tpj.14516.

Lu, E., 2021. Sustainability in the Cannabis Industry: Truly Going Green. Forbes.

Lubell, J.D., Brand, M.H., 2018. Foliar sprays of silver thiosulfate produce male flowers on female hemp plants. HortTechnology 28 (6), 743–747. https://doi.org/10.21273/HORTTECH04188-18.

Luo, X., Reiter, M.A., D'Espaux, L., Wong, J., Denby, C.M., Lechner, A., Zhang, Y., Grzybowski, A.T., Harth, S., Lin, W., Lee, H., Yu, C., Shin, J., Deng, K., Benites, V.T., Wang, G., Baidoo, E.E.K., Chen, Y., Dev, I., Keasling, J.D., 2019. Complete biosynthesis of cannabinoids and their unnatural analogues in yeast. Nature 567 (7746), 123–126. https://doi.org/10.1038/s41586-019-0978-9.

Lynch, R.C., Vergara, D., Tittes, S., White, K., Schwartz, C.J., Gibbs, M.J., Ruthenburg, T.C., DeCesare, K., Land, D.P., Kane, N.C., 2016. Genomic and chemical diversity in Cannabis. Crit. Rev. Plant Sci. 35 (5–6), 349–363. https://doi.org/10.1080/07352689.2016.1265363.

Mandolino, G., Carboni, A., Forapani, S., Faeti, V., Ranalli, P., 1999. Identification of DNA markers linked to the male sex in dioecious hemp (*Cannabis sativa* L.). TAG Theor. Appl. Genet. 98 (1), 86–92. https://doi.org/10.1007/s001220051043.

Matsuda, L.A., Lolait, S.J., Brownstein, M.J., Young, A.C., Bonner, T.I., 1990. Structure of a cannabinoid receptor and functional expression of the cloned cDNA. Nature 346 (6284), 561–564. https://doi.org/10.1038/346561a0.

Maybin, S., 2019. Uruguay: The world's Marijuana Pioneer. BBC News.

McKenzie, B., 2020. Plant Variety Rights Summary.

McPartland, J.M., 2018. Cannabis systematics at the levels of family, genus, and species. Cannabis Cannabinoid Res. 3 (1), 203–212. https://doi.org/10.1089/can.2018.0039.

McPartland, J.M., Small, E., 2020. A classification of endangered high-THC cannabis (Cannabissativasubsp.indica) domesticates and their wild relatives. PhytoKeys 144, 81. https://doi.org/10.3897/PHYTOKEYS.144.46700.

McPartland, J.M., Hegman, W., Long, T., 2019. Cannabis in Asia: its center of origin and early cultivation, based on a synthesis of subfossil pollen and archaeobotanical studies. Veg. Hist. Archaeobotany 28 (6), 691–702. https://doi.org/10.1007/s00334-019-00731-8.

Mechoulam, R., Hanuš, L.O., Pertwee, R., Howlett, A.C., 2014. Early phytocannabinoid chemistry to endocannabinoids and beyond. Nat. Rev. Neurosci. 15. https://doi.org/10.1038/nrn3811.

Mihalyov, P.D., Garfinkel, A.R., 2021. Discovery and genetic mapping of PM1, a powdery mildew resistance gene in *Cannabis sativa* L. Front. Agron. 3, 66. https://doi.org/10.3389/FAGRO.2021.720215.

Mitchell, J., 2020. The Business Case for Sustainability. Cannabis Doing Good. https://cannabisdoinggood.com/the-business-case-for-sustainability/.

Mohan Ram, H.Y., Sett, R., 1982. Induction of fertile male flowers in genetically female *Cannabis sativa* plants by silver nitrate and silver thiosulphate anionic complex. Theor. Appl. Genet. 62, 369–375. https://doi.org/10.1007/BF00275107.

Montone, C.M., Cerrato, A., Botta, B., Cannazza, G., Capriotti, A.L., Cavaliere, C., Citti, C., Ghirga, F., Piovesana, S., Laganà, A., 2020. Improved identification of phytocannabinoids using a dedicated structure-based workflow. Talanta 219, 121310. https://doi.org/10.1016/J.TALANTA.2020.121310.

Mudge, E.M., Brown, P.N., Murch, S.J., 2019. The terroir of Cannabis: terpene metabolomics as a tool to understand *Cannabis sativa* selections. Planta Med. 85 (9–10), 781–796. https://doi.org/10.1055/a-0915-2550.

Munro, S., Thomas, K.L., Abu-Shaar, M., 1993. Molecular characterization of a peripheral receptor for cannabinoids. Nature 365 (6441), 61–65. https://doi.org/10.1038/365061a0.

Naim-Feil, E., Pembleton, L.W., Spooner, L.E., Malthouse, A.L., Miner, A., Quinn, M., Polotnianka, R.M., Baillie, R.C., Spangenberg, G.C., Cogan, N.O.I., 2021. The characterization of key physiological traits of medicinal cannabis (*Cannabis sativa* L.) as a tool for precision breeding. BMC Plant Biol. 21 (1), 294. https://doi.org/10.1186/S12870-021-03079-2.

Norrgard, K., 2008. Forensics, DNA fingerprinting, and CODIS | learn science at scitable. Nat. Educ. 1 (1), 35.

Okorafor, I.C., Chen, M., Tang, Y., 2021. High-titer production of olivetolic acid and analogs in engineered fungal host using a nonplant biosynthetic pathway. ACS Synth. Biol. 10 (9), 2159–2166. https://doi.org/10.1021/acssynbio.1c00309.

Ormancey, M., Guillotin, B., San Clemente, H., Thuleau, P., Plaza, S., Combier, J.P., 2021. Use of microRNA-encoded peptides to improve agronomic traits. Plant Biotechnol. J. 19 (9), 1687–1689. https://doi.org/10.1111/PBI.13654.

Peckenpaugh, D.J., 2020. Taking a look at today's Cannabis consumer. In: Food Engineering. https://www.foodengineeringmag.com/articles/98969-taking-a-look-at-todays-cannabis-consumer.

Pehota, J., 2020. Using Data Analytics to Understand Consumer Preferences. Cannabis Retailer. https://cannabisretailer.ca/2020/using-data-analytics-to-understand-consumer-preferences/.

Pépin, N., Hebert, F.O., Joly, D.L., 2021. Genome-wide characterization of the MLO gene family in Cannabis sativa reveals two genes as strong candidates for powdery mildew susceptibility. BioRxiv. https://doi.org/10.1101/2021.07.16.452661. 2021.07.16.452661.

Pertwee, R.G., 2006. Cannabinoid pharmacology: the first 66 years. Br. J. Pharmacol. 147 (S1), S163–S171. https://doi.org/10.1038/sj.bjp.0706406.

Petit, J., Salentijn, E.M.J., Paulo, M.-J., Denneboom, C., Trindade, L.M., 2020. Genetic architecture of flowering time and sex determination in hemp (*Cannabis sativa* L.): a genome-wide association study. Front. Plant Sci. 11, 1704. https://doi.org/10.3389/fpls.2020.569958.

Pisanti, S., Bifulco, M., 2017. Modern history of medical Cannabis: from widespread use to prohibitionism and back. Trends Pharmacol. Sci. 38 (3), 195–198. https://doi.org/10.1016/J.TIPS.2016.12.002.

Prentout, D., Stajner, N., Cerenak, A., Tricou, T., Brochier-Armanet, C., Jakse, J., Käfer, J., Marais, G.A.B., 2021. Plant genera Cannabis and Humulus share the same pair of well-differentiated sex chromosomes. New Phytol. 231 (4), 1599–1611. https://doi.org/10.1111/NPH.17456.

Punja, Z.K., 2018. Flower and foliage-infecting pathogens of marijuana (*Cannabis sativa* L.) plants. Can. J. Plant Pathol. 40 (4), 514–527. https://doi.org/10.1080/07060661.2018.1535467.

Punja, Z.K., Holmes, J.E., 2020. Hermaphroditism in marijuana (*Cannabis sativa* L.) inflorescences—impact on floral morphology, seed formation, progeny sex ratios, and genetic variation. Front. Plant Sci., 718. https://doi.org/10.3389/FPLS.2020.00718.

Pusiak, R.J., Cox, C., Harris, C.S., 2021. Growing pains: an overview of cannabis quality control and quality assurance in Canada. Int. J. Drug Policy 93, 103111. https://doi.org/10.1016/J.DRUGPO.2021.103111.

Qiao, L., Lan, C., Capriotti, L., Ah-Fong, A., Sanchez, J.N., Hamby, R., Heller, J., Zhao, H., Glass, N.L., Judelson, H.S., Mezzetti, B., Niu, D., Jin, H., 2021. Spray-induced gene silencing for disease control is dependent on the efficiency of pathogen RNA uptake. BioRxiv. https://doi.org/10.1101/2021.02.01.429265.

Reimann-Philipp, U., Speck, M., Orser, C., Johnson, S., Hilyard, A., Turner, H., Stokes, A.J., Small-Howard, A.L., 2020. Cannabis chemovar nomenclature misrepresents chemical and genetic diversity; survey of variations in chemical profiles and genetic markers in Nevada Medical Cannabis samples. Cannabis Cannabinoid Res. 5 (3), 215–230. https://doi.org/10.1089/CAN.2018.0063.

Ren, M., Tang, Z., Wu, X., Spengler, R., Jiang, H., Yang, Y., Boivin, N., 2019. The origins of cannabis smoking: chemical residue evidence from the first millennium BCE in the Pamirs. Sci. Adv. 5 (6), eaaw1391. 12 https://doi.org/10.1126/SCIADV.AAW1391.

ResearchAndMarkets.com, 2021. Global Industrial Hemp Market Report and Forecast 2021–2026: Favourable Regulations Aid Growth with 22.5% CAGR Forecast Between 2021 and 2026. Business Wire.

Richins, R.D., Rodriguez-Uribe, L., Lowe, K., Ferral, R., O'Connell, M.A., 2018. Accumulation of bioactive metabolites in cultivated medical Cannabis. PLoS One 13 (7), e0201119. https://doi.org/10.1371/journal.pone.0201119.

Riggs, T.J., 1988. Breeding F1 hybrid varieties of vegetables. J. Horticult. Sci. 63 (3), 369–382. https://doi.org/10.1080/14620316.1988.11515871.

Righetti, L., Paris, R., Ratti, C., Calassanzio, M., Onofri, C., Calzolari, D., Menzel, W., Knierim, D., Magagnini, G., Pacifico, D., Grassi, G., 2018. Not the one, but the only one: about Cannabis cryptic virus in plants showing 'hemp streak' disease symptoms. Eur. J. Plant Pathol. 150 (3), 575–588. https://doi.org/10.1007/s10658-017-1301-y.

Riskin, D., 2021. The Fast-Growing cannabis Industry Has a Big Pollution Problem | CTV News. CTV News.

Rodriguez, C.E.B., Ouyang, L., Kandasamy, R., 2021. Antinociceptive effects of minor cannabinoids, terpenes and flavonoids in Cannabis. Behav. Pharmacol. https://doi.org/10.1097/fbp.0000000000000627.

Roscow, R.F., 2017. Plants and Methods for Increasing and Decreasing Synthesis of Cannabinoids (Patent No. WO 2018/035450 Al). World Intellectual Property Organization.

Royal Queen Seeds, 2021. The Home Growers Guide to Cannabis Breeding. RQS Blog. https://www.royalqueenseeds.com/blog-breeding-and-preserving-cannabis-genetics-at-home-n511.

Russo, E.B., 2011. Taming THC: potential entourage effects. Br. J. Pharmacol. 163, 1344–1364. https://doi.org/10.1111/j.1476-5381.2011.01238.x.

Russo, E.B., Jiang, H.-E., Li, X., Sutton, A., Carboni, A., Del Bianco, F., Mandolino, G., Potter, D.J., Zhao, Y.-X., Bera, S., Zhang, Y.-B., Lü, E.-G., Ferguson, D.K., Hueber, F., Zhao, L.-C., Liu, C.-J., Wang, Y.-F., Li, C.-S., 2008. Phytochemical and genetic analyses of ancient cannabis from Central Asia. J. Exp. Bot. 59 (15), 4171–4182. https://doi.org/10.1093/jxb/ern260.

Sawler, J., Stout, J.M., Gardner, K.M., Hudson, D., Vidmar, J., Butler, L., Page, J.E., Myles, S., 2015. The genetic structure of marijuana and hemp. PLoS One 10 (8). https://doi.org/10.1371/journal.pone.0133292.

Schwabe, A.L., McGlaughlin, M.E., 2019. Genetic tools weed out misconceptions of strain reliability in *Cannabis sativa*: implications for a budding industry. BioRxiv 1 (1), 3. https://doi.org/10.1101/332320.

Shao, H., Song, S.J., Clarke, R.C., 2003. Female-associated DNA polymorphisms of hemp (*Cannabis sativa* l.). J. Ind. Hemp 8 (1), 5–9. https://doi.org/10.1300/J237v08n01_02.

Simmons, J., 2020. Cannabis Viruses, Viroids, and Phytoplasmas. Indica Info. https://indicainfo.com/2020/03/15/cannabis-viruses-viroids-and-phytoplasmas/.

Small, E., Brookes, B., 2012. Temperature and moisture content for storage maintenance of germination capacity of seeds of industrial hemp, marijuana, and ditchweed forms of *Cannabis sativa*. J. Nat. Fibers 9 (4), 240–255. https://doi.org/10.1080/15440478.2012.737179.

Smith, C.J., Vergara, D., Keegan, B., Jikomes, N., 2021. The phytochemical diversity of commercial Cannabis in the United States. BioRxiv. https://doi.org/10.1101/2021.07.05.451212. 2021.07.05.451212.

Summers, H.M., Sproul, E., Quinn, J.C., 2021. The greenhouse gas emissions of indoor cannabis production in the United States. Nat. Sustain. 4 (7), 644–650. https://doi.org/10.1038/s41893-021-00691-w.

Tahir, M.N., Shahbazi, F., Rondeau-Gagné, S., Trant, J.F., 2021. The biosynthesis of the cannabinoids. J. Cannabis Res. 3 (1), 1–2.

Taliansky, M., Samarskaya, V., Zavriev, S.K., Fesenko, I., Kalinina, N.O., Love, A.J., 2021. RNA-based technologies for engineering plant virus resistance. Plan. Theory 10 (1). https://doi.org/10.3390/plants10010082.

Thomas, F.J., Kayser, O., 2019. Minor cannabinoids of *Cannabis sativa* L. J. Med. Sci. 88 (3), 141–149. https://doi.org/10.20883/JMS.367.

Thompson III, G., Tuscano, J., Dennis, M., Singapuri, A., Libertini, S., Gaudino, R., Torres, A., Delisle, J., Gillece, J., Schupp, J., Engelthaler, D., 2017. A microbiome assessment of medical marijuana, letter to the editor. Clin. Microbiol. Infect. 23, 269–270. https://doi.org/10.1016/j.cmi.2016.12.001.

Tomko, A.M., Whynot, E.G., Ellis, L.D., Dupré, D.J., 2020. Anti-cancer potential of cannabinoids, terpenes, and flavonoids present in Cannabis. Cancers 12 (7), 1985. https://doi.org/10.3390/CANCERS12071985.

Toth, J.A., Stack, G.M., Cala, A.R., Carlson, C.H., Wilk, R.L., Crawford, J.L., Viands, D.R., Philippe, G., Smart, C.D., Rose, J.K.C.C., Smart, L.B., 2020. Development and validation of genetic markers for sex and cannabinoid chemotype in *Cannabis sativa* L. GCB Bioenergy 12 (3), 213–222. https://doi.org/10.1111/gcbb.12667.

Truniger, V., Aranda, M.A., 2009. Recessive resistance to plant viruses. In: Advances in Virus Research. vol. 75. Elsevier, https://doi.org/10.1016/s0065-3527(09)07504-6. Issue 09.

United Nations Office on Drugs and Crime, 2021a. Single Convention on Narcotic Drugs. https://www.unodc.org/unodc/en/treaties/single-convention.html?ref=menuside.

United Nations Office on Drugs and Crime, 2021b. World Drug Report 2021. https://www.unodc.org/unodc/en/data-and-analysis/wdr2021.html.

van Bakel, H., Stout, J.M., Cote, A.G., Tallon, C.M., Sharpe, A.G., Hughes, T.R., Page, J.E., 2011. The draft genome and transcriptome of *Cannabis sativa*. Genome Biol. 12 (10), R102. https://doi.org/10.1186/gb-2011-12-10-r102.

Wang, M., Jin, H., 2017. Spray-induced gene silencing: a powerful innovative strategy for crop protection. Trends Microbiol. 25 (1), 4–6. https://doi.org/10.1016/j.tim.2016.11.011.

Wang, M., Thomas, N., Jin, H., 2017. Cross-kingdom RNA trafficking and environmental RNAi for powerful innovative pre- and post-harvest plant protection. Curr. Opin. Plant Biol. 38, 133–141. https://doi.org/10.1016/j.pbi.2017.05.003.

Wang, R., Zhao, S., Wang, Z., Koffas, M.A.G., 2020. Recent advances in modular co-culture engineering for synthesis of natural products. Curr. Opin. Biotechnol. 62, 65–71. https://doi.org/10.1016/j.copbio.2019.09.004.

Warren, G., 2015. Regulating pot to save the polar bear: energy and climate impacts of the marijuana industry. Columbia J. Environ. Law 40.

Warren, J., Mercado, J., Grace, D., 2019. The occurrence of hop latent viroid causing disease in Cannabis sativa in California. Plant Dis. https://doi.org/10.1094/PDIS-03-19-0530-PDN. PDIS-03-19-0530-PDN.

Wayne Labs, 2019. Cannabis Testing Is an Exact Science—Regulations Are Not. Food Engineering. https://www.foodengineeringmag.com/articles/98370-cannabis-testing-is-an-exact-science- - -regulations-are-not.

Weiblen, G.D., Wenger, J.P., Craft, K.J., Elsohly, M.A., Mehmedic, Z., Treiber, E.L., Marks, M.D., 2015. Gene duplication and divergence affecting drug content in *Cannabis sativa*. New Phytol. 208 (4), 1241–1250. https://doi.org/10.1111/nph.13562.

Welling, M.T., Shapter, T., Rose, T.J., Liu, L., Stanger, R., King, G.J., 2016. A belated green revolution for Cannabis: virtual genetic resources to fast-track cultivar development. Front. Plant Sci. 7, 1113. https://doi.org/10.3389/FPLS.2016.01113.

Welling, M.T., Liu, L., Kretzschmar, T., Mauleon, R., Ansari, O., King, G.J., 2020. An extreme-phenotype genome-wide association study identifies candidate cannabinoid pathway genes in Cannabis. Sci. Rep. 10 (1), 18643. https://doi.org/10.1038/S41598-020-75271-7.

Werner, B.T., Gaffar, F.Y., Schuemann, J., Biedenkopf, D., Koch, A.M., 2020. RNA-spray-mediated silencing of fusarium graminearum AGO and DCL genes improve barley disease resistance. Front. Plant Sci. 11, 476. https://doi.org/10.3389/fpls.2020.00476.

Wetzstein, H.Y., Porter, J.A., Janick, J., Ferreira, J.F.S., Mutui, T.M., 2018. Selection and clonal propagation of high artemisinin genotypes of *Artemisia annua*. Front. Plant Sci. 9, 358. https://doi.org/10.3389/FPLS.2018.00358.

World Health Organization, 2019. WHO expert committee on drug dependence: forty first report. In: World Health Organization—Technical Report Series. vol. 1018.

Zager, J.J., Lange, I., Srividya, N., Smith, A., Lange, B.M., 2019. Gene networks underlying cannabinoid and terpenoid accumulation in Cannabis. Plant Physiol. 180 (4), 1877–1897. https://doi.org/10.1104/pp.18.01506.

Zhang, Q., Chen, X., Guo, H., Trindade, L.M., Salentijn, E.M.J., Guo, R., Guo, M., Xu, Y., Yang, M., 2018. Latitudinal adaptation and genetic insights into the origins of *Cannabis sativa* L. Front. Plant Sci. 1876.

Zhang, S.L., Gao, H.Y., 1999. 荥阳青台遗址出土的丝麻品观察与研究 (Observation and study of silk and hemp recovered from Qingtai archaeological site, Xingyang). Zhōngyuán Wénwù 3, 10–16.

Zhao, Y., Yang, X., Zhou, G., Zhang, T., 2020. Engineering plant virus resistance: from RNA silencing to genome editing strategies. Plant Biotechnol. J. 18 (2), 328–336. https://doi.org/10.1111/PBI.13278.

Zhou, J.M., Zhang, Y., 2020. Plant immunity: danger perception and signaling. Cell 181 (5), 978–989. https://doi.org/10.1016/J.CELL.2020.04.028.

Zias, J., Stark, H., Seligman, J., Levy, R., Werker, E., Breuer, A.A., Mechoulam, R., Zias, J., Stark, H., Sellgman, J., Levy, R., Werker, E., Breuer, A.A., Mechoulam, R., 1993. Early medical use of cannabis. Nature 363 (6426), 215. https://doi.org/10.1038/363215A0.

Zimmer, K., 2020. Variation in Cannabis Testing Challenges a Young Industry. The Scientist.

Zirpel, B., Degenhardt, F., Martin, C., Kayser, O., Stehle, F., 2017. Engineering yeasts as platform organisms for cannabinoid biosynthesis. J. Biotechnol. 259, 204–212.

Zirpel, B., Kayser, O., Stehle, F., 2018. Elucidation of structure-function relationship of THCA and CBDA synthase from *Cannabis sativa* L. J. Biotechnol. 284, 17–26.

Zirpel, B., Stehle, F., Kayser, O., 2015. Production of Δ9-tetrahydrocannabinolic acid from cannabigerolic acid by whole cells of *Pichia* (*Komagataella*) *pastoris* expressing Δ9-tetrahydrocannabinolic acid synthase from *Cannabis sativa* L. Biotechnol. Lett. 37 (9), 1869–1875.

Chapter 10

Multiomics techniques for plant secondary metabolism engineering: Pathways to shape the bioeconomy

Minxuan Li[a,b], Sen Cai[a,c], Shijun You[d,e], and Yuanyuan Liu[a]
[a]*Basic Forestry and Proteomics Center, Haixia Institute of Science and Technology, State Key Laboratory of Ecological Pest Control for Fujian and Taiwan Crops, College of Forestry, Fujian Agriculture and Forestry University, Fuzhou, China,* [b]*College of Life Sciences, Fujian Agriculture and Forestry University, Fuzhou, China,* [c]*School of Life Sciences, Capital Normal University, Beijing, China,* [d]*Institute of Applied Ecology, Fujian Agriculture and Forestry University, Fuzhou, China,* [e]*Joint International Research Laboratory of Ecological Pest Control, Ministry of Education, Fuzhou, China*

Introduction

Plants produce an array of low-molecular-weight compounds classified as primary metabolites, secondary metabolites, or hormones (Erb and Kliebenstein, 2020). Primary metabolites are highly conserved and essential for plant growth and development (Fernie and Pichersky, 2015). By contrast, plant secondary metabolites (PSMs) are involved in the adaptation of plants to the biotic and abiotic environments but do not directly participate in plant growth, development, and reproduction (Yang et al., 2018). Considerable attention has been paid to PSMs because of their biological importance, such as major contribution to the specific odor, color, and taste of plant; regulation of plant defense, growth, and development; as well as pharmacological and toxicological effects on humans. Furthermore, PSMs serve as taxonomic markers because of their lineage-specific distribution in taxonomic groups (Hartmann, 2007). The interest in PSMs is not purely academic but also commercial, as PSMs are of interest to various industries, particularly the pharmaceutical, cosmetic, and food industries, that either directly commercialize natural compounds or develop derived products. PSMs contribute significantly to accomplish the SDGs, ranging from improving human health as new drugs and nutraceuticals to protecting the environment as sustainable chemicals. Because of the complexity of PSM

Genomics and the Global Bioeconomy. https://doi.org/10.1016/B978-0-323-91601-1.00013-4

biosynthesis systems, identifying the functions of associated enzyme-encoding genes is no longer satisfactory. For instance, the coordination and interactions between various metabolic pathways should also be elucidated. In this context, an integrative approach entailing large-scale experiments, the so-called omics approach, including genomics, transcriptomics, metabolomics, and proteomics, provides the complementary information required to decipher such coordination (Oksman-Caldentey and Saito, 2005; Yuan et al., 2008). Further, advances in high-throughput untargeted metabolomics have led to the generation of large datasets whose exploitation and interpretation generally require bioinformatics tools. The development of sequencing technology and the advancement of bioinformatics provide possibilities for genomics-based research methods to study the metabolic potential of the entire plant. Transcriptomic and proteomic tools used to discover PSMs couple genomic methods and metabolomic experiments. In this chapter, we provide an overview of PSMs, including their classification, biosynthetic pathways, and function, as well as the methodology, case studies, and challenges associated with multiomics data integration.

Classification of secondary metabolites

According to their biosynthetic pathways, PSMs are generally classified into four large molecular families: terpenoids, phenolic compounds, sulfur-containing compounds, and nitrogen-containing compounds (Fig. 10.1; Table 10.1).

Terpenoids

The class terpenoids contains more than 40,000 structurally diverse compounds and is the largest and most diverse group of PSMs (Bohlmann and Keeling, 2008). The terms "terpene" and "terpenoid" are used because the first member was isolated from turpentine oil (Rocke, 1985). Terpenoids play diverse functional roles in plants and humans (Table 10.1). Monoterpenes constitute important plant defenses against pathogens. For instance, pyrethrins, monoterpene esters (C10 terpenoids) produced by Chrysanthemum species (plants of the daisy family with brightly colored ornamental flowers), are neurotoxic and insecticidal for many insects, such as bees, wasps, beetles, and moths. Sesquiterpenes (C15 terpenoids) act as phytoalexins, antibiotic compounds produced during a microbial attack, and serve as antifeedants that deter herbivores. They are also principal constituents of essential oils and display a wide range of pharmaceutical activities, including effects on the central nervous system, antimicrobial and antitumor activities (Sousa, 2012). Caryophyllene, a sesquiterpene widely distributed in various plants, plays an important role in antipathogen defense. It also contributes to several physiological functions in plants, including antiinflammatory, antibiotic, antioxidant, anticarcinogenic, and anesthetic activities (Legault and Pichette, 2007; Huang et al., 2012). Abscisic acid, which promotes leaf abscission, is one of the most common

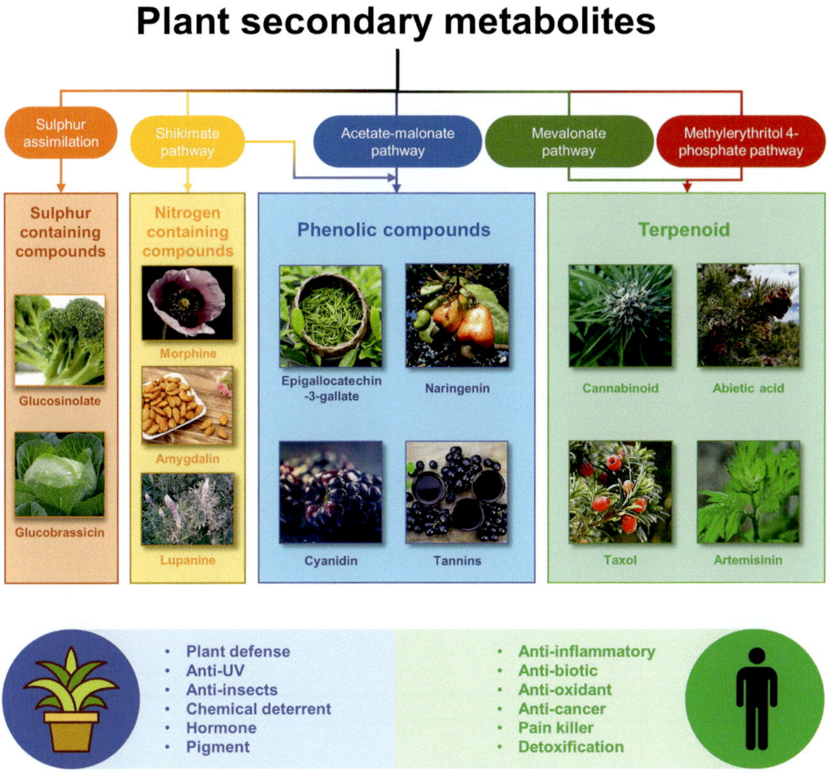

FIG. 10.1 Outline of plant secondary metabolites biosynthesis and effects of PSMs in plants and humans. *(Credit: Four main types of secondary metabolites are highlighted in different colors, namely, sulfur-containing compounds (*brown*), nitrogen-containing compounds (*orange*), phenolic compounds (*blue*), and terpenoids (*green*). Important pathways involved in metabolite biosynthesis are marked by* round rectangles. *The main effects of plant secondary metabolites in plants and humans are summarized below.)*

sesquiterpene phytohormones produced by all types of higher plants. Because of its ability to inhibit seed germination and enhance crop yield, as well as stress resilience-promoting activity, it also has agricultural applications (Fujii et al., 2007). Diterpenoids (C20 terpenoids) exhibit robust antimicrobial activities. Taxol (paclitaxel), a natural product first isolated from the bark of yew trees, is used as a potent anticancer agent because of its ability to "freeze" the microtubules and prevent them from separating chromosomes during cell division, which promotes the death of cancer cell (Band, 1992; Xiao et al., 2006). Triterpenes and sterols (C30) are isoprenoids synthesized via the cytoplasmic mevalonate (MVA) pathway. Triterpenes are involved in biological processes, such as embryonic development and cell and plant growth, while sterols are responsible for the construction of membranes and hormone signaling, which are

TABLE 10.1 Four main groups of plant secondary metabolites and some examples of each group.

Secondary	Metabolites	Origin	Structure	Type	Role of secondary metabolites in plants	Effect of secondary metabolites to human
Terpenoid	Pyrethrins	*Tanacetum cinerariaefolium* (Chrysanthemum)	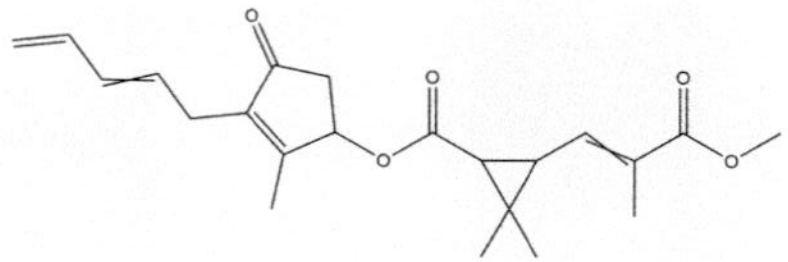	Monoterpene ester	Great insecticides, low mammalian toxicity, low persistence in the environment	Strong insecticides with low toxicity to human
	Pinene	Gymnosperms like pine and fir	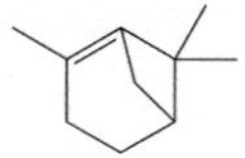	Monoterpene	Antiinflammatory compounds, insecticides	Antiinflammatory compounds, insecticides
	Caryophyllene	Widespread in plants	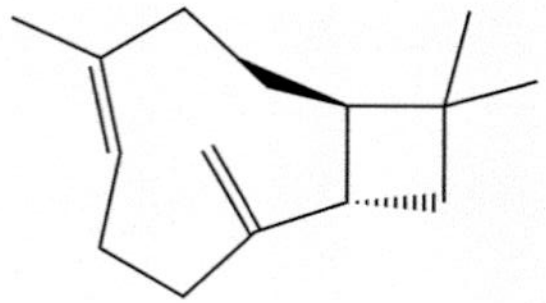	Sesquiterpene	Pathogen defense agents	Antiinflammatory, antibiotic, antioxidant, anticarcinogenic, and anesthetic activities
	Abscisic acid	Widespread in higher plants	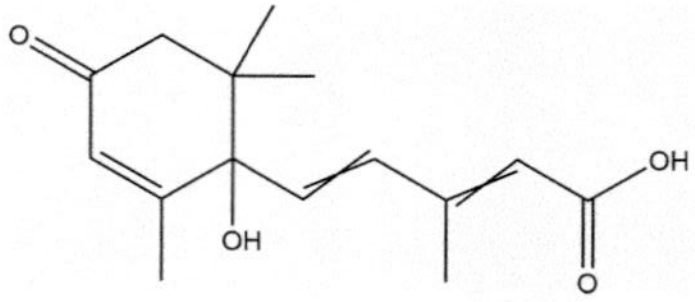	Sesquiterpene	Plant growth regulatory compounds, responsor to water stress, transcriptional activator, signaling compounds	Germination Inhibitor of seeds, enhance crop yield and stress resilience
	Abietic acid	Pines and leguminous tress	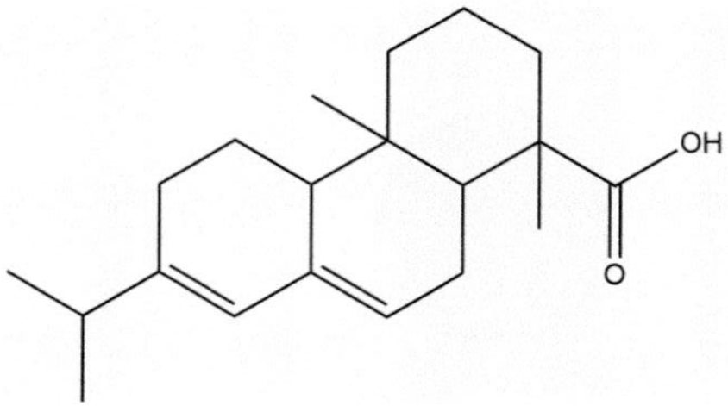	Diterpene	Chemical deterrent	Antiinflammatory, antioxidant, antimicrobial, and antibiofilm activities

	β-Carotene	Widespread in plants	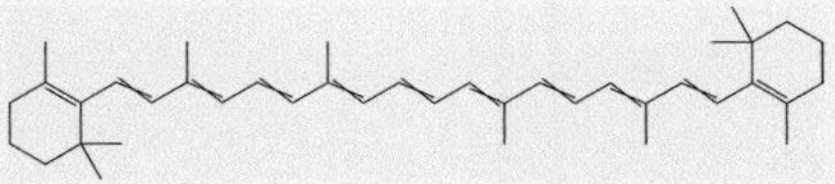	Tetraterpene	Pigment, antioxidant compound	Anticancer and antioxidant activities, potential role in preventing cardiovascular disease, precursor of vitamin A
	Delta-9-tetra-hydrocannabinol acid (Δ9-THCA)	*Cannabis sativa L.*	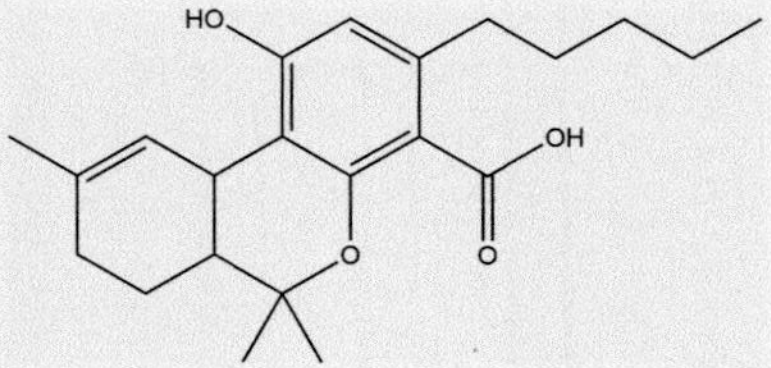	Terpene derivative	Protect against UV-B, antiinflammatory compounds	Antitumor, antihyperalgesia, antiinflammatory, and antimicrobial activities
	β-Sitosterol	Widespread in plants	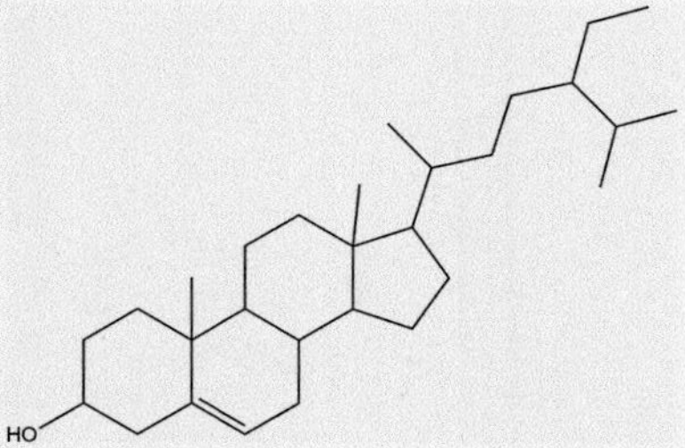	Phytosterol	Responsor to pathogens and water stress	Anxiolytic and sedative effects, analgesic, immunomodulatory, antimicrobial, anticancer, antiinflammatory, antioxidant, and antidiabetic activities
Phenolic compound	Cinnamic acid	Widespread in vascular plants	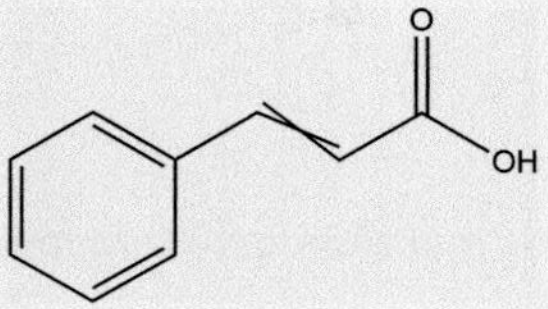	Phenylpropanoid	Plant defense compounds against insect herbivores and fungi, broad-spectrum antimicrobial compounds	Potential antidiabetic, antioxidant, and antimicrobial activities

Continued

TABLE 10.1 Four main groups of plant secondary metabolites and some examples of each group—cont'd

Secondary	Metabolites	Origin	Structure	Type	Role of secondary metabolites in plants	Effect of secondary metabolites to human
	Quercetin	Widespread in plants	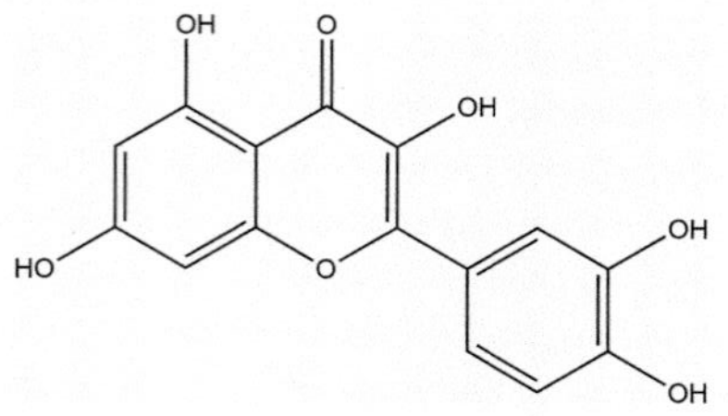	Flavonoid	Antimicrobic, antiinflammatory compound	Efficacy in diseases like Alzheimer's Disease(AD); antiinflammatory, antiprotozoal, antimicrobial, antiviral, and antioxidant effects
	(−)-Epigallocatechin-3-gallate	*Camellia sinensis* (green tea)	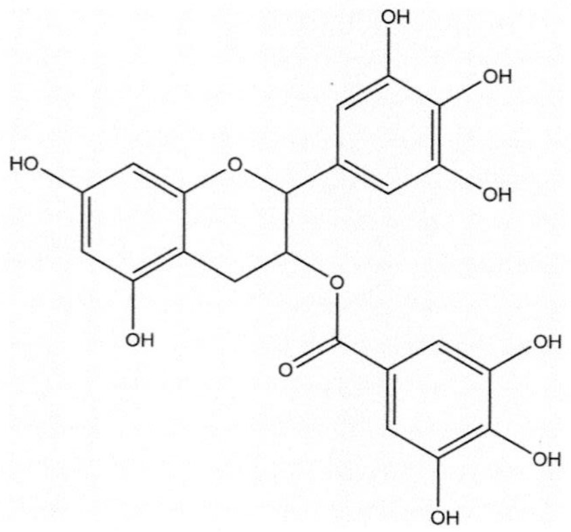	Flavanol	Antioxidant compounds, UV-B protection	Anticancer, antioxidant activities

	Naringenin	Citrus fruits, bergamot, tomatoes and other fruits	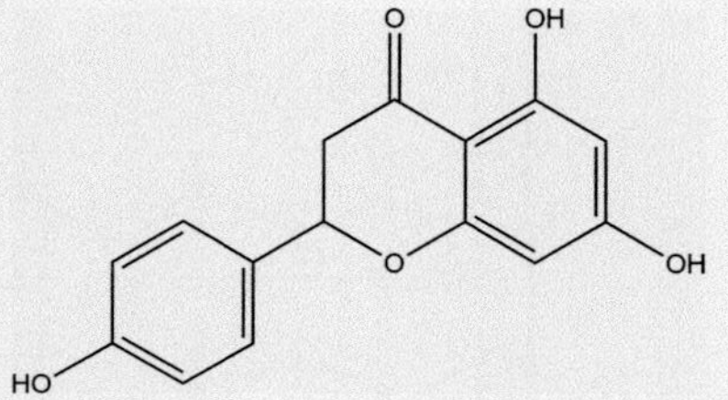	Flavanone	Plant developmental and defense response, promoting the formation of nitrogen fixing nodules by symbiotic rhizobia	Hepatoprotective, antioxidant, antidiabetic, antiatherogenic, antidepressant, immunomodulatory, antitumor, antiinflammatory, DNA-protective activities
	Cyanidin	Widespread in plants	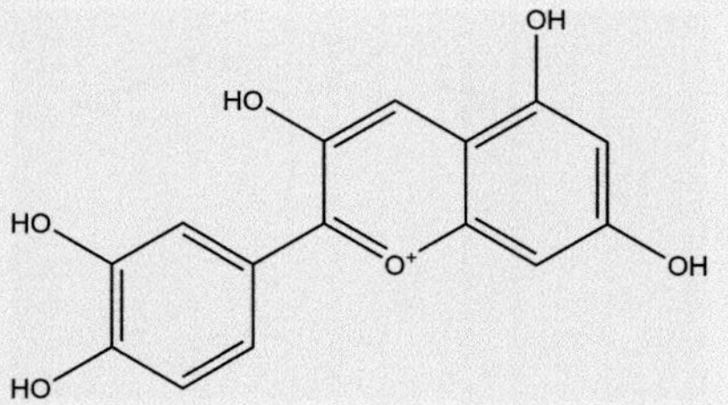	Flavonoid	Pigment, attract animal pollinators and seed dispersers, UV-B protection	Antioxidant, antitumor, antiaging effects, nervous system-protective activities
	Tannins	Grape, quebracho, oak, chestnut, tara, and galla	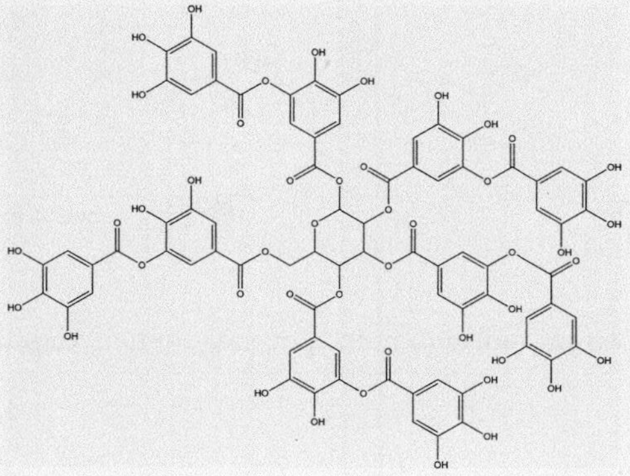	Polymer phenolic compound	General toxins that significantly reduce the growth and survivorship of many herbivores, feeding repellents to a great diversity of animals	Detoxification activity of snake venoms and bacterial toxins, antidiarrheic, antihemorrhagic, antioxidant activities

Continued

TABLE 10.1 Four main groups of plant secondary metabolites and some examples of each group—cont'd

Secondary	Metabolites	Origin	Structure	Type	Role of secondary metabolites in plants	Effect of secondary metabolites to human
Sulfur-containing compound	Glutathione	Widespread in plants	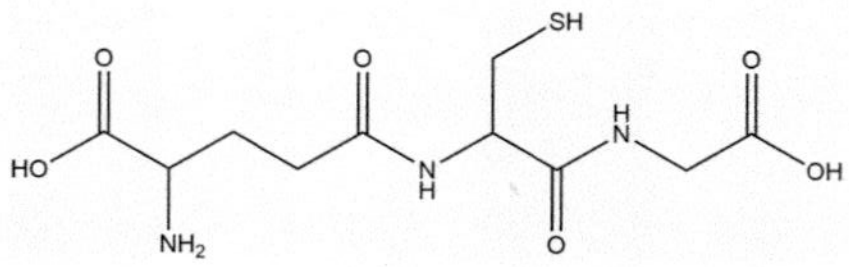	Peptide	Regulatory compounds of plant growth and development, cellular antioxidant in stress responses, defense compounds against reactive oxygen species(ROS) generated by O_3	Antioxidant, detoxification, antitumor activities
	Glucosinolate	Capparales and the genus *Drypetes*	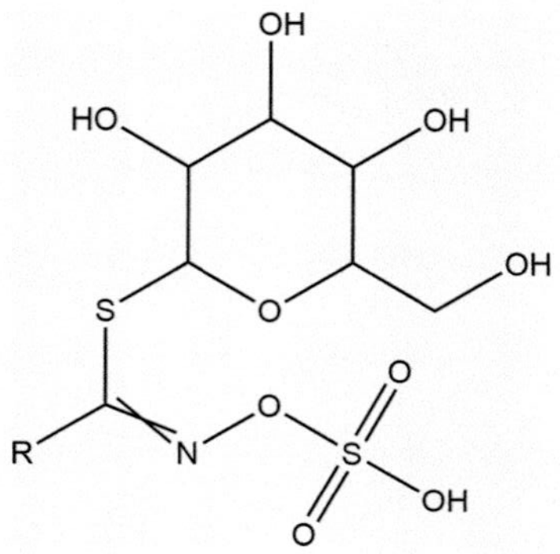	Thioglycoside	Defense compounds and attractants	Cancer-preventing agents, biopesticides, and flavor compounds

Nitrogen-containing compound	Morphine	*Papaver somniferum*	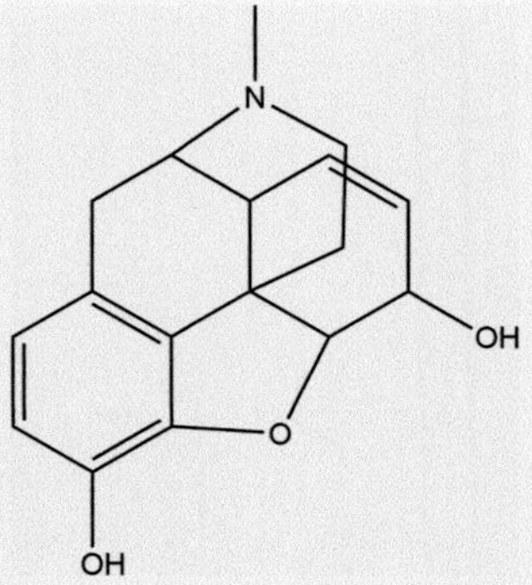	Alkaloid	Defensive elements against predators, narcotic analgesics	The most common anesthetics, efficient pain killer
	Nicotine	*Nicotiana tabacum*	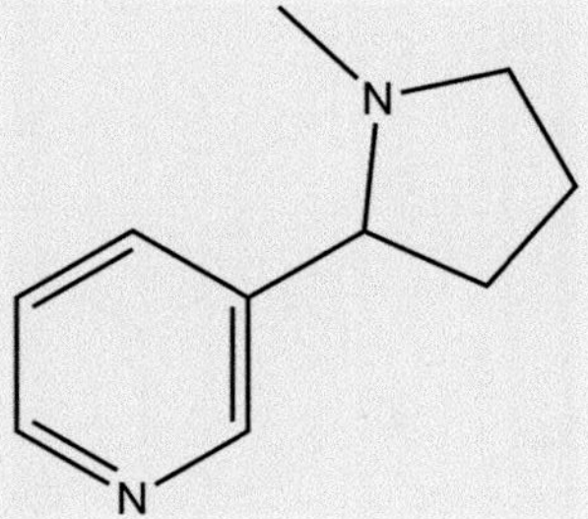	Alkaloid	Antibiotic compounds, insecticides, feeding deterrents	Cholinoceptor agonists, toxicity, addictive compound contributing to cancer and cardiovascular diseases
	Lupanine	*Lupinus polyphyllus*	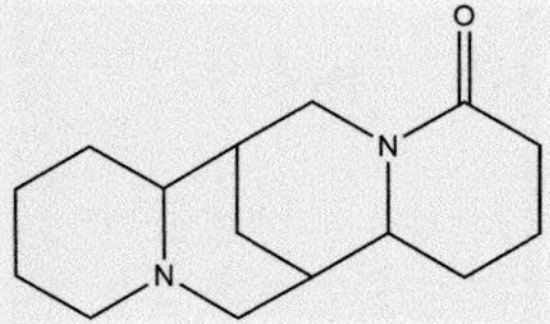	Quinolizidine alkaloid	Feeding deterrent	Preganglionic stimulation suppressor of the pneumogastric nerve, toxicity

Continued

TABLE 10.1 Four main groups of plant secondary metabolites and some examples of each group—cont'd

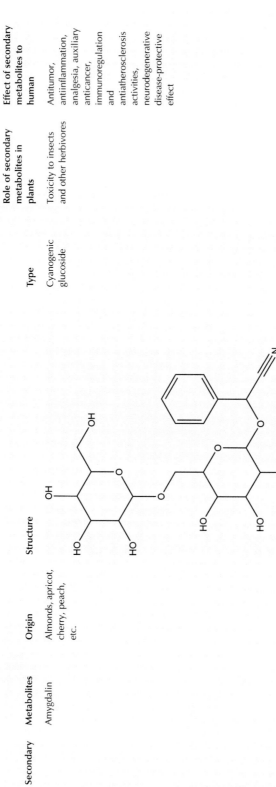

Secondary	Metabolites	Origin	Structure	Type	Role of secondary metabolites in plants	Effect of secondary metabolites to human
	Amygdalin	Almonds, apricot, cherry, peach, etc.	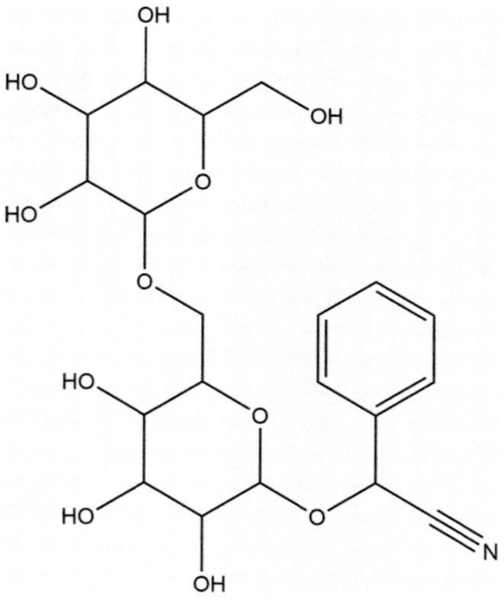	Cyanogenic glucoside	Toxicity to insects and other herbivores	Antitumor, antiinflammation, analgesia, auxiliary anticancer, immunoregulation and antiatherosclerosis activities, neurodegenerative disease-protective effect

crucial for normal development (Lindsey et al., 2003). Finally, tetraterpenes, such as lycopene and carotene, are naturally occurring pigments. Carotene, the precursor of vitamin A, is one of the most important compounds for humans and is widely used as a food supplement.

Phenolic compounds

Phenolic compounds are widely distributed in plants. While these compounds' chemical structures vary widely, they typically consist of at least one benzene ring (C_6H_6) with one or more acidic hydroxyl groups (—OH) attached (Nicholson and Hammerschmidt, 1992) (Table 10.1). Regular consumption of vegetables and beverages with a high level of phenolic compounds may reduce the risk of developing several diseases, including colon cancer, due to their antioxidant power (Yi et al., 2005; Lin et al., 2016). Cinnamic acid is a natural phenolic compound found in many plants, which plays an important role in plant defense against insect herbivores (Valiela et al., 1979). In humans, cinnamic acid and its derivatives have potential antidiabetic, antioxidant, and antimicrobial activities (Kasetti et al., 2012; Sova, 2012). Coumarins and coumarin derivatives are widely existing in plants. Plant-derived coumarin derivatives play an important role in plants and soil microbial communities (Niro et al., 2016). They also show potent antioxidant, antiinflammatory, and antiplatelet effects, which can prevent the aggregation of platelets to treat cardiovascular diseases (Teng et al., 1992; Witaicenis et al., 2014). Flavonoids and flavonoid derivatives are probably the largest class of PSMs with polyphenolic structures and are widely distributed in several parts of various plant species (Wang et al., 2018a; Jin et al., 2020). In the tea plant, (−)-epigallocatechin-3-gallate (EGCG), a bioactive flavonol compound, protects the plant from harmful UV-B light. In humans, EGCG shows excellent biological activities, including antiviral, antiinflammatory, antifungal, and antiobesity activities, and confers protection against fatal diseases, such as cancer (Du et al., 2012; Liu and Yan, 2019). Cyanidins, another group of flavonoid compounds, contribute to the pigmentation of many flowers, fruits, vegetables, and grains. Almost all colors (except green) can be ascribed to cyanidins. The colors depend on the number of hydroxyl groups and glycosylation, as well as pH, which affects the compound structure. Besides acting as natural dyes, cyanidins have attracted increasing interest as functional food coloring compounds and potent antioxidant and antiinflammatory agents (Noda et al., 2002; Stintzing et al., 2002). The levels of phenolic compounds tannin and tannic acid are high in Chinese sumac (traditional Chinese medicine) and the grape's skin and exhibit general toxicity that greatly reduces the growth and survival of many herbivores. On the contrary, they also have several bioactivities, including antidiarrheic, antihemorrhagic, and antioxidant activities. Of note, tannin detoxifies snake venom and bacterial toxins via the formation of tannin-protein complexes (Frazier et al., 2010).

Nitrogen-containing compounds

As the name implies, nitrogen-containing PSMs, including cyanogenic glycosides, glucosinolates, and alkaloids, are characterized by the presence of nitrogen molecules in their structure (Table 10.1). Amino acids, such as lysine, tyrosine, and tryptophan, act as precursors of their biosynthesis (Fig. 10.1). Alkaloids are a large group of nitrogen-containing compounds. Based on their chemical structure, alkaloids can be subdivided into more than 60 subclasses. Morphine is believed to be one of the most common alkaloids with multiple functions. Morphine was first isolated from opium poppy by Germen chemist Serturner in 1805 though the history of using opium poppy can date back to 600 BCE (Brook et al., 2017). For the opium plant, morphine defends the plant host against predators. As the most widely used painkiller and narcotic analgesic, morphine is of great importance in clinical medicine despite its addictive side effects. Amygdalin is one of the major active compounds in the seed of plants, such as peach, apple, and bayberry. It is toxic to insects and other herbivores, as well as to humans at sufficiently high concentrations. The pharmacological activity of amygdalin, including antitumor, antifibrosis, antiinflammation, pain-relieving, and antiatherosclerosis effects, has been recently reviewed (He et al., 2020).

Sulfur-containing compounds

Although sulfur-containing secondary metabolites are a relatively small group of plant compounds, they play a vital role in plant-insect interactions and antimicrobial pathogen defense in a variety of different plant families and act as the major chemical defenses in Brassicaceae, Alliaceae, and Asteraceae (Table 10.1). Sulfur-containing secondary metabolites present in Brassicaceae are a highly diversified group of phytochemicals, currently represented by almost 200 structures, including glutathione (GSH), glycosphingolipids (GSL), glucosinolates, phytoalexins, defensins, and allicin (Czerniawski and Bednarek, 2018). GSH is a small intracellular thiol molecule. The key functions of GSH in plants include detoxification, plant defense, antioxidant biochemistry, redox signaling, and abiotic stress tolerance (Foyer et al., 2001; Xiang et al., 2001; Yadav, 2010). GSH also exhibits antioxidant, detoxification, and antitumor activities. The glucosinolate-rich plant material is mixed with agricultural soils to deter pathogens, nematodes, and weeds (Zasada and Ferris, 2004; Vaughn et al., 2005). Additionally, plant-derived sulfur-containing secondary metabolites show promising pharmacological properties, including neuroprotective properties, for humans. For example, glucosinolates are candidates for tumor prevention. Further, cysteine sulfoxides found in garlic are considered to be potent antimicrobial agents. Other garlic sulfur-containing secondary metabolites provide neuroprotection via their direct and indirect antioxidant properties, modulation of apoptosis mediators, and inhibition of amyloid protein formation.

PSM as an important strategy in accomplishing SDGs

To tackle the global challenges in the 21st century, the United Nations announced 17 SDGs. PSM plays an essential role in accomplishing SDGs (Fig. 10.2). To achieve no poverty (SDG 1) and zero hunger (SDG 2), PSM has contributed to facilitating crop production by enhancing stress resilience. It is reported that flavonoids and polyphenols can help plants induce drought resistance by scavenging reactive oxygen species (ROS) produced following drought-induced oxidative stress (Treml and Šmejkal, 2016). In salvia, rosemary, and sage, diterpene provides drought resistance by a similar antioxidative protection mechanism (Munné-Bosch et al., 2001). Rice is one of the most important crops in Asia. It is suggested that the concentration of phenolic

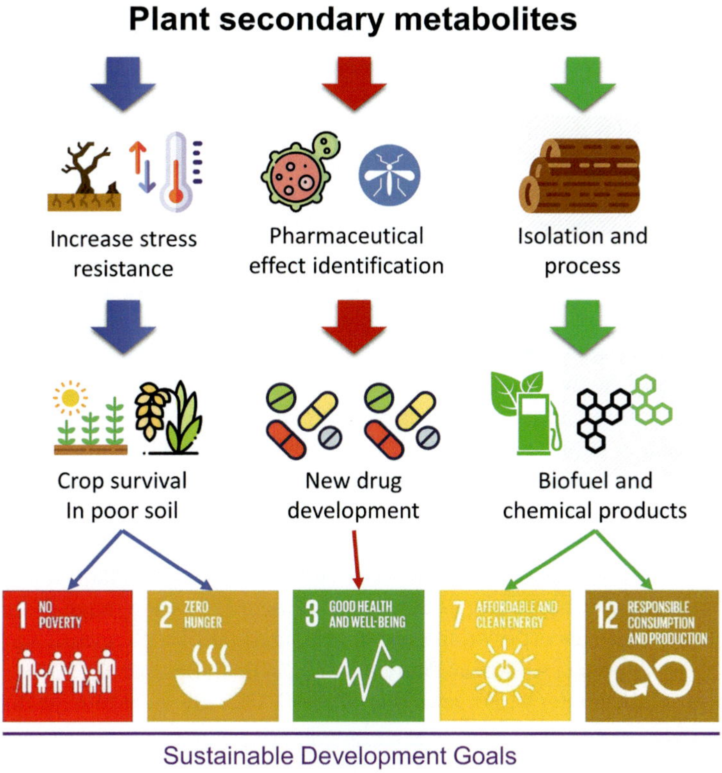

FIG. 10.2 Outline of the relationship between PSMs and sustainable development goals (SDGs). *(Credit: PSMs have a close relationship with SDGs. PSMs increase the resistance of abiotic and biotic stresses, which are closely related to SDG1 (No poverty) and SDG2 (Zero hunger) (*blue arrow*). The pharmaceutical effects of PSMs play key roles in public health (SDG3, good health, and well-being,* red arrow*). PSMs also contribute to the biofuel, biomaterial, and chemical products, aligning with the environmental carbon cycle and respecting green chemistry principles (SDG7, affordable and clean energy, and SDG12, responsible consumption and production,* green arrow*).)*

compounds such as gallic acid and rutin in rice strongly positively correlates with drought resilience (Quan et al., 2016). Another primary food crop, maize, exhibits a significant increase in several sulfur-containing compounds like glutathione and methionine with the effect of drought stress, implying these PSMs' potential drought resistance role (Yang et al., 2018). The biosynthesis of drought resistance PSMs in different plants increases the survival rate under water shortage conditions, allowing crop breeding and planting in the dry soil, where poverty and famine often occur. Other stress resistances, including abiotic and biotic stress resistance, are closely related to PSMs (Akula and Ravishankar, 2011; Yang et al., 2018).

Other than food problems, good health and well-being (SDG 3) can be achieved by employing PSMs to treat diseases. Of the most prominent examples is artemisinin, a well-known antimalarial drug from the Chinese herb Artemisia (Qinghao in Chinese). Malaria is a life-threatening disease caused by parasites, one of the most severe public health issues worldwide. According to WHO's World malaria report 2020, an estimated 409,000 people died from malaria in 2019, even though the total funding for malaria control and elimination reached 3 billion dollars in the same year (https://www.who.int/publications/i/item/9789240015791). At the end of the last century, researchers headed by Tu investigated more than 2000 Chinese herb preparations. Finally, the Artemisia extract was targeted due to its significant inhibition of parasite growth (Tu, 2011). Furthermore, the potent antimalarial effect of artemisinin has been well characterized (de Vries and Dien, 1996; O'Neill et al., 2010; Tu, 2016).

Cancer is one of the world's largest health problems. The Global Burden of Disease (GBD) study estimates that 9.56 million people died prematurely due to cancer in 2017. In the United States, lung and bronchus cancer, prostate cancer, colon and rectum cancer, as well as breast cancer are leading life-threatening cancers (Siegel et al., 2020). The development of anticancer drug helps fight cancer and promotes human health (Holmes et al., 1991; Forastiere et al., 1993; Murphy et al., 1993; Smit et al., 1998). More than 3000 plant species have reportedly been used in the treatment of cancer (Cragg and Newman, 2005); several notable plant-derived anticancer agents include the taxol, a powerful and complex anticancer agent used as the treatment for a wide range of cancers, including breast cancer, ovarian cancer, nonsmall-cell lung cancer (NSCLC), and Kaposi sarcoma (Ajikumar Parayil et al., 2010; Gallego et al., 2017); vinca alkaloids, vinblastine (VLB), and vincristine (VCR) are used to treat diabetes and a variety of cancers, including leukemia, lymphomas, advanced testicular cancer, breast and lung cancers, and Kaposi's sarcoma (Guéritte and Fahy, 2005; Shoeb, 2006).

PSM has contributed to the pursuit of responsible consumption and production (SDG 12), affordable and clean energy (SDG 7), and climate action (SDG 13) in several different ways. In 2008, about one-tenth of the global primary energy was contributed by biomass, which is expected to increase up to 60% in 2050 (Rabaçal et al., 2017). Plant-derived biomass such as cellulose, hemicellulose, and lignin are widely used as renewable resources to produce gasoline, biodiesel, and jet fuel (Alzagameem et al., 2018). These natural biodegradable polymers can be biodegraded and recycled once disposed of, which aids the environmental carbon cycle and respects green chemistry principles.

Multiomics in PSM studies

Understanding the complex biosynthetic processes of PSMs is crucial for realizing the full potential of PSMs as a path for a sustainable bioeconomy. A detailed understanding of PSM biosynthesis requires interpretation of the molecular intricacies and variations at multiple levels, such as the genome, transcriptome, metabolome, and proteome levels (Fig. 10.2).

Genomics

Genome sequencing provides an opportunity for the identification of genes and pathways involved in PSM biosynthesis. By the end of 2020, the genomes of 435 plant species had been published in the NCBI Genome Assembly Database (www.ncbi.nlm.nih.gov/assembly/; Fig. 10.4). In recent years, increasingly rapid and more cost-effective high-throughput genome sequencing technologies, coupled with advanced computational power, have yielded a large amount of plant genomic data, revealing the great potential for the biosynthesis of diverse and novel PSMs (Fig. 10.3).

Since specialized PSMs are highly plant species-specific, comparative genomics studies are used to identify candidate genes associated with their biosynthetic pathways for further functional characterization. For example, comparative genomic analysis of 25 plant species revealed orthologous genes (homologous genes found in different species which have evolved through speciation from the same ancestral gene) associated with terpenoid, alkaloid, phenolic, and glucosinolate metabolism (Tohge et al., 2014). The species-specific biosynthetic pathways such as the glucosinolate pathway in Brassica species were investigated, and comparative genome analysis as a basis for predicting metabolite types within a plant species was proposed. The approach of identifying biosynthetic genes by using genome mining (Bachmann et al., 2014), which is currently successfully applied to microorganisms with small genomes, is also exploited in plants with large genomes. A genome mining approach can be automated and has been used for exploring Arabidopsis genomic information to identify novel terpenoids (Fazio et al., 2004). This approach also revealed rich lyciumin genotypes and chemotypes that are widespread in flowering plants (Kersten and Weng, 2018). Recent studies involving genome mining in different plant species have revealed the existence of metabolic gene clusters associated with PSM biosynthesis. More than 20 clusters associated with secondary metabolism, including diterpene, triterpene, polyketide, alkaloid, and cyanogenic glycoside metabolism, have been reported. Examples include gene

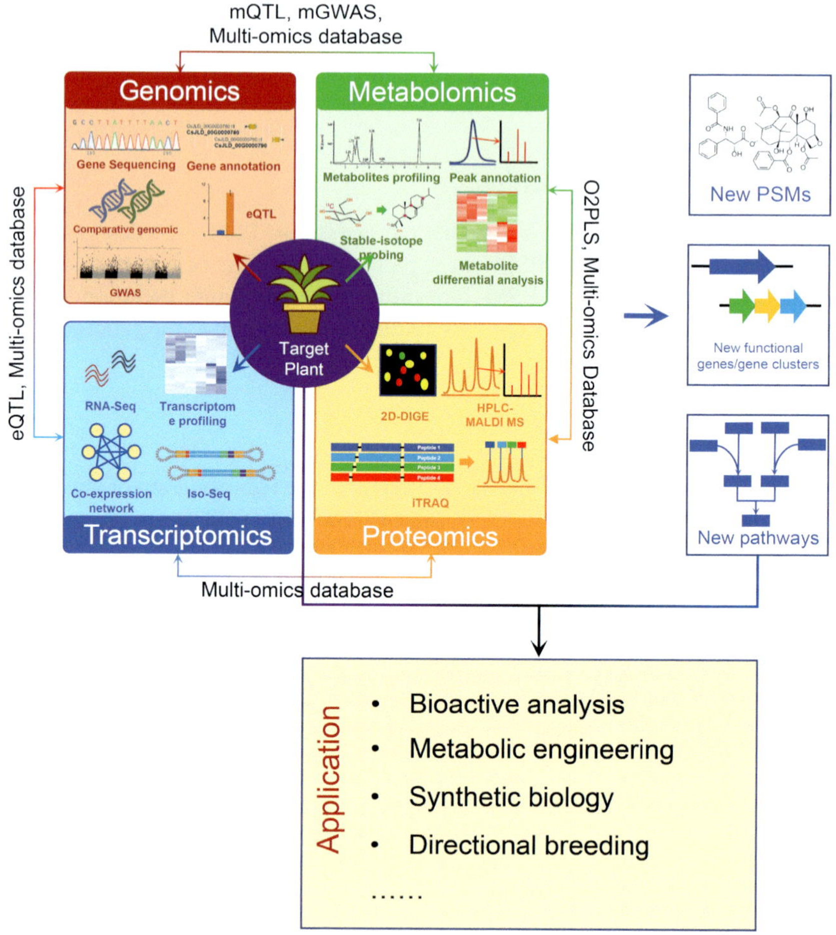

FIG. 10.3 Schematic representation of the workflow for applying multiomics methods in the study of PSMs. *(Credit: Research approaches for each -omics method are displayed, and integrated -omics methods are also shown. The multiomics approaches can be used to discover new PSMs, identify new functional genes, and mine new PSM biosynthesis pathway. This knowledge furthers bioactive and medical analyses, metabolic engineering, biosynthesis and fermentation in heterologous expression systems, and directional breeding attempts.)*

cluster for the synthesis of avenacin, a triterpenoid, in oat (Qi et al., 2004, 2006; Mugford et al., 2013); terpene synthase gene cluster in tomato (Falara et al., 2011; Matsuba et al., 2013; Zhou and Pichersky, 2020); gene cluster involved in the thalianol pathway in Arabidopsis and oat (Field and Osbourn, 2008); gene cluster for cyanogenic glucoside biosynthesis in sorghum (Takos et al., 2011); diterpenoid biosynthetic gene cluster in rice (Swaminathan et al., 2009); α-chaconine/α-solanine biosynthesis gene cluster in potato (Itkin et al., 2013); cucurbitacin gene cluster in cucumber (Shang et al., 2014; Boutanaev et al., 2015);

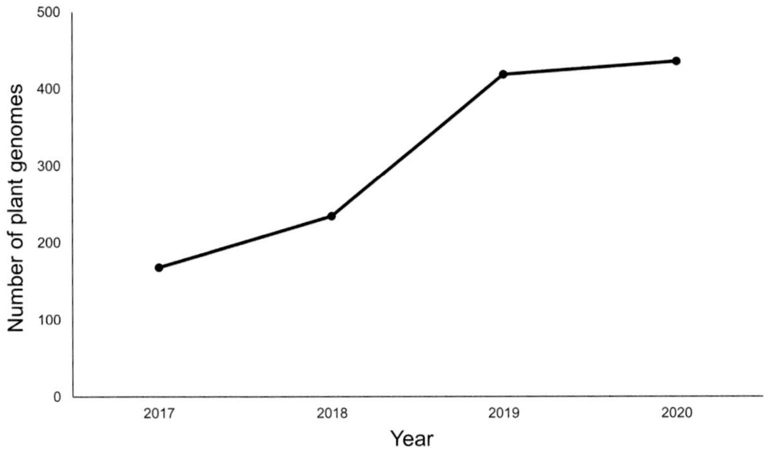

FIG. 10.4 Number of published plant genomes with time. *(Credit: One hundred and sixty-eight plant genomes had been published by 2017, 234 by 2018, 418 by 2019, and 435 by 2020.)*

gene cluster linked to a novel noscapine (an antitumor alkaloid) biosynthesis pathway in opium poppy (Winzer et al., 2012); and gene cluster for known enzymes of the caffeine biosynthetic pathway in coffee (Denoeud et al., 2014).

Other genomics approaches for the discovery of metabolite pathways, including genome-wide association studies (GWAS) and expression quantitative trait loci (eQTLs), are important scientific tools and have been successfully used in the model plant Arabidopsis and crops, such as rice, maize, and tomato (Brachi et al., 2011; Ingvarsson and Street, 2011; Pasam et al., 2012). GWAS and eQTL approaches, combined with metabolomics, facilitate the identification of genomic regions necessary for the production of the desired PSMs (Lu et al., 2003; Atwell et al., 2010; Bergelson and Roux, 2010; Huang et al., 2010; Miura et al., 2011; Abe et al., 2012). Together with computing abilities, such as multiple-sequence alignment and deep learning, and crystallographic analysis tools, researchers are able to predict and identify heretofore-unknown biosynthesis genes and elucidate the functions and characteristics of key enzymes involved in the assembly and tailoring of PSMs.

Transcriptomics

Transcriptomics is a set of high-throughput technologies used to analyze a complete set of RNA transcripts produced by an organism or cell under specific conditions. According to the central genetic dogma, messenger RNA (mRNA) is the molecule that transfers biological information between DNA and protein; other RNAs like microRNA also play important roles in regulating transcription (Filipowicz et al., 2008). Transcriptome researches are the basis and starting point of the study of gene function and structure. Therefore, exploring the transcriptome is essential to interpret the genome's functional elements and reveal the molecular composition of cells and tissues.

Tag-based sequencing and hybridization-based approaches were the technology of choice for large-scale gene expression studies a decade ago (Wang et al., 2009), usually based on previously identified target genes. RNA sequencing (RNA-Seq), a recently developed approach, uses deep-sequencing technology to delineate cell transcriptome at a given time point providing information on the gene sequence in addition to their expression level (Behera et al., 2021). RNA-Seq also provides a more accurate measurement of transcript levels and transcript isoforms than other methods (Wang et al., 2009). In addition to quantifying gene expression, the data generated by RNA-Seq can help discover novel transcripts, identify differential mRNA splicing, and detect allele-specific expression. In addition to mRNA transcripts, RNA-Seq can also be used to study different RNA sets, including total RNA, pre-mRNA, and noncoding RNA (ncRNA), such as microRNA and long ncRNA (Chen et al., 2015; Chiang et al., 2016; Song et al., 2016; Xiong et al., 2019). MicroRNA play an important role in plant development and other physiological processes such as hormone response, cell division, flowering, and environmental or stress response (Palatnik et al., 2003; Schmid et al., 2003; Achard et al., 2004; Jones-Rhoades and Bartel, 2004). While long ncRNA functions as microRNA noncleavable target mimic, chromatin modifiers (enzymes that catalyze the chemical conversion of cytosine residues in DNA, or lysine, arginine, tyrosine, and serine residues in histone proteins) recruitment guide and molecular cargo for protein relocalization, serving as key regulators of a variety of biological processes (Zhu and Wang, 2012).

The last few years has witnessed an exponential rise in the number of RNA-Seq-based transcriptome studies conducted by plant biologists (Rai et al., 2017). RNA-Seq-based transcriptome profiling, de novo transcriptome assembly (reconstruction of the transcriptome without referring to a reference genome), coexpression, and comparative transcriptome analyses of various plant species have changed the view of the scope and complexity of PSM biosynthesis (Wisecaver et al., 2017; Rao and Dixon, 2019). Since fully sequenced genomes are not yet available for many plant species, de novo transcriptome assembly is the basis for understanding molecular and functional mechanisms of secondary metabolite biosynthesis. For example, the establishment and characterization of de novo transcriptome assembly resource for purple gromwell, a valuable source of bioactive metabolites with medicinal and industrial values, revealed several genes involved in the biosynthesis of shikonin, which is the major component of purple gromwell and a novel class of inhibitors of tumor PKM2 and human immunodeficiency virus (HIV) type 1 (HIV-1) (Chen et al., 2011; Rai et al., 2018). In addition, analysis of allele-specific expression in tea plant (*Camellia sinensis* L.) suggests that genes related to metabolism and biosynthesis of alkaloids and aromatic chemicals undergo strong selection contributing to the most important feature of tea plants, the flavor characteristics

(Zhang et al., 2021). Further, tissue-specific RNA-Seq revealed active transcription of genes involved in the caffeine biosynthetic pathway in the developing endosperm of coffee (Denoeud et al., 2014).

Coexpression network analysis is one of the most powerful approaches for interpreting large transcriptomic datasets. It correlates RNA-Seq results and metabolites of interest. It is constructed using weighted gene coexpression network analysis (WGCNA), enabling the identification of coexpressed gene modules involved in the same metabolic pathway and prediction of biological functions of unknown genes (Gaudinier et al., 2018). Similar behavior of mRNA levels of genes within the same metabolic pathway has been observed (Wei et al., 2006). Terpenes, like monoterpenes and sesquiterpenes in cannabis, are major aromatic flavor substances. Different terpenes are combined in a certain ratio to give various flavors like fruit, spicy, or wood. So, it is important but difficult to identify the terpenes responsible for the flavors and the TPS involved in these compounds' biosynthesis. In addition, the entourage effect of cannabinoids combined with terpenes has long been a hot spot in cannabis research because of a better pharmaceutical effect compared to pure cannabinoids (Nahler et al., 2019; Ferber et al., 2020). Although most gene sequences acquired by genomic analysis are functionally ambiguous, the construction of a coexpression network of genes related to the synthesis of cannabinoids and terpenoids enabled the identification of the main terpene synthase genes involved in the biosynthesis of major monoterpenes and sesquiterpenes in cannabis (Zager et al., 2019). Combining RNA-Seq-based transcriptome profiling and coexpression analysis of opium poppy extracts containing high levels of morphine, thebaine, and noscapine, ten genes that were highly expressed and specific to the noscapine-producing plant samples were identified (Winzer et al., 2012).

Recently, third-generation sequencing approaches, including Oxford Nanopore Technologies (ONT), Pacific Biosciences (PacBio) single-molecule real-time (SMRT) sequencing, and long-read isoform sequencing (Iso-Seq), have been widely used for transcriptome sequencing because they greatly facilitate de novo transcriptome assembly, isoform expression quantification, and in-depth RNA species analysis by providing particularly long reads in a high-throughput manner (Wang et al., 2016). For example, Qiao et al. (2019) used TGS transcriptome analysis to reveal that alternative splicing plays a crucial role in the regulation of secondary metabolism, such as that of flavonoids, theanine, and caffeine biosynthesis, in tea plant (Qiao et al., 2019). In another study, researchers used TGS transcriptome and coexpression analysis to identify 22 genes potentially involved in taxol biosynthesis in *Taxus cuspidata* (Kuang et al., 2019).

Metabolomics

Metabolomics studies small molecules (metabolites), including the substrates and products of enzymatic reactions, in a given organism or cell type. Plant secondary metabolite is one of the most intuitive phenotypes rooted downstream of the entire physiological process. The concentration and structure diversity of metabolites can directly or indirectly reflect the difference in genome level (gene sequence difference) and transcription level (gene expression level). Because the concentrations of metabolites can be considered the final response of a biological system to genetic or environmental change, the metabolome can serve as the ultimate output of the biochemical and physiological status of a cell, directly linking mechanistic biochemistry to cellular phenotype (Patti et al., 2012). With the latest improvements in high-throughput technologies, such as high-performance liquid chromatography (HPLC) and gas chromatography (GC) coupled with mass spectrometry (MS) and nuclear magnetic resonance (NMR), it is now possible to sample hundreds to thousands of unique metabolite peaks, translate these peaks into putative molecular formulas, and calculate their native intracellular concentrations, all from a small starting quantity of the test material (Bennett et al., 2009; Tautenhahn et al., 2012). Because of these recent innovations, metabolomics has become a major contributor to the identification of diverse PSMs, including taxol, a highly effective anticancer diterpenoid obtained from the stem bark of the Pacific yew tree (Holmes et al., 1991); morphine, a robust pain-relieving alkaloid isolated from the opium poppy (Huang and Kutchan, 2000); and artemisinin, a remarkable sesquiterpene antimalarial therapeutic, originating from *Artemisia annua* (Mannan et al., 2010). In these studies, metabolomic approaches were applied to identify the specific and temporal distribution of target PSMs and to identify the related compounds that could be intermediates of the biosynthetic pathway of interest or alternative products of promiscuous enzymes that contribute to the biosynthesis of target PSMs. To discover new metabolic functions and pathways, researchers combine metabolomics with stable isotope probing to track the metabolic fate of target compounds and their flow through specific pathways (Prosser et al., 2014). In this approach, plants are fed labeled substrates or labeled carbon sources, leading to the formation of labeled metabolites that can be distinguished from contaminants and background noise. In addition, metabolic flux and metabolic pathways can be elucidated by detecting and analyzing the structure of labeled intermediate metabolites and end-products. For example, the biosynthesis of cannabinoids, a group of bioactive terpene derivatives specifically produced by Cannabis, was studied using mixtures of unlabeled glucose and two differently labeled glucose molecules harboring 13-C at different positions (Fellermeier et al., 2001). Tracing ^{13}C-labeling glucose revealed that the C10-terpenoid moiety is biosynthesized via the deoxyxylulose phosphate pathway. The phenolic moiety is generated by a polyketide-type reaction, shedding insights on the biosynthesis pathway of cannabinoids for the first time and laying the foundation for further research.

Proteomics

According to the central genetic dogma, proteins are formed through mRNA translation after the transcription of DNA into RNA. The protein level is highly associated with the gene copy numbers and gene expression level. Therefore, we can investigate the protein level, analyze the amino acid sequences of proteins, and trace back the functional genes through a reverse deduction to understanding a certain physiological process in another dimension. Here comes proteomics. Proteomics is the study of the composition and abundance of proteins in tissues, cells, or organelles (Wasinger et al., 1995). Three major types of proteins are involved in secondary metabolism: enzymes that catalyze sequential reactions, transport proteins responsible for the translocation of natural compounds in plant organs, and transcription factors involved in the regulation of biosynthetic pathways (Yazaki, 2006; Cao et al., 2020). The multiple and important properties of these secondary metabolism-related proteins make the proteome an essential focus of metabolic pathway research.

The analysis of proteins mainly depends on the purification technology used to separate complex protein mixtures and MS, which converts a protein to an amino sequence for further annotation and identification. Traditionally, gel-based two-dimensional polyacrylamide gel electrophoresis (2D-PAGE) has been used to resolve proteins. For comparative analysis, two-dimensional fluorescence difference gel electrophoresis (2D-DIGE) was used to capture differential protein expression. These gel-based classical methods have made a great contribution to the study of PSM biosynthesis in wheat (*Triticum durum*) (Caruso et al., 2009), sunflower (*Helianthus annuus* L.) (da Silva and Arruda, 2012), the medicinal plant java plum (*Syzygium cumini*) (Binita et al., 2014), and rosy periwinkle (*Catharanthus roseus*) (Jacobs et al., 2005).

With technological developments, advanced approaches, such as HPLC that incorporates protein analysis, isobaric tags for relative and absolute quantitation (iTRAQ), and matrix-assisted laser desorption/ionization tandem MS (MALDI MS/MS), are becoming widely used. Grapevine is an economically important fruit crop. In one study, developmental analysis of grapevine was performed using the iTRAQ method to explore the proteomic profile changes during plant maturation (Martínez-Esteso et al., 2013). The detected dynamic changes in protein levels provided insight into biological processes and metabolic pathways involved in plant development and metabolite storage, giving a new method to decode the processes controlling grape berry ripening and lay the foundation for the breed improvement and yield increasing. A similar analysis was performed to investigate the biosynthetic network of flavonoid biosynthesis in tea plant (*Camellia sinensis* L.) (Wu et al., 2019), and anthocyanin biosynthesis pathway in black rice (*Oryza sativa* L.) (Chen et al., 2016a) and Chrysanthemum (*Chrysanthemum* × *morifolium*) (Hong et al., 2019). iTRAQ enables higher throughput, with high accuracy and resolution, for quantitative and relative qualitative protein analyses than traditional differential proteomics analysis.

Multiomics data integration, interpretation, and its application

Genomics, transcriptomics, metabolomics, and proteomics have been invaluable to characterize PSM but the use of multiomic approaches will enable to reach the big picture and see the forest for the trees. Multiomics approaches have generated an enormous amount of information critical to expanding our understanding of plant secondary metabolism. Each multiomics approach has its advantages and disadvantages, which are discussed in Table 10.2. Sequential or simultaneous integration of multiomics data provides different layers of information on PSM biosynthesis, enabling understanding of the interplay between various molecules (Fig. 10.2) (Ni et al., 2021; Yang et al., 2021). Integrated approaches facilitate the flow of information from one-omics level to a multiomics level, thus helping to bridge the gap between genotype and phenotype. However, because of the inherent differences in the data generated by different approaches, integrating multiple-omics platforms remains an ongoing challenge for many researchers. Below, we discuss databases, software tools, and methods for multiomics integration, along with their applications.

Multiomics data repositories

Since new developments and scientific discoveries are generally based on the predecessors' achievements, researchers must systematically collect and store different data types. On the other hand, as an important part of research results, high-quality data inherits the analysis methods and research ideas of different scientists in the same field, increasing the impact of research results. Therefore, data sharing under the protection of intellectual property rights creates a virtuous circle that promotes research even in different fields with an expanding impact. Data repository, as a container for data storage and platform for data sharing, standardizes the format and use of data. Open-access databases, conducive to data dissemination and data acquirement, are generally accepted by the scientific community, laying the foundation for the unified data standard formation in the field. Databases construct an information flow of scientific value and provide an impetus for the further development of science.

A number of public databases are available, listed in Table 10.3, which house multiomics datasets for plant scientists. AtMAD, a public repository for multiomics integrated large-scale investigation of the model plant Arabidopsis, serves as a powerful tool for various analyses, such as identifying eQTL and phenotype-pathway links, GWAS, and transcriptome-wide association studies (Lan et al., 2021). Maize GDB collects maize reference sequences, stocks, and phenotypic and genotypic data, as well as tools, such as SNPversity, Pedigree viewer, and qTeller, for further data mining (Portwood II et al., 2019). With the help of the SNPversity tool, a candidate gene relating to type C cytoplasmic male sterility (CMS-C) and fertility restoration was identified in maize hybrids. In addition, the F87Y substitution of the gene was demonstrated to play key roles in the fertility restoration *in planta* (Jaqueth et al., 2020). More recently, ZEAMAP, a comprehensive maize multiomics database incorporating genomic, transcriptomic, phenotypic, metabolic, and epigenomic data, was established. The large scale of data containing SNP information of 3000 rice

TABLE 10.2 Advantages and disadvantages of multiomics methods in PSM study.

Omics technologies	Advantages	Disadvantages
Genomics	➤ Large amount of dataset harboring limitless information ➤ Studies on gene polymorphisms reveal key loci relating to PSM biosynthesis	➤ Deviation in gene function annotation ➤ Difficulty to predict the final biological effect of DNA by only genome analysis ➤ Limited reference genome sequenced
Transcriptomics	➤ Identification of PSM biosynthesis genes ➤ Exploration of gene expression pattern ➤ Good repeatability for laboratory studies	➤ Influence of posttranslational modification ➤ Complex network of changes in the transcriptome
Proteomics	➤ Direct analysis of structure and concentration of functional enzymes ➤ Close relationship with PSMs	➤ High cost, insensitivity to low-copy proteins ➤ Influence of posttranslational modification ➤ Difficulty in sample preparation
Metabolites	➤ Fewer amounts comparing to genes, transcripts, and proteins, less available data to be interpreted ➤ Most direct analysis method of the PSM biosynthesis pathways end-products ➤ Technical improvement for high-sensitivity and resolution metabolites detection and identification	➤ Lack of standard compounds for metabolites identification ➤ High dependence on the precious and delicate instruments ➤ Numerous factors affecting the accuracy of results

varieties dramatically facilitated the association analysis (Gui et al., 2020). Soybase is an extensive repository for genetics, genomics, and related data resources for soybean, with a recent update including new features for visualization of gene expression and soya pan-genome (Brown et al., 2021). RICE SNP-seek is a unique resource and tool for rice genomics and allele mining (Mansueto et al., 2017). A recent study on retrotransposons in rice genome carried out a GWAS for transposable elements (TE) copy number (as a phenotype)

TABLE 10.3 List of multiomics databases for plant secondary metabolism study.

Database	Web link	Species	Type of omics data available	Reference
AtMAD	http://www.megabionet.org/atmad	*Arabidopsis thaliana*	Genome, transcriptome, methylome, pathway, phenotype	Lan et al. (2021)
maizeGDB	https://www.maizegdb.org/	*Zea mays*	Genome, transcriptome, phenotype	Portwood II et al. (2019)
ZEAMAP	http://www.zeamap.com/	*Z. mays*	Genome, transcriptome, phenotype, metabolome, Epigenome, GWAS, QTL	Gui et al. (2020)
Soybase	https://www.soybase.org/	*Glycine* max	Genome, transcriptome, phenotype, GWAS, QTL	Brown et al. (2021)
Rice SNP-seek	https://snp-seek.irri.org	*Oryza sativa*	Genome, transcriptome, phenotype, SNP	Mansueto et al. (2017)
GourdBase	http://www.gourdbase.cn/	*Lagenaria siceraria*	Genome, transcriptome, phenotype, QTL	Wang et al. (2018b)
Sol Genomics Network	https://solgenomics.net/	*Solanum lycopersicum, Solanum tuberosum, Capsicum annuum*	Genome, transcriptome, phenotype	Fernandez-Pozo et al. (2015)
MDSi	http://foxtail-millet.biocloud.net/home	*Setaria italica*	Genome, transcriptome, phenotype, SNP	Yang et al. (2020)
CannabisGDB	https://gdb.supercann.net	*Cannabis sativa*	Genome, transcriptome, metabolome, proteome	Cai et al. (2021)

with a 404 K coreSNP dataset obtained from the Rice SNP-Seek Database combining with a linear-mixed model, demonstrating associations between TE copy number and the SNP (Carpentier et al., 2019). GourdBase, a user-friendly multiomics database, provides genome and phenome resources and molecular markers, and facilitates marker-assisted breeding of bottle gourd (*Lagenaria siceraria*) (Wang et al., 2018b). Sol Genomics Network, a large-scale genome-centered database for the Solanaceae family and close relatives, hosts whole-genome data and useful tools, including BLAST, GBrowse, JBrowse, QTL analysis, and the recently developed virus-induced gene silencing (VIGS) tool (Mueller et al., 2005). MDSi, a multiomics database for *Setaria italica*, provides a reference genome, spatial and temporal-scale transcriptome, phenotype, and SNP data, as well as several powerful tools to facilitate the functional studies of *Setaria italica* and other C4 plants (Yang et al., 2020). Our team developed the first integrated functional genomics database for Cannabis (CannabisGDB, https://gdb.supercann.net), including the complete eight cultivated cannabis plant genomes, gene structures, and comprehensively annotated genes from different perspectives including gene ontology (GO), KEGG orthology, and gene family, as well as widely collected gene and protein expression information and chemical phenotype data. Furthermore, CannabisGDB provides built-in tools, such as JBrowse, BLAST, and SeqExtractor, to perform further comparative, evolutionary, and functional analysis. We also designed multiple dynamic charts to present metabolites in 190 varieties. By integrating the genomics, transcriptomics, proteomics, and metabolomics data, CannabisGDB can be used to assemble valuable information to facilitate basic, translational, and applied research in cannabis (Cai et al., 2021).

Multiomics integration software tools and web applications

Multiomics data integration and analysis are highly complicated. The integration and postintegration analysis of multiomics data put forward higher requirements for researchers. The association study of cross-omics data requires researchers to have a broader scientific research perspective together with a more systematic and abstract understanding of research topics to explain multidimensional biological phenomena. Besides researchers' perspectives toward research projects, the integration and postintegration analysis of multiomics data generates an even higher demand on analysis methods. In the era of high throughput, the computational pressure caused by the huge data scale has to be compensated by the robustness of the analytical method. A comprehensive applicability analysis method can meet a variety of analysis needs and can quickly and accurately obtain analysis results to promote the integration and development of multiomics data.

Several widely applicable methods exist; for example, the R package, designed to improve the function and efficiency of R language by the community, is a group of powerful methods and tools for multiomics data analysis and display. In addition, ClusterProfiler provides functional enrichment analysis for thousands of species based on gene ontology (Yu et al., 2012). MixOmics (Rohart et al., 2017) provides semisupervised methods to integrate multiomics datasets focusing on variable selection. Based on the two-way orthogonal partial least-squares (O2PLS) analysis, OmicsPLS offers an efficient solution for two-dimensional or multidimensional data processing (Bouhaddani et al., 2018). WGCNA is widely utilized for association analyses of highly correlated gene sets and external phenotypic features (Langfelder and Horvath, 2008).

Besides R packages, there are a large number of data visualization tools that play important roles in data display, data interaction, information acquisition, and data communication. For example, Cell Illustrator and SimCell can graphically display cell models and biological pathways (Wishart et al., 2005; Nagasaki et al., 2010) (Fig. 10.5). Web applications, PathVisio and SynVisio, are useful visualization tools for metabolic pathways and genome collinearity, respectively (Kutmon et al., 2015).

In addition to statistical language R-based methods and visualization tools, other tools specialized in integrated multiomics analysis have also been developed, facilitating the establishment of correlation models, networks, and pathways with integrated datasets. MetScape, a Cytoscape plug-in, provides functional enrichment and other external analysis functions, completing the analysis results of Cytoscape (Gao et al., 2010; Karnovsky et al., 2012). 3Omics, KaPPA-View, and MiBiOmics provide user-friendly analysis interfaces for transcriptome, proteome, and metabolome data (Tokimatsu et al., 2005; Kuo et al., 2013; Zoppi et al., 2021). MarVis, MassTrix, and MetaboLights focus on metabolomics, relying on clustering algorithms, metabolite annotation, and other cross-species and cross-technology methods to guide the association analyses of metabolomics and other omics (Suhre and Schmitt-Kopplin, 2008; Kaever et al., 2009; Haug et al., 2013). We provide a summary of multiomics integration software tools and web applications in Table 10.4.

Data analysis for multiomics integration

Recent integrated multiomics studies involve two mainstream workflows: integrated post multiomics data analysis and multiomics integrated data analysis. In the former workflow, individual-omics analysis methods are first used to acquire massive amounts of different dimension information. This is followed by data integration approaches that extract related valuable information from each -omics dataset and bridge different data types based on linkage of different physiological activities corresponding to each -omics approach, combined with pathway models or other features. This workflow has been widely used for the analysis of PSM biosynthesis. For example, Yu et al. (2020) performed untargeted metabolomic and proteomic landscape analyses to explore the secondary metabolites and proteins in different *Taxus* stem tissues. The authors thus

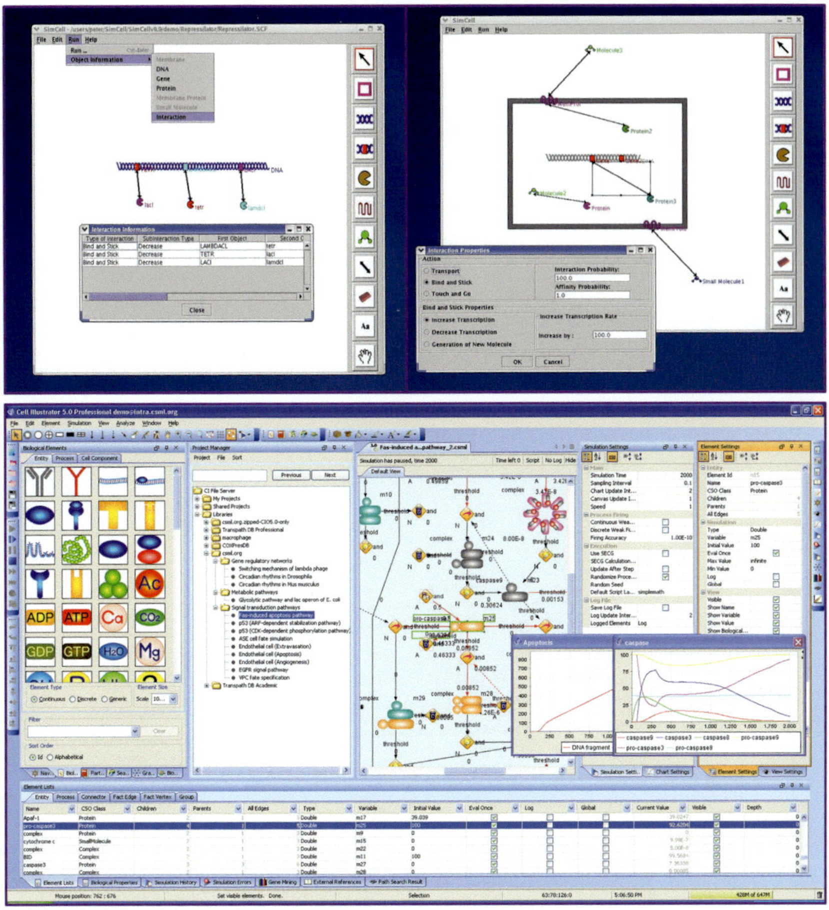

FIG. 10.5 Screenshots of Simcell (above) and Cell Illustrator (below) user interface. *(Credit: Cell Illustrator and SimCell can graphically display cell models and biological pathways with a great accessibility.)*

identified stem tissue-specific taxoid metabolic pathways involving paclitaxel biosynthesis-related enzymes (Yu et al., 2020). The other workflow, multiomics integrated data analysis, is an efficient approach to screening data using specific analytical methods and tools from different -omics datasets merged before further data analysis and interpretation (Kuo et al., 2013). Among analytical approaches, metabolite-based GWAS (mGWAS) associates specific genetic variations with particular plant metabolic phenotypes to dissect the correlation between genetic and biochemical features and plant metabolism (Luo, 2015). In *Populus*, mGWAS was used to analyze three caffeoylquinic acid and caffeoyl conjugate traits based on >8.2 million SNPs, and nucleotide insertions and deletions (indels) from 917 accessions (Zhang et al., 2018). The analysis

TABLE 10.4 Summary of multiomics integration software tools and web applications.

				Types of omics data					
Classification	Software/ web site	Link	Functionality	*Genomics*	*Transcriptomics*	*Metabolomics*	*Proteomics*	*Phenomics*	Reference
R package	clusterProfiler	https://bioconductor.org/packages/release/bioc/html/clusterProfiler.html	R-package for statistical analysis and visualization of functional profiles for genes and gene clusters	✓					Wu et al. (2021)
	mixOmics	https://bioconductor.org/packages/release/bioc/html/mixOmics.html	R-package for omics feature selection and multiple data integration		✓	✓	✓		Rohart et al. (2017)
	OmicsPLS	https://cran.r-project.org/web/packages/OmicsPLS/index.html	R-package with an open-source implementation of two-way orthogonal PLS		✓	✓	✓		Bouhaddani et al. (2018)
	WGCNA	https://cran.r-project.org/web/packages/WGCNA/index.html	R-package for weighted correlation network analysis		✓			✓	Langfelder and Horvath (2008)

Visualization tools	Cell Illustrator	http://www.cellillustrator.com/home	Visualization of biological pathway	✓	✓		✓		Nagasaki et al. (2010)
	Cytoscape	https://cytoscape.org/	Visualizing complex networks and integrating these with any type of attribute data		✓	✓	✓		Shannon et al. (2003)
	PathVisio	https://github.com/PathVisio/pathvisio	Visualize omics data based on common data nodes and interactions in the pathway		✓	✓	✓		Kutmon et al. (2015)
	SimCell	http://wishart.biology.ualberta.ca/SimCell/index.html	Graphical modeling tool		✓	✓	✓		Wishart et al. (2005)
	SynVisio	https://github.com/kiranbandi/synvisio	A visualization tool for McScanX and DAGChainer	✓					Bandi et al. (2022)
Other	MarVis	http://marvis.gobics.de/	Filtering, combination, clustering, visualization, and functional analysis of datasets containing intensity-based profile vectors		✓	✓			Kaever et al., 2015

Continued

TABLE 10.4 Summary of multiomics integration software tools and web applications—cont'd

Classification	Software/ web site	Link	Functionality	Types of omics data					Reference
				Genomics	*Transcriptomics*	*Metabolomics*	*Proteomics*	*Phenomics*	
	MassTrix	http://masstrix3.helmholtz-muenchen.de/masstrix3/	Generation of colored pathway maps KEGG data analysis		✓	✓			Wägele et al. (2012)
	MetaboLights	https://www.ebi.ac.uk/metabolights/index	Storing metabolomics experiments and derived information			✓			Haug et al. (2020)
	KaPPA-View	http://kpv.kazusa.or.jp/kpv4/index.action	Integrates transcriptomics and metabolomics data to map pathways		✓	✓			Tokimatsu et al. (2005)
	MetScape	http://metscape.ncibi.org/index.html	Integrates data from KEGG and EHMN databases		✓	✓			Karnovsky et al. (2012)
	MiBiOmics	https://shiny-bird.univ-nantes.fr/app/Mibiomics	Integrative multiomics analyses through an intuitive and interactive interface	✓	✓	✓	✓	✓	Zoppi et al. (2021)

revealed the potential key gene, shikimate hydroxycinnamoyl transferase 2 (PtHCT2), on chromosome 18 as significantly associated with the abundance of chlorogenic acid and other partially characterized caffeoyl conjugates. Other analytical approaches, such as two-way orthogonal partial least squares (O2PLS, a data integration method for two datasets, such as metabolomics dataset and transcriptomics dataset, yielding joint and data-specific parts for each dataset) and its variant version, OnPLS, are also used for integrated analysis, to extract crucial linked information from two or multiomics datasets (Bylesjö et al., 2007; Turner et al., 2016).

Case studies—Application of multiomics approaches in research

To further understand the application of multiomics methods in the study of PSMs, two case studies are discussed in detail below, with additional case studies presented in Table 10.5.

Discovery of biosynthesis pathways and evolutionary routes for triterpene acids in loquat (Su et al., 2021)

Background and aim. Loquat is a subtropical fruit that originated in southcentral China with a long cultivation history. The fruits of loquat are widely used in the treatment of coughing and throat pain. Natural small molecules, such as triterpene acids, e.g., ursolic acid (UA) and corosolic acid (CA) produced by loquat tree, are potential bioactive compounds with antiinflammatory, antidiabetic, and anticancer activities, indicating the potential in the medical application. To elucidate the molecular evolutionary mechanism of triterpene acid accumulation, methods such as genomic, metabolomic, transcriptomic, and synthetic biology approaches were used to systematically analyze the synthesis pathways, expression patterns, and evolutionary routes of the main triterpene acids in loquat and the apple tribe plants.

Genomic analysis. First, the chromosome-level genomes of loquat (cv. "Jiefangzhong") and bowman's root (the allied species of loquat) were sequenced and assembled by using PacBio, Illumina, and Hi-C technology. Molecular clock analysis describes the relationship between evolutionary rate and time across species. Genomic syntenic analysis investigating the genome collinearity comparing the genome of loquat with other species indicated a whole-genome duplication event (WGD), creating an organism with additional copies of the entire genome of a species that occurred in a common ancestor of the apple tribe.

Metabolomic analysis. Genomic analysis played a fundamental role for further omics researches, providing large amounts of genomic information for the following metabolites and gene function identification and characterization. In addition, the discovery of the WGD event also implies the duplication and diversification of functional genes, providing clues for the study of terpene

TABLE 10.5 Recent cases of PSM biosynthesis study using multiomics approaches.

Plant species	Key PSMs	Multiomics	Analytical platform	Analytical approaches	Reference
Populus trichocarpa	Lignin	Transcriptomic, proteomic, fluxomic, and phenomic	Illumina GAIIx, PC-IDMS, 2D NMR	RNA-Seq, comparative transcriptome analysis, targeted protein quantification, metabolic fluxes analysis, comprehensively targeted metabolites analysis, physical properties analysis, integrative systems analysis	Wang et al. (2018a, b)
Prunus L.	Phenylpropanoids	Genomic, transcriptomic, metabolomic	Illumina, PacBio, Hi-C, HPLC-MS/MS	Genome sequencing, assembly and annotation, phylogenetic analysis, genomic feature analysis, comparative genome analysis, comparative metabolites analysis, metabolite profiling and analysis, mGWAS, PCA, GO enriching, methylome analyses, short interspersed nuclear elements (SINEs) detection	Wang et al. (2021)
Carthamus tinctorius L.	Flavonoids	Transcriptomic, metabolomic	UHPLC-MS/MS, Illumina Hi-Seq	Comparative metabolites analysis, RNA-Seq, GO/KEGG enriching, expression level analysis	Chen et al. (2020)
Taxus L.	Paclitaxel	Genomic, transcriptomic, metabolomic	Illumina, PacBio, Hi-C, PlantiSMASH Database, GC-MS, *Escherichia coli* expression system	Genome sequencing, assembly and annotation, genomic feature analysis, phylogenomic analysis, expression-level analysis, in vitro functional identification	Xiong et al. (2021)

Taxus media	Paclitaxel	Metabolomic, proteomic	UPLC-MS/MS (Metabolomic), HPLC-LC-MS/MS system (Proteomic, HPLC for fractionation, LC-MS/MS for analysis)	Targeted and untargeted metabolites analysis, comparative metabolites analysis, comparative proteomics analysis, GO/KEGG, integrated metabolomic and proteomic analysis, chromosome walking, rapid amplification of cDNA ends (RACE), electrophoretic mobility shift assay (EMSA), chromatin immunoprecipitation(ChIP)-quantitative (q)PCR, dual-luciferase analysis	Yu et al. (2020)
Camptotheca acuminata	Camptothecin	Genomic, transcriptomic	Illumina, PacBio, Hi-C, *Saccharomyces cerevisiae* expression system, UPLC-MS/MS	Genome sequencing, assembly and annotation, RNA-Seq, phylogenetic analysis, expression-level analysis, coexpression analysis, in vitro functional identification, protein structure modeling and structure analysis	Kang et al. (2021)
Senna tora	Anthraquinone	Genomic, transcriptomic	Illumina, PacBio, Hi-C, *E. coli* expression system, StoraCyc Database, UPLC-MS/MS	Genome sequencing, assembly and annotation, comparative genomic analysis, KEGG/GO, gene expansion analysis, targeted metabolites quantification, transcript expression profiles and expression-level analysis, in vitro functional identification	Kang et al. (2020)

Continued

TABLE 10.5 Recent cases of PSM biosynthesis study using multiomics approaches—cont'd

Plant species	Key PSMs	Multiomics	Analytical platform	Analytical approaches	Reference
Populus trichocarpa	Caffeoylquinic acid	Transcriptomic, metabolomic	Illumina Hi-Seq, GC-MS	Targeted metabolites quantification, mGWAS analysis, expression-level analysis, eQTL analysis, coexpression analysis, stress-induced expression pattern analysis, protein structure modeling and structure analysis	Zhang et al. (2018)
Salvia miltiorrhiza	Salvianolic acid	Metabolomic, proteomic, transcriptomic	UPLC-MS/MS (Metabolomic), HPLC-LC-MS/MS system (Proteomic, HPLC for fractionation, LC-MS/MS for analysis), Sequencing company	Comparative metabolites analysis, iTRAQ analysis, metabolomic-proteomic correlation analysis, expression-level analysis, gene overexpressing and silencing analysis, comparative transcriptome analysis	Yin et al. (2020)
Andrographis paniculata	Neoandrographolide	Genomic, transcriptomic	Illumina, PacBio, Hi-C, UPLC-MS/MS, GC-MS, *E. coli* expression system	Genome sequencing, assembly and annotation, RNA-Seq, phylogenetic and evolution analysis, targeted metabolites analysis, in vitro functional identification, phytohormone-induced expression-level analysis	Sun et al. (2019)

metabolic pathways in the apple tribe. To determine the difference in triterpene acid accumulation in loquat and related species, the research team systematically analyzed the differences in the composition and total concentration of triterpene acids in the leaves of peach, bowman's root, loquat, and apple at different developmental stages. As indicated by targeted metabolomic analysis, oleanolic acid (OA) and UA were commonly detected in the metabolite profiles of leaves from all species tested. Interestingly, the total triterpene acid content in loquat leaves was ten times higher than that in other species, and predominantly consisted of ursane-type triterpene acids. In addition, UA largely accumulated in the middle development stage, while CA (triterpene acid derived from UA) was synthesized in the mature stage.

Transcriptomic analysis. To identify the key biosynthetic enzymes of ursane-type triterpene acid, sequences for oxidosqualene cyclases (OSCs), which catalyze the first committed step of triterpene biosynthesis in plants, were systematically mined in the genomes of loquat, pear, and apple (Maleae), as well as peach and bowman's root (outgroup species). Using tissue-specific gene expression pattern analysis, two OSCs, named *EjOSC1* and *EJOSC2*, were identified as AA and BA (UA and OA precursors, respectively) synthases because of their close relationship with OSCs from apple, which had been previously characterized as having the same function. To identify other downstream genes, WGCNA was used, and genes highly correlated with *EjOSC1* and *EJOSC2* were identified. Combined with phylogenetic analysis, four CYP716 genes were identified as triterpene-modifying P450 genes and may participate in the biosynthesis of ursane-type triterpene acids in loquat.

Characterization of synthetic biology. Though the function of a certain gene can be speculated by advanced omics methods, the sequence or structure similarity of genes does not equal the function similarity to a great extent. Therefore, it is believed that gene function characterization in vitro, when research is performed outside of a living organism (in microorganism, insect, or mammalian cell lines or plant platforms) or in vivo, which refers to research done with or within an entire living organism (gene overexpression, gene silence, or gene knock out) is the most direct and reliable way to decipher a gene's function, serving as the complement and validation to the former omics analysis. The key genes mentioned above were transiently expressed in tobacco to confirm the formation route of major ursane-type triterpene acids in loquat. The experiment indicated the existence of at least two complete ursane-type triterpene acid synthesis pathways in loquat: the EjOSC1-EjCYP716A1-EjCYP716C1 pathway and the EjOSC2-EjCYP716A2-EjCYP716C2 pathway.

Conclusion. By integrating multiple-omics approaches, such as genomics, metabolomics, and transcriptomics, two possible ursane-type triterpene acid biosynthesis pathways in loquat were delineated and characterized. Furthermore, the analysis indicated that GWD events might play a major role in the expansion of metabolic pathways, thereby improving the biosynthesis rate of major bioactive compounds. The study provides a genomic foundation for the medicinal properties of loquat, promotes comparative genomics and evolutionary biology in Rosaceae, supports loquat molecular breeding, and contributes to the use of synthetic biology to develop highly valuable natural small molecules.

Discovery of physiological effects and enzyme evolution for flavonoids in rice (Peng et al., 2017)

Background and aim. Flavonoids, a type of phenolic compound widely distributed in the Plant Kingdom, are well known on account of their various bioactivities in humans and physiological effects in plants. They are usually modified by transferases, such as UDP-dependent glycosyltransferases (UGTs), a large group of enzymes having the ability to transfer glycosyl and "stick" the glycosyl to various acceptors in a different position. Various UGT types accept different flavonoids as substrates to form large numbers of glycosylated flavonoids with different decoration positions and moieties, but the roles of these PSMs remain unknown. Consequently, natural genetic variation of flavones in rice (*O. sativa*) and the biological function of glycosylated flavones in plants were studied using the multiomics approach.

Metabolomic analysis. Flavonoid compounds from 14 plant species spanning the different plant clades were detected and quantified using widely targeted LC-MS. Overall, 85 modified flavonoids were detected and clustered. The highest amounts of most of the flavonoids were accumulated in rice, a model monocot and an important crop for humans, which was then selected for further study.

Genomics analysis. mGWAS analysis was performed to investigate the genetic variation of the major flavonoids in rice. The association analysis revealed four major loci, and 21 putative UGT-encoding genes were identified as candidates responsible for the PSM variety. To determine the function of these UGTs, a phylogenetic tree of 29 previously reported flavonoid UGTs and 21 candidates was constructed to show the evolutionary interrelations of these UGTs. Eighteen putative UGTs formed a distinct clade, which had not been previously characterized, indicating novel functions of the identified candidates.

Synthetic biology and transgenic characterization. To conduct functional characterization in vitro, the candidate genes were expressed in *Escherichia coli*, and enzyme activity was assayed. Among these candidates, nine UGTs exhibited specific activity with flavonol aglycones, while OsUGT706D1 and OsUGT707A2 were identified and characterized as the two major flavone UGTs responsible for glycosylation of the 7-OH and 5-OH groups on rice flavones, respectively. To perform functional characterization in vivo, OsUGT706D1 and OsUGT707A2 were independently overexpressed. Both UGT transgenic lines utilized only flavones, with a dramatically increased flavone 5-*O*-glucoside accumulation, suggesting that OsUGT707A2 (termed flavone

5-*O*-glucosyltransferase, F5GlcT) and OsUGT706D1 (termed flavone 7-*O*--glucosyltransferase, F7GlcT) are the major enzymes contributing to the natural variation of flavones in rice. Further genetic and transcription analyses suggested that the levels of flavone 5-*O*-glucosides and flavone 7-*O*-glucosides in different rice varieties are controlled by several key allelic variations, affecting the transcription and enzyme activity levels.

Functional characterization in the laboratory. Flavanols have long been reported to be involved in UV protection in many plant species. Hence, UV-B resistance of transgenic rice plants was tested by direct exposure to UV-B light. Compared with the wild type, the transgenic plants survived, with the leaves remaining green after the exposure, while the wild-type plants became dramatically withered. This indicated that the UGT transgenic plants, which accumulated more flavone glycosides than the wild type, exhibited improved UV endurance.

Function characterization under natural conditions. To confirm whether flavone glycosides play a similar role under natural conditions, the geographical distribution of plants with different flavone glycoside-producing genotypes that accumulate different levels of flavone glycosides was related to the UV-B radiation level based on the latitude data. The analysis suggested that the abundance of flavone *O*-glucosides might be related to the adaptation to UV-B irradiation, indicating the biological function of these glycosylated flavonoids in rice.

Conclusion. Taking the variation in the metabolic profiles of different types of plants as the starting point of analysis, key variation loci were identified using the metabolomics-genomics-integrated approach mGWAS, and the related UGTs were characterized both in vitro and in vivo. The GWAS analysis allowed the identification of trait-related variation loci in the whole genome, accelerating the study progress by generating high-resolution data. mGWAS is more accurate for correlation analyses than the traditional GWAS, especially for studying traits related to flavor, taste, secondary metabolites, etc., and has become a popular approach in PSM studies (Dong et al., 2015; Chen et al., 2016b).

Conclusion and future prospects

In this chapter, we summarized the diverse PSMs, their biosynthetic pathways, “omics” -based approaches used to delineate these biosynthetic pathways, and the integrative approach of using multiomics data to derive the related mechanistic details. The enormous number of -omics datasets generated in the last decade poses a considerable challenge to plant biologists in terms of data storage, management, and comparison among multiple platforms and/or laboratories. To tackle these challenges, computational efficiency should be increased, and a universally accepted international standard for data deposition in databases should be devised. Quantitative-omics data must be generated instead of qualitative or semiquantitative data. This will allow for a consistent and accurate integration of multiomics data, where the measurements are typically performed using multiple platforms, across multiple laboratories, and over a long period. Future multiomics data integration algorithms should make use of the "big data" resources, and take advantage of the emergence of machine learning and artificial intelligence algorithms. Further, based on the development of multiomics database and quantitative -omics data, a Plant Cell Atlas (PCA), which serves as a high-resolution community resource to comprehensively and precisely display the state of various types of plant cells supplying with multidimensional information in a cellular, subcellular, or even molecular scale, becomes possible and essential (Rhee et al., 2019). Lastly, significant findings should be validated using traditional wet laboratory techniques to confirm the results and avoid false positives.

References

Abe, A., et al., 2012. Genome sequencing reveals agronomically important loci in rice using MutMap. Nat. Biotechnol. 30 (2), 174–178. https://doi.org/10.1038/nbt.2095.

Achard, P., Herr, A., Baulcombe, D.C., Harberd, N.P., 2004. Modulation of floral development by a gibberellin-regulated microRNA. Development 131 (14), 3357–3365. https://doi.org/10.1242/dev.01206.

Ajikumar Parayil, K., Xiao, W.-H., Tyo Keith, E.J., Wang, Y., Simeon, F., Leonard, E., Mucha, O., Phon Too, H., Pfeifer, B., Stephanopoulos, G., 2010. Isoprenoid pathway optimization for Taxol precursor overproduction in *Escherichia coli*. Science 330 (6000), 70–74. https://doi.org/10.1126/science.1191652.

Akula, R., Ravishankar, G.A., 2011. Influence of abiotic stress signals on secondary metabolites in plants. Plant Signal. Behav. 6 (11), 1720–1731. https://doi.org/10.4161/psb.6.11.17613.

Alzagameem, A., El Khaldi-Hansen, B., Kamm, B., Schulze, M., 2018. Lignocellulosic biomass for energy, biofuels, biomaterials, and chemicals. In: Vaz Jr., S. (Ed.), Biomass and Green Chemistry: Building a Renewable Pathway. Springer International Publishing, Cham, ISBN: 978-3-319-66736-2, pp. 95–132.

Atwell, S., et al., 2010. Genome-wide association study of 107 phenotypes in *Arabidopsis thaliana* inbred lines. Nature 465 (7298), 627–631. https://doi.org/10.1038/nature08800.

Bachmann, B.O., Van Lanen, S.G., Baltz, R.H., 2014. Microbial genome mining for accelerated natural products discovery: is a renaissance in the making? J. Ind. Microbiol. Biotechnol. 41 (2), 175–184. https://doi.org/10.1007/s10295-013-1389-9.

Band, H.S., 1992. Mechanism of action of taxol. Trends Pharmacol. Sci. 13, 134–136. https://doi.org/10.1016/0165-6147(92)90048-B.

Bandi, V., Gutwin, C., Siri, J.N., Neufeld, E., Sharpe, A., Parkin, I., 2022. Visualization tools for genomic conservation. Methods Mol. Biol. 2443, 285–308. https://doi.org/10.1007/978-1-0716-2067-0_16.

Behera, S., Voshall, A., Moriyama, E., 2021. Plant Transcriptome Assembly: Review and Benchmarking. Exon Publications, pp. 109–130.

Bennett, B.D., Kimball, E.H., Gao, M., Osterhout, R., Van Dien, S.J., Rabinowitz, J.D., 2009. Absolute metabolite concentrations and implied enzyme active site occupancy in *Escherichia coli*. Nat. Chem. Biol. 5 (8), 593–599. https://doi.org/10.1038/nchembio.186.

Bergelson, J., Roux, F., 2010. Towards identifying genes underlying ecologically relevant traits in *Arabidopsis thaliana*. Nat. Rev. Genet. 11 (12), 867–879. https://doi.org/10.1038/nrg2896.

Binita, K., Kumar, S., Sharma, V.K., Sharma, V., Yadav, S., 2014. Proteomic identification of *Syzygium cumini* seed extracts by MALDI-TOF/MS. Appl. Biochem. Biotechnol. 172 (4), 2091–2105. https://doi.org/10.1007/s12010-013-0660-x.

Bohlmann, J., Keeling, C.I., 2008. Terpenoid biomaterials. Plant J. 54 (4), 656–669. https://doi.org/10.1111/j.1365-313X.2008.03449.x.

Bouhaddani, S., Uh, H.-W., Jongbloed, G., Hayward, C., Klarić, L., Kiełbasa, S.M., Houwing-Duistermaat, J., 2018. Integrating omics datasets with the OmicsPLS package. BMC Bioinf. 19 (1), 371. https://doi.org/10.1186/s12859-018-2371-3.

Boutanaev, A.M., Moses, T., Zi, J., Nelson, D.R., Mugford, S.T., Peters, R.J., Osbourn, A., 2015. Investigation of terpene diversification across multiple sequenced plant genomes. Proc. Natl. Acad. Sci. U. S. A. 112 (1), E81. https://doi.org/10.1073/pnas.1419547112.

Brachi, B., Morris, G.P., Borevitz, J.O., 2011. Genome-wide association studies in plants: the missing heritability is in the field. Genome Biol. 12 (10), 232. https://doi.org/10.1186/gb-2011-12-10-232.

Brook, K., Bennett, J., Desai, S.P., 2017. The chemical history of morphine: an 8000-year journey, from resin to de-novo synthesis. J. Anesth. Hist. 3 (2), 50–55.

Brown, A.V., Conners, S.I., Huang, W., Wilkey, A.P., Grant, D., Weeks, N.T., Cannon, S.B., Graham, M.A., Nelson, R.T., 2021. A new decade and new data at SoyBase, the USDA-ARS soybean genetics and genomics database. Nucleic Acids Res. 49 (D1), D1496–D1501. https://doi.org/10.1093/nar/gkaa1107.

Bylesjö, M., Eriksson, D., Kusano, M., Moritz, T., Trygg, J., 2007. Data integration in plant biology: the O2PLS method for combined modeling of transcript and metabolite data. Plant J. 52 (6), 1181–1191. https://doi.org/10.1111/j.1365-313X.2007.03293.x.

Cai, S., et al., 2021. CannabisGDB: a comprehensive genomic database for *Cannabis sativa* L. Plant Biotechnol. J. 19 (5), 857–859. https://doi.org/10.1111/pbi.13548.

Cao, Y., Li, K., Li, Y., Zhao, X., Wang, L., 2020. MYB transcription factors as regulators of secondary metabolism in plants. Biology 9 (3). https://doi.org/10.3390/biology9030061.

Carpentier, M.-C., Manfroi, E., Wei, F.-J., Wu, H.-P., Lasserre, E., Llauro, C., Debladis, E., Akakpo, R., Hsing, Y.-I., Panaud, O., 2019. Retrotranspositional landscape of Asian rice revealed by 3000 genomes. Nat. Commun. 10 (1), 24. https://doi.org/10.1038/s41467-018-07974-5.

Caruso, G., Cavaliere, C., Foglia, P., Gubbiotti, R., Samperi, R., Laganà, A., 2009. Analysis of drought responsive proteins in wheat (Triticum durum) by 2D-PAGE and MALDI-TOF mass spectrometry. Plant Sci. 177 (6), 570–576. https://doi.org/10.1016/j.plantsci.2009.08.007.

Chen, J., Wang, J., Wang, R., Xian, B., Ren, C., Liu, Q., Wu, Q., Pei, J., 2020. Integrated metabolomics and transcriptome analysis on flavonoid biosynthesis in safflower (*Carthamus tinctorius* L.) under MeJA treatment. BMC Plant Biol. 20 (1), 353. https://doi.org/10.1186/s12870-020-02554-6.

Chen, J., Xie, J., Jiang, Z., Wang, B., Wang, Y., Hu, X., 2011. Shikonin and its analogs inhibit cancer cell glycolysis by targeting tumor pyruvate kinase-M2. Oncogene 30 (42), 4297–4306. https://doi.org/10.1038/onc.2011.137.

Chen, J., Quan, M., Zhang, D., 2015. Genome-wide identification of novel long non-coding RNAs in Populus tomentosa tension wood, opposite wood and normal wood xylem by RNA-seq. Planta 241 (1), 125–143. https://doi.org/10.1007/s00425-014-2168-1.

Chen, L., Huang, Y., Xu, M., Cheng, Z., Zhang, D., Zheng, J., 2016a. iTRAQ-based quantitative proteomics analysis of black rice grain development reveals metabolic pathways associated with anthocyanin biosynthesis. PLoS One 11 (7), e0159238.

Chen, W., et al., 2016b. Comparative and parallel genome-wide association studies for metabolic and agronomic traits in cereals. Nat. Commun. 7 (1), 12767. https://doi.org/10.1038/ncomms12767.

Chiang, C.-P., Yim, W.C., Sun, Y.-H., Ohnishi, M., Mimura, T., Cushman, J.C., Yen, H.E., 2016. Identification of ice plant (*Mesembryanthemum crystallinum* L.) MicroRNAs using RNA-Seq and their putative roles in high salinity responses in seedlings. Front. Plant Sci. 7 (1143). https://doi.org/10.3389/fpls.2016.01143.

Cragg, G.M., Newman, D.J., 2005. Plants as a source of anti-cancer agents. J. Ethnopharmacol. 100 (1), 72–79. https://doi.org/10.1016/j.jep.2005.05.011.

Czerniawski, P., Bednarek, P., 2018. Glutathione S-transferases in the biosynthesis of sulfur-containing secondary metabolites in Brassicaceae plants. Front. Plant Sci. 9, 1639.

da Silva, M.A.O., Arruda, M.A.Z., 2012. Identification of selenium in the leaf protein of sunflowers by a combination of 2D-PAGE and laser ablation ICP-MS. Microchim. Acta 176 (1–2), 131–136.

de Vries, P.J., Dien, T.K., 1996. Clinical pharmacology and therapeutic potential of artemisinin and its derivatives in the treatment of malaria. Drugs 52 (6), 818–836. https://doi.org/10.2165/00003495-199652060-00004.

Denoeud, F., et al., 2014. The coffee genome provides insight into the convergent evolution of caffeine biosynthesis. Science 345 (6201), 1181. https://doi.org/10.1126/science.1255274.

Dong, X., Gao, Y., Chen, W., Wang, W., Gong, L., Liu, X., Luo, J., 2015. Spatiotemporal distribution of phenolamides and the genetics of natural variation of hydroxycinnamoyl spermidine in rice. Mol. Plant 8 (1), 111–121. https://doi.org/10.1016/j.molp.2014.11.003.

Du, G.-J., Zhang, Z., Wen, X.-D., Yu, C., Calway, T., Yuan, C.-S., Wang, C.-Z., 2012. Epigallocatechin Gallate (EGCG) is the most effective cancer chemopreventive polyphenol in green tea. Nutrients 4 (11), 1679–1691.

Erb, M., Kliebenstein, D.J., 2020. Plant secondary metabolites as defenses, regulators, and primary metabolites: the blurred functional trichotomy. Plant Physiol. 184 (1), 39–52. https://doi.org/10.1104/pp.20.00433.

Falara, V., et al., 2011. The tomato terpene synthase gene family. Plant Physiol. 157 (2), 770–789. https://doi.org/10.1104/pp.111.179648.

Fazio, G.C., Xu, R., Matsuda, S.P.T., 2004. Genome mining to identify new plant triterpenoids. J. Am. Chem. Soc. 126 (18), 5678–5679. https://doi.org/10.1021/ja0318784.

Fellermeier, M., Eisenreich, W., Bacher, A., Zenk, M.H., 2001. Biosynthesis of cannabinoids: incorporation experiments with 13C-labeled glucoses. Eur. J. Biochem. 268 (6), 1596–1604.

Ferber, S.G., Namdar, D., Hen-Shoval, D., Eger, G., Koltai, H., Shoval, G., Shbiro, L., Weller, A., 2020. The "entourage effect": terpenes coupled with cannabinoids for the treatment of mood disorders and anxiety disorders. Curr. Neuropharmacol. 18 (2), 87–96.

Fernandez-Pozo, N., Menda, N., Edwards, J.D., Saha, S., Tecle, I.Y., Strickler, S.R., Bombarely, A., Fisher-York, T., Pujar, A., Foerster, H., Yan, A., Mueller, L.A., 2015. The Sol Genomics Network (SGN)—from genotype to phenotype to breeding. Nucleic Acids Res. 43 (D1), D1036–D1041. https://doi.org/10.1093/nar/gku1195.

Fernie, A.R., Pichersky, E., 2015. Focus issue on metabolism: metabolites, metabolites everywhere. Plant Physiol. 169 (3), 1421–1423. https://doi.org/10.1104/pp.15.01499.

Field, B., Osbourn, A.E., 2008. Metabolic diversification—independent assembly of operon-like gene clusters in different plants. Science 320 (5875), 543. https://doi.org/10.1126/science.1154990.

Filipowicz, W., Bhattacharyya, S.N., Sonenberg, N., 2008. Mechanisms of post-transcriptional regulation by microRNAs: are the answers in sight? Nat. Rev. Genet. 9 (2), 102–114. https://doi.org/10.1038/nrg2290.

Forastiere, A.A., Neuberg, D., Taylor, S.G., DeConti, R., Adams, G., 1993. Phase II evaluation of taxol in advanced head and neck cancer: an eastern cooperative oncology group trial. J. Natl. Cancer Inst. Monogr. (15), 181–184.

Foyer, C.H., Theodoulou, F.L., Delrot, S., 2001. The functions of inter- and intracellular glutathione transport systems in plants. Trends Plant Sci. 6 (10), 486–492. https://doi.org/10.1016/s1360-1385(01)02086-6.

Frazier, R.A., Deaville, E.R., Green, R.J., Stringano, E., Willoughby, I., Plant, J., Mueller-Harvey, I., 2010. Interactions of tea tannins and condensed tannins with proteins. J. Pharm. Biomed. Anal. 51 (2), 490–495.

Fujii, H., Verslues, P.E., Zhu, J.-K., 2007. Identification of two protein kinases required for abscisic acid regulation of seed germination, root growth, and gene expression in Arabidopsis. Plant Cell 19 (2), 485–494.

Gallego, A., Malik, S., Yousefzadi, M., Makhzoum, A., Tremouillaux-Guiller, J., Bonfill, M., 2017. Taxol from *Corylus avellana*: paving the way for a new source of this anti-cancer drug. Plant Cell Tissue Organ Cult. 129 (1), 1–16.

Gao, J., et al., 2010. Metscape: a cytoscape plug-in for visualizing and interpreting metabolomic data in the context of human metabolic networks. Bioinformatics 26 (7), 971–973. https://doi.org/10.1093/bioinformatics/btq048.

Gaudinier, A., et al., 2018. Transcriptional regulation of nitrogen-associated metabolism and growth. Nature 563 (7730), 259–264. https://doi.org/10.1038/s41586-018-0656-3.

Guéritte, F., Fahy, J., 2005. The vinca alkaloids. In: Anticancer Agents from Natural Products., https://doi.org/10.1201/b11185-8.

Gui, S., et al., 2020. ZEAMAP, a comprehensive database adapted to the maize multi-omics era. iScience 23 (6), 101241. https://doi.org/10.1016/j.isci.2020.101241.

Hartmann, T., 2007. From waste products to ecochemicals: fifty years research of plant secondary metabolism. Phytochemistry 68 (22–24), 2831–2846. https://doi.org/10.1016/j.phytochem.2007.09.017.

Haug, K., et al., 2013. MetaboLights—an open-access general-purpose repository for metabolomics studies and associated meta-data. Nucleic Acids Res. 41 (D1), D781–D786. https://doi.org/10.1093/nar/gks1004.

Haug, K., Cochrane, K., Nainala, V.C., Williams, M., Chang, J., Jayaseelan, K.V., O'Donovan, C., 2020. MetaboLights: a resource evolving in response to the needs of its scientific community. Nucleic Acids Res. 48 (D1), D440–D444. https://doi.org/10.1093/nar/gkz1019.

He, X.Y., Wu, L.J., Wang, W.X., Xie, P.J., Chen, Y.H., Wang, F., 2020. Amygdalin—a pharmacological and toxicological review. J. Ethnopharmacol. 254, 112717. https://doi.org/10.1016/j.jep.2020.112717.

Holmes, F.A., Walters, R.S., Theriault, R.L., Buzdar, A.U., Frye, D.K., Hortobagyi, G.N., Forman, A.D., Newton, L.K., Raber, M.N., 1991. Phase II trial of taxol, an active drug in the treatment of metastatic breast cancer. J. Natl. Cancer Inst. 83 (24), 1797–1805. https://doi.org/10.1093/jnci/83.24.1797.

Hong, Y., Li, M., Dai, S., 2019. iTRAQ-based protein profiling provides insights into the mechanism of light-induced anthocyanin biosynthesis in Chrysanthemum (Chrysanthemum × morifolium). Genes 10 (12). https://doi.org/10.3390/genes10121024.

Huang, F.-C., Kutchan, T.M., 2000. Distribution of morphinan and benzo[c]phenanthridine alkaloid gene transcript accumulation in *Papaver somniferum*. Phytochemistry 53 (5), 555–564. https://doi.org/10.1016/S0031-9422(99)00600-7.

Huang, X., et al., 2010. Genome-wide association studies of 14 agronomic traits in rice landraces. Nat. Genet. 42 (11), 961–967. https://doi.org/10.1038/ng.695.

Huang, M., Sanchez-Moreiras, A.M., Abel, C., Sohrabi, R., Lee, S., Gershenzon, J., Tholl, D., 2012. The major volatile organic compound emitted from Arabidopsis thaliana flowers, the sesquiterpene (E)-beta-caryophyllene, is a defense against a bacterial pathogen. New Phytol. 193 (4), 997–1008. https://doi.org/10.1111/j.1469-8137.2011.04001.x.

Ingvarsson, P.K., Street, N.R., 2011. Association genetics of complex traits in plants. New Phytol. 189 (4), 909–922. https://doi.org/10.1111/j.1469-8137.2010.03593.x.

Itkin, M., et al., 2013. Biosynthesis of antinutritional alkaloids in solanaceous crops is mediated by clustered genes. Science 341 (6142), 175. https://doi.org/10.1126/science.1240230.

Jacobs, D.I., Gaspari, M., van der Greef, J., van der Heijden, R., Verpoorte, R., 2005. Proteome analysis of the medicinal plant *Catharanthus roseus*. Planta 221 (5), 690–704. https://doi.org/10.1007/s00425-004-1474-4.

Jaqueth, J.S., Hou, Z., Zheng, P., Ren, R., Nagel, B.A., Cutter, G., Niu, X., Vollbrecht, E., Greene, T. W., Kumpatla, S.P., 2020. Fertility restoration of maize CMS-C altered by a single amino acid substitution within the Rf4 bHLH transcription factor. Plant J. 101 (1), 101–111. https://doi.org/10.1111/tpj.14521.

Jin, D., Dai, K., Xie, Z., Chen, J., 2020. Secondary metabolites profiled in Cannabis inflorescences, leaves, stem barks, and roots for medicinal purposes. Sci. Rep. 10 (1), 3309. https://doi.org/10.1038/s41598-020-60172-6.

Jones-Rhoades, M.W., Bartel, D.P., 2004. Computational identification of plant microRNAs and their targets, including a stress-induced miRNA. Mol. Cell 14 (6), 787–799. https://doi.org/10.1016/j.molcel.2004.05.027.

Kaever, A., Landesfeind, M., Feussner, K., Mosblech, A., Heilmann, I., Morgenstern, B., Feussner, I., Meinicke, P., 2015. MarVis-Pathway: integrative and exploratory pathway analysis of non-targeted metabolomics data. Metabolomics 11 (3), 764–777. https://doi.org/10.1007/s11306-014-0734-y.

Kaever, A., Lingner, T., Feussner, K., Göbel, C., Feussner, I., Meinicke, P., 2009. MarVis: a tool for clustering and visualization of metabolic biomarkers. BMC Bioinf. 10 (1), 92. https://doi.org/10.1186/1471-2105-10-92.

Kang, M., Fu, R., Zhang, P., Lou, S., Yang, X., Chen, Y., Ma, T., Zhang, Y., Xi, Z., Liu, J., 2021. A chromosome-level *Camptotheca acuminata* genome assembly provides insights into the evolutionary origin of camptothecin biosynthesis. Nat. Commun. 12 (1), 3531. https://doi.org/10.1038/s41467-021-23872-9.

Kang, S.-H., Pandey, R.P., Lee, C.-M., Sim, J.-S., Jeong, J.-T., Choi, B.-S., Jung, M., Ginzburg, D., Zhao, K., Won, S.Y., Oh, T.-J., Yu, Y., Kim, N.-H., Lee, O.R., Lee, T.-H., Bashyal, P., Kim, T.-S., Lee, W.-H., Hawkins, C., Kim, C.-K., Kim, J.S., Ahn, B.O., Rhee, S.Y., Sohng, J.K., 2020. Genome-enabled discovery of anthraquinone biosynthesis in *Senna tora*. Nat. Commun. 11 (1), 5875. https://doi.org/10.1038/s41467-020-19681-1.

Karnovsky, A., et al., 2012. Metscape 2 bioinformatics tool for the analysis and visualization of metabolomics and gene expression data. Bioinformatics 28 (3), 373–380. https://doi.org/10.1093/bioinformatics/btr661.

Kasetti, R.B., Nabi, S.A., Swapna, S., Apparao, C., 2012. Cinnamic acid as one of the antidiabetic active principle(s) from the seeds of Syzygium alternifolium. Food Chem. Toxicol. 50 (5), 1425–1431. https://doi.org/10.1016/j.fct.2012.02.003.

Kersten, R.D., Weng, J.-K., 2018. Gene-guided discovery and engineering of branched cyclic peptides in plants. Proc. Natl. Acad. Sci. U. S. A. 115 (46), E10961. https://doi.org/10.1073/pnas.1813993115.

Kuang, X., Sun, S., Wei, J., Li, Y., Sun, C., 2019. Iso-Seq analysis of the *Taxus cuspidata* transcriptome reveals the complexity of taxol biosynthesis. BMC Plant Biol. 19 (1), 210. https://doi.org/10.1186/s12870-019-1809-8.

Kuo, T.-C., Tian, T.-F., Tseng, Y.J., 2013. 3Omics: a web-based systems biology tool for analysis, integration and visualization of human transcriptomic, proteomic and metabolomic data. BMC Syst. Biol. 7 (1), 64. https://doi.org/10.1186/1752-0509-7-64.

Kutmon, M., van Iersel, M.P., Bohler, A., Kelder, T., Nunes, N., Pico, A.R., Evelo, C.T., 2015. PathVisio 3: an extendable pathway analysis toolbox. PLoS Comput. Biol. 11 (2), e1004085. https://doi.org/10.1371/journal.pcbi.1004085.

Lan, Y., Sun, R., Ouyang, J., Ding, W., Kim, M.-J., Wu, J., Li, Y., Shi, T., 2021. AtMAD: Arabidopsis thaliana multi-omics association database. Nucleic Acids Res. 49 (D1), D1445–D1451. https://doi.org/10.1093/nar/gkaa1042.

Langfelder, P., Horvath, S., 2008. WGCNA: an R package for weighted correlation network analysis. BMC Bioinf. 9 (1), 559. https://doi.org/10.1186/1471-2105-9-559.

Legault, J., Pichette, A., 2007. Potentiating effect of β-caryophyllene on anticancer activity of α-humulene, isocaryophyllene and paclitaxel. J. Pharm. Pharmacol. 59 (12), 1643–1647. https://doi.org/10.1211/jpp.59.12.0005.

Lin, D., et al., 2016. An overview of plant phenolic compounds and their importance in human nutrition and management of type 2 diabetes. Molecules 21 (10). https://doi.org/10.3390/molecules21101374.

Lindsey, K., Pullen, M.L., Topping, J.F., 2003. Importance of plant sterols in pattern formation and hormone signalling. Trends Plant Sci. 8 (11), 521–525. https://doi.org/10.1016/j.tplants.2003.09.012.

Liu, B., Yan, W., 2019. Lipophilization of EGCG and effects on antioxidant activities. Food Chem. 272, 663–669. https://doi.org/10.1016/j.foodchem.2018.08.086.

Lu, H., Rate, D.N., Song, J.T., Greenberg, J.T., 2003. ACD6, a novel ankyrin protein, is a regulator and an effector of salicylic acid signaling in the Arabidopsis defense response. Plant Cell 15 (10), 2408–2420. https://doi.org/10.1105/tpc.015412.

Luo, J., 2015. Metabolite-based genome-wide association studies in plants. Curr. Opin. Plant Biol. 24, 31–38.

Mannan, A., Ahmed, I., Arshad, W., Asim, M.F., Qureshi, R.A., Hussain, I., Mirza, B., 2010. Survey of artemisinin production by diverse Artemisia species in northern Pakistan. Malar. J. 9 (1), 310. https://doi.org/10.1186/1475-2875-9-310.

Mansueto, L., et al., 2017. Rice SNP-seek database update: new SNPs, indels, and queries. Nucleic Acids Res. 45 (D1), D1075–D1081. https://doi.org/10.1093/nar/gkw1135.

Martínez-Esteso, M.J., Vilella-Antón, M.T., Pedreño, M.Á., Valero, M.L., Bru-Martínez, R., 2013. iTRAQ-based protein profiling provides insights into the central metabolism changes driving grape berry development and ripening. BMC Plant Biol. 13 (1), 167. https://doi.org/10.1186/1471-2229-13-167.

Matsuba, Y., et al., 2013. Evolution of a complex locus for terpene biosynthesis in solanum. Plant Cell 25 (6), 2022–2036. https://doi.org/10.1105/tpc.113.111013.

Miura, K., Ashikari, M., Matsuoka, M., 2011. The role of QTLs in the breeding of high-yielding rice. Trends Plant Sci. 16 (6), 319–326.

Mueller, L.A., et al., 2005. The SOL genomics network. A comparative resource for Solanaceae biology and beyond. Plant Physiol. 138 (3), 1310–1317. https://doi.org/10.1104/pp.105.060707.

Mugford, S.T., et al., 2013. Modularity of plant metabolic gene clusters: a trio of linked genes that are collectively required for acylation of triterpenes in oat. Plant Cell 25 (3), 1078–1092. https://doi.org/10.1105/tpc.113.110551.

Munné-Bosch, S., Mueller, M., Schwarz, K., Alegre, L., 2001. Diterpenes and antioxidative protection in drought-stressed *Salvia officinalis* plants. J. Plant Physiol. 158 (11), 1431–1437. https://doi.org/10.1078/0176-1617-00578.

Murphy, W.K., et al., 1993. Phase II study of taxol in patients with untreated advanced non-small-cell lung cancer. J. Natl. Cancer Inst. 85 (5), 384–388. https://doi.org/10.1093/jnci/85.5.384.

Nagasaki, M., Saito, A., Jeong, E., Li, C., Kojima, K., Ikeda, E., Miyano, S., 2010. Cell illustrator 4.0: a computational platform for systems biology. In Silico Biol. 10, 5–26. https://doi.org/10.3233/ISB-2010-0415.

Nahler, G., Jones, T., Russo, E., 2019. Cannabidiol and contributions of major hemp phytocompounds to the "entourage effect"; possible mechanisms. J. Altern. Complement. Integr. Med. 5.

Ni, X., Jin, C., Liu, A., Chen, Y., Hu, Y., 2021. Physiological and transcriptomic analyses to reveal underlying phenolic acid action in consecutive monoculture problem of Polygonatum odoratum. BMC Plant Biol. 21 (1), 362. https://doi.org/10.1186/s12870-021-03135-x.

Nicholson, R.L., Hammerschmidt, R., 1992. Phenolic compounds and their role in disease resistance. Annu. Rev. Phytopathol. 30 (1), 369–389.

Niro, E., Marzaioli, R., De Crescenzo, S., D'Abrosca, B., Castaldi, S., Esposito, A., Fiorentino, A., Rutigliano, F.A., 2016. Effects of the allelochemical coumarin on plants and soil microbial community. Soil Biol. Biochem. 95, 30–39. https://doi.org/10.1016/j.soilbio.2015.11.028.

Noda, Y., Kaneyuki, T., Mori, A., Packer, L., 2002. Antioxidant activities of pomegranate fruit extract and its Anthocyanidins: Delphinidin, Cyanidin, and Pelargonidin. J. Agric. Food Chem. 50 (1), 166–171. https://doi.org/10.1021/jf0108765.

O'Neill, P.M., Barton, V.E., Ward, S.A., 2010. The molecular mechanism of action of artemisinin—the debate continues. Molecules 15 (3). https://doi.org/10.3390/molecules15031705.

Oksman-Caldentey, K.M., Saito, K., 2005. Integrating genomics and metabolomics for engineering plant metabolic pathways. Curr. Opin. Biotechnol. 16 (2), 174–179. https://doi.org/10.1016/j.copbio.2005.02.007.

Palatnik, J.F., Allen, E., Wu, X., Schommer, C., Schwab, R., Carrington, J.C., Weigel, D., 2003. Control of leaf morphogenesis by microRNAs. Nature 425 (6955), 257–263. https://doi.org/10.1038/nature01958.

Pasam, R.K., Sharma, R., Malosetti, M., van Eeuwijk, F.A., Haseneyer, G., Kilian, B., Graner, A., 2012. Genome-wide association studies for agronomical traits in a world wide spring barley collection. BMC Plant Biol. 12 (1), 16. https://doi.org/10.1186/1471-2229-12-16.

Patti, G.J., Yanes, O., Siuzdak, G., 2012. Metabolomics: the apogee of the omics trilogy. Nat. Rev. Mol. Cell Biol. 13 (4), 263–269. https://doi.org/10.1038/nrm3314.

Peng, M., et al., 2017. Differentially evolved glucosyltransferases determine natural variation of rice flavone accumulation and UV-tolerance. Nat. Commun. 8 (1), 1975. https://doi.org/10.1038/s41467-017-02168-x.

Portwood II, J.L., et al., 2019. MaizeGDB 2018: the maize multi-genome genetics and genomics database. Nucleic Acids Res. 47 (D1), D1146–D1154. https://doi.org/10.1093/nar/gky1046.

Prosser, G.A., Larrouy-Maumus, G., de Carvalho, L.P.S., 2014. Metabolomic strategies for the identification of new enzyme functions and metabolic pathways. EMBO Rep. 15 (6), 657–669. https://doi.org/10.15252/embr.201338283.

Qi, X., Bakht, S., Leggett, M., Maxwell, C., Melton, R., Osbourn, A., 2004. A gene cluster for secondary metabolism in oat: implications for the evolution of metabolic diversity in plants. Proc. Natl. Acad. Sci. U. S. A. 101 (21), 8233. https://doi.org/10.1073/pnas.0401301101.

Qi, X., et al., 2006. A different function for a member of an ancient and highly conserved cytochrome P450 family: from essential sterols to plant defense. Proc. Natl. Acad. Sci. U. S. A. 103 (49), 18848. https://doi.org/10.1073/pnas.0607849103.

Qiao, D., Yang, C., Chen, J., Guo, Y., Li, Y., Niu, S., Cao, K., Chen, Z., 2019. Comprehensive identification of the full-length transcripts and alternative splicing related to the secondary metabolism pathways in the tea plant (*Camellia sinensis*). Sci. Rep. 9 (1), 2709. https://doi.org/10.1038/s41598-019-39286-z.

Quan, N.T., et al., 2016. Involvement of secondary metabolites in response to drought stress of rice (*Oryza sativa* L.). Agriculture 6 (2). https://doi.org/10.3390/agriculture6020023.

Rabaçal, M., Ferreira, A.F., Silva, C.A., Costa, M., 2017. Biorefineries: Targeting Energy, High Value Products and Waste Valorisation. Springer.

Rai, A., Saito, K., Yamazaki, M., 2017. Integrated omics analysis of specialized metabolism in medicinal plants. Plant J. 90 (4), 764–787. https://doi.org/10.1111/tpj.13485.

Rai, A., Nakaya, T., Shimizu, Y., Rai, M., Nakamura, M., Suzuki, H., Saito, K., Yamazaki, M., 2018. De novo transcriptome assembly and characterization of *Lithospermum officinale* to discover putative genes involved in specialized metabolites biosynthesis. Planta Med. 84 (12/13), 920–934.

Rao, X., Dixon, R.A., 2019. Co-expression networks for plant biology: why and how. Acta Biochim. Biophys. Sin. 51 (10), 981–988. https://doi.org/10.1093/abbs/gmz080.

Rhee, S.Y., Birnbaum, K.D., Ehrhardt, D.W., 2019. Towards building a plant cell atlas. Trends Plant Sci. 24 (4), 303–310. https://doi.org/10.1016/j.tplants.2019.01.006.

Rocke, A.J., 1985. Hypothesis and experiment in the early development of Kekulé's benzene theory. Ann. Sci. 42 (4), 355–381. https://doi.org/10.1080/00033798500200411.

Rohart, F., Gautier, B., Singh, A., Lê Cao, K.-A., 2017. mixOmics: an R package for 'omics feature selection and multiple data integration. PLoS Comput. Biol. 13 (11), e1005752. https://doi.org/10.1371/journal.pcbi.1005752.

Schmid, M., Uhlenhaut, N.H., Fo, G., Demar, M., Bressan, R., Weigel, D., Lohmann, J.U., 2003. Dissection of floral induction pathways using global expression analysis. Development 130 (24), 6001–6012. https://doi.org/10.1242/dev.00842.

Shang, Y., et al., 2014. Biosynthesis, regulation, and domestication of bitterness in cucumber. Science 346 (6213), 1084. https://doi.org/10.1126/science.1259215.

Shannon, P., Markiel, A., Ozier, O., Baliga, N.S., Wang, J.T., Ramage, D., Amin, N., Schwikowski, B., Ideker, T., 2003. Cytoscape: a software environment for integrated models of biomolecular interaction networks. Genome Res. 13 (11), 2498–2504. https://doi.org/10.1101/gr.1239303.

Shoeb, M., 2006. Anti-cancer agents from medicinal plants. Bangladesh J. Pharmacol. 1 (2), 35–41. https://doi.org/10.3329/bjp.v1i2.486.

Siegel, R.L., Miller, K.D., Jemal, A., 2020. Cancer statistics, 2020. CA Cancer J. Clin. 70 (1), 7–30. https://doi.org/10.3322/caac.21590.

Smit, E.F., Fokkema, E., Biesma, B., Groen, H.J.M., Snoek, W., Postmus, P.E., 1998. A phase II study of paclitaxel in heavily pretreated patients with small-cell lung cancer. Br. J. Cancer 77 (2), 347–351. https://doi.org/10.1038/bjc.1998.54.

Song, X., Sun, L., Luo, H., Ma, Q., Zhao, Y., Pei, D., 2016. Genome-wide identification and characterization of long non-coding RNAs from mulberry (Morus notabilis) RNA-seq data. Genes 7 (3). https://doi.org/10.3390/genes7030011.

Sousa, D., 2012. Medicinal Essential Oils: Chemical, Pharmacological and Therapeutic Aspects. New York, pp. 1–236.

Sova, M., 2012. Antioxidant and antimicrobial activities of cinnamic acid derivatives. Mini-Rev. Med. Chem. 12 (8), 749–767.

Stintzing, F.C., Stintzing, A.S., Carle, R., Frei, B., Wrolstad, R.E., 2002. Color and antioxidant properties of Cyanidin-based anthocyanin pigments. J. Agric. Food Chem. 50 (21), 6172–6181. https://doi.org/10.1021/jf0204811.

Su, W., et al., 2021. Polyploidy underlies co-option and diversification of biosynthetic triterpene pathways in the apple tribe. Proc. Natl. Acad. Sci. U. S. A. 118 (20), e2101767118. https://doi.org/10.1073/pnas.2101767118.

Suhre, K., Schmitt-Kopplin, P., 2008. MassTRIX: mass translator into pathways. Nucleic Acids Res. 36 (suppl_2), W481–W484. https://doi.org/10.1093/nar/gkn194.

Sun, W., Leng, L., Yin, Q., Xu, M., Huang, M., Xu, Z., Zhang, Y., Yao, H., Wang, C., Xiong, C., Chen, S., Jiang, C., Xie, N., Zheng, X., Wang, Y., Song, C., Peters, R.J., Chen, S., 2019. The genome of the medicinal plant *Andrographis paniculata* provides insight into the biosynthesis of the bioactive diterpenoid neoandrographolide. Plant J. 97 (5), 841–857. https://doi.org/10.1111/tpj.14162.

Swaminathan, S., Morrone, D., Wang, Q., Fulton, D.B., Peters, R.J., 2009. CYP76M7 is an ent-cassadiene C11α-hydroxylase defining a second multifunctional diterpenoid biosynthetic gene cluster in rice. Plant Cell 21 (10), 3315–3325. https://doi.org/10.1105/tpc.108.063677.

Takos, A.M., et al., 2011. Genomic clustering of cyanogenic glucoside biosynthetic genes aids their identification in Lotus japonicus and suggests the repeated evolution of this chemical defence pathway. Plant J. 68 (2), 273–286. https://doi.org/10.1111/j.1365-313X.2011.04685.x.

Tautenhahn, R., Cho, K., Uritboonthai, W., Zhu, Z., Patti, G.J., Siuzdak, G., 2012. An accelerated workflow for untargeted metabolomics using the METLIN database. Nat. Biotechnol. 30 (9), 826–828. https://doi.org/10.1038/nbt.2348.

Teng, C.-M., Li, H.-L., Wu, T.-S., Huang, S.-C., Huang, T.-F., 1992. Antiplatelet actions of some coumarin compounds isolated from plant sources. Thromb. Res. 66 (5), 549–557. https://doi.org/10.1016/0049-3848(92)90309-X.

Tohge, T., de Souza, L.P., Fernie, A.R., 2014. Genome-enabled plant metabolomics. J. Chromatogr. B 966, 7–20.

Tokimatsu, T., Sakurai, N., Suzuki, H., Ohta, H., Nishitani, K., Koyama, T., Umezawa, T., Misawa, N., Saito, K., Shibata, D., 2005. KaPPA-view. A web-based analysis tool for integration of transcript and metabolite data on Plant metabolic pathway maps. Plant Physiol. 138 (3), 1289–1300. https://doi.org/10.1104/pp.105.060525.

Treml, J., Šmejkal, K., 2016. Flavonoids as potent scavengers of hydroxyl radicals. Compr. Rev. Food Sci. Food Saf. 15 (4), 720–738. https://doi.org/10.1111/1541-4337.12204.

Tu, Y., 2011. The discovery of artemisinin (qinghaosu) and gifts from Chinese medicine. Nat. Med. 17 (10), 1217–1220. https://doi.org/10.1038/nm.2471.

Tu, Y., 2016. Artemisinin—a gift from traditional Chinese medicine to the world (Nobel lecture). Angew. Chem. Int. Ed. 55 (35), 10210–10226. https://doi.org/10.1002/anie.201601967.

Turner, M.F., Heuberger, A.L., Kirkwood, J.S., Collins, C.C., Wolfrum, E.J., Broeckling, C.D., Prenni, J.E., Jahn, C.E., 2016. Non-targeted metabolomics in diverse sorghum breeding lines indicates primary and secondary metabolite profiles are associated with plant biomass accumulation and photosynthesis. Front. Plant Sci. 7, 953.

Valiela, I., Koumjian, L., Swain, T., Teal, J.M., Hobbie, J.E., 1979. Cinnamic acid inhibition of detritus feeding. Nature 280 (5717), 55–57.

Vaughn, S.F., Isbell, T.A., Weisleder, D., Berhow, M.A., 2005. Biofumigant compounds released by field pennycress (*Thlaspi arvense*) seedmeal. J. Chem. Ecol. 31 (1), 167–177.

Wägele, B., Witting, M., Schmitt-Kopplin, P., Suhre, K., 2012. MassTRIX reloaded: combined analysis and visualization of transcriptome and metabolome data. PLoS One 7 (7), e39860. https://doi.org/10.1371/journal.pone.0039860.

Wang, Z., Gerstein, M., Snyder, M., 2009. RNA-Seq: a revolutionary tool for transcriptomics. Nat. Rev. Genet. 10 (1), 57–63. https://doi.org/10.1038/nrg2484.

Wang, X., Liu, S., Zuo, H., Zheng, W., Zhang, S., Huang, Y., Pingcuo, G., Ying, H., Zhao, F., Li, Y., Liu, J., Yi, T.-S., Zan, Y., Larkin, R.M., Deng, X., Zeng, X., Xu, Q., 2021. Genomic basis of high-altitude adaptation in Tibetan *Prunus* fruit trees. Curr. Biol. 31 (17), 3848–3860.e8. https://doi.org/10.1016/j.cub.2021.06.062.

Wang, B., Tseng, E., Regulski, M., Clark, T.A., Hon, T., Jiao, Y., Lu, Z., Olson, A., Stein, J.C., Ware, D., 2016. Unveiling the complexity of the maize transcriptome by single-molecule long-read sequencing. Nat. Commun. 7 (1), 11708. https://doi.org/10.1038/ncomms11708.

Wang, T.-Y., Li, Q., K-S, B., 2018a. Bioactive flavonoids in medicinal plants: structure, activity and biological fate. Asian J. Pharm. Sci. 13 (1), 12–23.

Wang, Y., Xu, P., Wu, X., Wu, X., Wang, B., Huang, Y., Hu, Y., Lin, J., Lu, Z., Li, G., 2018b. GourdBase: a genome-centered multi-omics database for the bottle gourd (*Lagenaria siceraria*), an economically important cucurbit crop. Sci. Rep. 8 (1), 3604. https://doi.org/10.1038/s41598-018-22007-3.

Wasinger, V.C., Cordwell, S.J., Cerpa-Poljak, A., Yan, J.X., Gooley, A.A., Wilkins, M.R., Duncan, M.W., Harris, R., Williams, K.L., Humphery-Smith, I., 1995. Progress with gene-product mapping of the Mollicutes: *Mycoplasma genitalium*. Electrophoresis 16 (1), 1090–1094. https://doi.org/10.1002/elps.11501601185.

Wei, H., Persson, S., Mehta, T., Srinivasasainagendra, V., Chen, L., Page, G.P., Somerville, C., Loraine, A., 2006. Transcriptional coordination of the metabolic network in Arabidopsis. Plant Physiol. 142 (2), 762–774. https://doi.org/10.1104/pp.106.080358.

Winzer, T., et al., 2012. A *Papaver somniferum* 10-gene cluster for synthesis of the anticancer alkaloid noscapine. Science 336 (6089), 1704. https://doi.org/10.1126/science.1220757.

Wisecaver, J.H., Borowsky, A.T., Tzin, V., Jander, G., Kliebenstein, D.J., Rokas, A., 2017. A global coexpression network approach for connecting genes to specialized metabolic pathways in plants. Plant Cell 29 (5), 944–959. https://doi.org/10.1105/tpc.17.00009.

Wishart, D., Yang, R., Arndt, D., Tang, P., Cruz, J., 2005. Dynamic cellular automata: an alternative approach to cellular simulation. In Silico Biol. 5, 139–161.

Witaicenis, A., Seito, L.N., da Silveira, C.A., de Almeida Jr., L.D., Luchini, A.C., Rodrigues-Orsi, P., Cestari, S.H., Di Stasi, L.C., 2014. Antioxidant and intestinal anti-inflammatory effects of plant-derived coumarin derivatives. Phytomedicine 21 (3), 240–246. https://doi.org/10.1016/j.phymed.2013.09.001.

Wu, L.-Y., Fang, Z.-T., Lin, J.-K., Sun, Y., Du, Z.-Z., Guo, Y.-L., Liu, J.-H., Liang, Y.-R., Ye, J.-H., 2019. Complementary iTRAQ proteomic and transcriptomic analyses of leaves in tea plant (*Camellia sinensis* L.) with different maturity and regulatory network of flavonoid biosynthesis. J. Proteome Res. 18 (1), 252–264. https://doi.org/10.1021/acs.jproteome.8b00578.

Wu, T., Hu, E., Xu, S., Chen, M., Guo, P., Dai, Z., Feng, T., Zhou, L., Tang, W., Zhan, L., Fu, X., Liu, S., Bo, X., Yu, G., 2021. clusterProfiler 4.0: a universal enrichment tool for interpreting omics data. Innovation (Camb) 2 (3), 100141. https://doi.org/10.1016/j.xinn.2021.100141.

Xiang, C., Werner, B.L., Christensen, E.M., Oliver, D.J., 2001. The biological functions of glutathione revisited in arabidopsis transgenic plants with altered glutathione levels. Plant Physiol. 126 (2), 564–574. https://doi.org/10.1104/pp.126.2.564.

Xiao, H., Verdier-Pinard, P., Fernandez-Fuentes, N., Burd, B., Angeletti, R., Fiser, A., Horwitz, S. B., Orr, G.A., 2006. Insights into the mechanism of microtubule stabilization by taxol. Proc. Natl. Acad. Sci. U. S. A. 103 (27), 10166. https://doi.org/10.1073/pnas.0603704103.

Xiong, X., Gou, J., Liao, Q., Li, Y., Zhou, Q., Bi, G., Li, C., Du, R., Wang, X., Sun, T., Guo, L., Liang, H., Lu, P., Wu, Y., Zhang, Z., Ro, D.-K., Shang, Y., Huang, S., Yan, J., 2021. The Taxus genome provides insights into paclitaxel biosynthesis. Nat. Plants 7 (8), 1026–1036. https://doi.org/10.1038/s41477-021-00963-5.

Xiong, F., Ren, J.-J., Yu, Q., Wang, Y.-Y., Kong, L.-J., Otegui, M.S., Wang, X.-L., 2019. AtBUD13 affects pre-mRNA splicing and is essential for embryo development in Arabidopsis. Plant J. 98 (4), 714–726. https://doi.org/10.1111/tpj.14268.

Yadav, S., 2010. Heavy metals toxicity in plants: an overview on the role of glutathione and phytochelatins in heavy metal stress tolerance of plants. S. Afr. J. Bot. 76 (2), 167–179.

Yang, L., Wen, K.-S., Ruan, X., Zhao, Y.-X., Wei, F., Wang, Q., 2018. Response of plant secondary metabolites to environmental factors. Molecules 23 (4). https://doi.org/10.3390/molecules23040762. ARTN 762.

Yang, Z., et al., 2020. A mini foxtail millet with an Arabidopsis-like life cycle as a C4 model system. Nat. Plants 6 (9), 1167–1178. https://doi.org/10.1038/s41477-020-0747-7.

Yang, F.-X., et al., 2021. The genome of Cymbidium sinense revealed the evolution of orchid traits. Plant Biotechnol. J. https://doi.org/10.1111/pbi.13676.

Yazaki, K., 2006. ABC transporters involved in the transport of plant secondary metabolites. FEBS Lett. 580 (4), 1183–1191. https://doi.org/10.1016/j.febslet.2005.12.009.

Yi, W., Fischer, J., Krewer, G., Akoh, C.C., 2005. Phenolic compounds from blueberries can inhibit colon cancer cell proliferation and induce apoptosis. J. Agric. Food Chem. 53 (18), 7320–7329. https://doi.org/10.1021/jf051333o.

Yu, G., Wang, L.-G., Han, Y., He, Q.-Y., 2012. clusterProfiler: an R package for comparing biological themes among gene clusters. OMICS: J. Integr. Biol. 16 (5), 284–287. https://doi.org/10.1089/omi.2011.0118.

Yin, X., Fan, H., Chen, Y., Li, L.-Z., Song, W., Fan, Y., Zhou, W., Ma, G., Alolga, R.N., Li, W., Zhang, B., Li, P., Tran, L.-S.P., Lu, X., Qi, L.-W., 2020. Integrative omic and transgenic analyses reveal the positive effect of ultraviolet-B irradiation on salvianolic acid biosynthesis through upregulation of SmNAC1. Plant J. 104 (3), 781–799. https://doi.org/10.1111/tpj.14952.

Yu, C., et al., 2020. Tissue-specific study across the stem of taxus media identifies a phloem-specific TmMYB3 involved in the transcriptional regulation of paclitaxel biosynthesis. Plant J. 103 (1), 95–110. https://doi.org/10.1111/tpj.14710.

Yuan, J.S., Galbraith, D.W., Dai, S.Y., Griffin, P., Stewart, C.N., 2008. Plant systems biology comes of age. Trends Plant Sci. 13 (4), 165–171. https://doi.org/10.1016/j.tplants.2008.02.003.

Zager, J.J., Lange, I., Srividya, N., Smith, A., Lange, B.M., 2019. Gene networks underlying cannabinoid and terpenoid accumulation in Cannabis. Plant Physiol. 180 (4), 1877–1897. https://doi.org/10.1104/pp.18.01506.

Zasada, I., Ferris, H., 2004. Nematode suppression with brassicaceous amendments: application based upon glucosinolate profiles. Soil Biol. Biochem. 36 (7), 1017–1024.

Zhang, J., et al., 2018. Genome-wide association studies and expression-based quantitative trait loci analyses reveal roles of HCT2 in caffeoylquinic acid biosynthesis and its regulation by defense-responsive transcription factors in Populus. New Phytol. 220 (2), 502–516. https://doi.org/10.1111/nph.15297.

Zhang, X., et al., 2021. Haplotype-resolved genome assembly provides insights into evolutionary history of the tea plant *Camellia sinensis*. Nat. Genet. 53 (8), 1250–1259. https://doi.org/10.1038/s41588-021-00895-y.

Zhou, F., Pichersky, E., 2020. The complete functional characterisation of the terpene synthase family in tomato. New Phytol. 226 (5), 1341–1360. https://doi.org/10.1111/nph.16431.

Zhu, Q.H., Wang, M.B., 2012. Molecular functions of long non-coding RNAs in plants. Genes (Basel) 3 (1), 176–190. https://doi.org/10.3390/genes3010176.

Zoppi, J., Guillaume, J.-F., Neunlist, M., Chaffron, S., 2021. MiBiOmics: an interactive web application for multi-omics data exploration and integration. BMC Bioinf. 22 (1), 6. https://doi.org/10.1186/s12859-020-03921-8.

Part IV

Why regulation and policy matter to advance the bioeconomy

Chapter 11

Regulatory frameworks applicable to food products of genome editing and synthetic biology in the United States, Canada, and the European Union

Emily Marden[a,b], Deepti Kulkarni[c], Eileen M. McMahon[d], Melanie Sharman Rowand[d], and Karin Verzijden[e]
[a]*Sidley Austin LLP, Palo Alto, CA, United States,* [b]*University of British Columbia, Vancouver, BC, Canada,* [c]*Sidley Austin LLP, Washington, DC, United States,* [d]*Torys LLP, Toronto, ON, Canada,* [e]*Axon Lawyers, Amsterdam, Netherlands*

Introduction

Advances in genomics and the development of techniques such as genome editing and synthetic biology have opened the door to a range of innovations in food and agricultural products. These innovations are particularly important as they have the potential to help address Sustainable Development Goals ("SDGs") 2 ("End hunger, achieve food security and improved nutrition and promotion of sustainable agriculture"), 12 ("Ensure sustainable consumption and production patterns"), and 13 ("Take urgent action to combat climate change and its impacts"). In part as a reflection of the importance of these innovations, regulatory authorities around the world are in the process of considering food and agriculture products that have utilized genome editing and synthetic biology and are addressing questions about how such products are or will be regulated. Given the rapid developments in science and technology, this area of regulatory oversight is the subject of much discussion and remains in flux.

This chapter provides an overview of the current state of regulatory oversight of food and agricultural products produced using genome editing and synthetic biology in the United States, Canada, and the European Union. For each

Genomics and the Global Bioeconomy. https://doi.org/10.1016/B978-0-323-91601-1.00001-8

of these jurisdictions, this chapter describes the regulatory framework for foods and agricultural products and then examines how products produced using genome editing and synthetic biology are specifically being considered. This chapter also addresses recent developments in each jurisdiction. Collectively, we note that while similar innovations are taking place around the globe, each of the jurisdictions considered negotiates them through a distinct regulatory framework. Thus, while there are certain parallels between the jurisdictions, it is still necessary for innovators to consider the regulatory framework carefully for each jurisdiction in which they intend to market a product.

In addition to the regulatory considerations addressed herein, it is important to be aware that there may also be environmental oversight in each jurisdiction. In addition, while much of the regulatory attention to date has focused on food and agriculture products that rely on genome editing and synthetic biology, there is also renewed regulatory attention in many jurisdictions on bioengineered animals and nonfood products of genome editing and synthetic biology. Given space considerations, these developments are not specifically addressed herein. Finally, we note that this chapter is purely descriptive in nature and should in no way be understood as legal advice.

US regulatory framework for foods

Overview of the US Coordinated Framework for regulation of biotechnology products

In the United States, products of biotechnology are regulated under the "Coordinated Framework for the Regulation of Biotechnology." Established in 1986 and subsequently updated in 1992 and 2017, the Coordinated Framework seeks to strike a balance between regulation that protects public health and the environment, while maintaining flexibility to avoid impeding innovation. The Coordinated Framework considers biotechnology to be an extension of traditional methods of production, and concludes that therefore existing laws administered by federal agencies are adequate to provide regulatory oversight based upon product uses. The exercise of such regulatory oversight, according to the Coordinated Framework, "should be based on the risk posed by the introduction and should not turn on the fact that an organism has been modified by a particular process or technique" (OSTP, 1986, 1992).

There have been major advancements in science and technology in the last decade, including genome editing and synthetic biology, that were not previously contemplated in the Coordinated Framework. Therefore, in 2017, the Coordinated Framework was updated; the goal was to acknowledge these advancements, facilitate appropriate regulatory oversight and increase transparency, while continuing to advance innovation (USDA, FDA, EPA, 2017). In addition to updating the Coordinated Framework, the federal regulatory agencies with oversight of biotechnology products issued a strategy document, titled

National Strategy for Modernizing the Regulatory System for Biotechnology Products ("National Strategy"), identifying future steps to ensure the regulatory system appropriately addresses novel products developed through future advances in science and technology (OSTP, 2016). Of note, the National Strategy calls upon these agencies to conduct horizon-scanning assessments of new biotechnology products, ensure that product evaluations are risk based and grounded in the best available science, and identify changes to regulatory requirements to more efficiently assess potential risks in a predictable and transparent manner.

The net result is that the regulatory landscape continues to evolve with the aim of better characterizing and addressing risks associated with genome editing and synthetic biology, as well as other emerging forms of biotechnology. Recent developments in the United States shed some light on the continuing evolution of the regulatory paradigm.

Regulatory approach

The primary agencies responsible for US oversight of products of biotechnology are the US Food and Drug Administration ("FDA"), the US Department of Agriculture ("USDA"), and the US Environmental Protection Agency ("EPA"). In this chapter, we focus on the current FDA and USDA oversight of food products developed using newer methods, such as genome editing and synthetic biology, under the Coordinated Framework.

FDA regulates foods developed using biotechnology to ensure human and animal food is safe and properly labeled pursuant to its expansive authority under the Federal Food, Drug, and Cosmetic Act ("FDCA"). Historically, the vast majority of these foods have been derived from genetically engineered ("GE") plants and microorganisms. FDA also has evaluated foods derived from GE animals, though to date has only approved two applications relating to such foods, including, for example, a line of domestic pigs intended to eliminate alpha-gal sugar on the surface of the pigs' cells as well as a genetically engineered Atlantic salmon (FDA, 2020f). While the area of animal biotechnology presents interesting scientific and regulatory issues, such issues are outside the immediate scope of this chapter and thus will not be addressed.

USDA, to date, primarily has regulated GE plants, particularly food crops, such as corn, soy, and canola, and microorganisms pursuant to the agency's interpretation and authority under the Plant Protection Act ("PPA"), to protect plants and plant products (USDA, FDA, EPA, 2017).

Despite the objectives of the Coordinated Framework, regulatory oversight by FDA and USDA is not always solely product-based, and products can be subject to multiple and sometimes overlapping authorities. For example, a genome edited plant could be subject to USDA oversight with regard to the plant pest risk for agriculture, as well as FDA jurisdiction with regard to the safety of the novel ingredient for human (or animal) food. In addition, regulatory oversight can appear confusing because there is no shared set of definitions between agencies pertaining to terms used to refer to the products of biotechnology, including "biotechnology," "genetic engineering," and "recombinant DNA techniques." For the purpose of this chapter, it is important to note that FDA and USDA have indicated that they consider "biotechnology" and "genetic engineering," as currently used in regulations, to include genome editing and synthetic biology (FDA, 2020e).

Food derived from genetically engineered plants and microorganisms

Development of plants and microorganisms for introduction into the environment

Genetically engineered plants and microbes are subject to the authority of USDA's Animal and Plant Health Inspection Service ("APHIS") under the PPA and APHIS's implementing regulations (7 U.S.C. § 7701 *et seq*; 7 C.F. R. part 340). As a general matter, these regulations require a permit before the introduction of certain GE plants and microorganisms if there is a potential "plant pest" risk. In this context, APHIS has interpreted the statute broadly to give it regulatory authority over products of biotechnology that could pose a risk to plant health or agriculture.

Historically, APHIS has taken the position that a broad range of plants and microbes, including those used for agricultural and food (for humans or animals) purposes, are subject to a permit requirement if they were developed using biotechnology. However, in an attempt to make regulation more efficient and risk-based, APHIS revised its approach in 2020 in a final rule, called the "Sustainable, Ecological, Consistent, Uniform, Responsible, Efficient Rule" (i.e., the "SECURE Rule") (APHIS, 2020a). This rule significantly alters the scope of APHIS's regulatory oversight, particularly for plants produced using genome editing and synthetic biology.

The SECURE Rule takes as its starting point a recognition of the following three principles:

1. Plants created through conventional breeding have a history of safe use related to plant pest risk;
2. The types of plants that qualify for these exemptions can also be created through conventional breeding; and.
3. There is no evidence that the use of recombinant deoxyribonucleic acid (DNA) or genome editing techniques necessarily and in and of itself introduces plant pest risk, irrespective of the technique employed (APHIS, 2020a).

Based on these principles, APHIS concluded that plants modified in a manner that could "otherwise be achieved through conventional breeding" would no longer be subject to a permit requirement. In its rulemaking process, APHIS therefore conducted a review of the scientific literature, and identified a list of genetic modifications that take place in conventional breeding. Based on the principles above, these types of modifications are specifically exempted from regulation by the SECURE Rule even if implemented through the use of genetic techniques:

- Is a change resulting from cellular repair of a targeted DNA break without an externally provided repair template;
- Is a targeted single base pair substitution; or
- Introduces a gene known to occur in the plant's gene pool, or makes a change in a targeted sequence to correspond to a known allele of such a gene or to a known structural variation present in the gene pool (7 C.F.R. part 340).

It is worth noting that APHIS recently proposed further exemptions, reflecting additional categories of genetic changes that also take place in conventional breeding. These additional exemptions would apply to other categories of genome edited products (APHIS, 2021). In addition, the SECURE Rule sets forth a streamlined "regulatory status review" process in which developers can establish whether novel GE plant/trait/mechanism of action combinations will require a permit. The goal of the SECURE Rule is to require permitting only for plants or organisms posing a "plausible plant pest risk" (APHIS, 2020a).

Importantly, the express exemptions from the permit requirement under the SECURE Rule do not apply to plants utilized to produce pharmaceuticals or other substances for industrial use, or to microorganisms. For these categories, APHIS has taken the position that it will, at least at the outset, require permits on the basis that there could be plausible plant pest risks associated with such products (APHIS, 2020a).

The net result is that as the SECURE Rule is implemented, a growing number of plants produced using genome editing and synthetic biology would no longer be subject to USDA oversight.

Recent developments: Genome editing

Even before the SECURE Rule went into effect, APHIS took the position that at least 69 genome edited plants did not pose a plant pest risk under the PPA and therefore were not subject to the permit requirement. These prior case-by-case determinations under the previous "Am I Regulated?" process provide a window into APHIS' evaluation process. APHIS granted exemptions for, among others:

- A genome edited white button mushroom with an antibrowning phenotype; the modification was achieved through "small deletions (1–14 base pairs) in a specific polyphenol oxidase gene," without the integration of foreign DNA into the mushroom's genome (APHIS, 2016)
- A genome-edited tomato containing deletions resulting in a change that permits the tomato fruit to come cleanly off the vine without the stem attached to the fruit when picked; the plants were backcrossed to wild-type plants and

progeny contained the deletions but lacked *Agrobacterium* T-DNA used to deliver the CRISPR-Cas9 protein (APHIS, 2018);

- A genome-edited high oleic low linolenic soybean; the soybean was a null-segregant of soybean lines developed by the transient expression of a transcription activator-like effector nuclease (TALEN) that resulted in the targeted knockout of five genes but none of the inserted DNA remained in the final soybean plant (APHIS, 2020b); and
- A genome-edited pea designed to have an improved flavor profile; the pea was a null-segregant of the transformed pea line in which the editing machinery was expressed, but the pea plants are self-pollinated to segregate the introduced DNA from the targeted deletion (APHIS, 2020c).

In each case, APHIS: (1) reviewed the changes to the genome described by the developer and issued a letter in response; (2) considered whether the final selected lines contained any introduced DNA or otherwise raised plant pest or weed potential; and (3) stated that the agency had no reason to believe the product was a plant pest. These determinations as well as APHIS's experience with GE plants over the course of the last three decades have undoubtedly paved the way for the approach set forth in the SECURE Rule, and should continue to serve as benchmarks moving forward. Importantly, the SECURE Rule was not developed with the aim of achieving harmonization between jurisdictions. As discussed in more detail below, the United States approach differs significantly from the EU approach and in some respect from the approach taken in Canada. Thus, it remains important for developers of these products to carefully review the applicable regulations in each jurisdiction.

Foods derived from genetically engineered plants and microorganisms

As a general matter, FDA relies on its "food additive" and "adulteration" authorities under the FDCA to ensure the safety of foods from GE plants and microorganisms. These authorities generally require that, prior to commercialization, a new food ingredient either be the subject of a food additive petition or a determination that the substance is generally recognized as safe ("GRAS") for its intended use (21 U.S.C. 321(s)).

With respect to GE plants, FDA takes the position that a substance that is expected to become a component of food as a result of genetic engineering and whose composition is or has been altered such that the substance is not GRAS or otherwise exempt is subject to premarket approval as a "food additive" (FDA, 1992). With respect to transferred genetic material (nucleic acids), FDA maintains that such material is presumed to be GRAS and thus would not itself be subject to food additive regulation (FDA, 1992). The intended expression product, such as a protein, carbohydrate, or fat, also would be *presumed* GRAS if the substance is "already present at generally comparable or greater levels in currently consumed foods" (FDA, 1992). If, however, the intended expression product differs significantly in structure, function, or composition from substances found currently in food, the substance may not be GRAS and may require regulation as a food additive or otherwise be prohibited (FDA, 1992).

To enable developers of foods from GE plants to ensure that such foods are safe prior to marketing, FDA provides a voluntary premarket consultation program through which agency scientists advise on the appropriate approach to assess the food's safety, and evaluate safety assessments performed by the developer. Under this program, FDA has evaluated the safety of over 180 GE plant varieties (FDA, 2018b, 2020e).

In line with the goals of the 2017 Update to the Coordinated Framework and National Strategy, FDA has asked for stakeholder input on the appropriate approach to evaluate the safety of foods derived from genome-edited plants, including whether there are specific risks that should be considered and whether there are categories of products that may pose less risk than others (FDA, 2017). In its request, FDA does not identify specific risks that should be considered, but rather asks the question in general terms; the agency also asks specifically whether any risks identified are the same as or different from those associated with other plant development methods (e.g., hybridization, chemical or radiation-induced mutagenesis, and nontargeted genetic modifications using in vitro recombinant DNA technologies). In addition, FDA has stated that it intends to publish guidance to clarify its regulatory approach for such foods and prevent unnecessary barriers (FDA, 2019b). To date, FDA has not yet issued such guidance. FDA, however, has emphasized that its premarket consultation program is open to foods derived from genome-edited plant varieties, and, to date, has completed a consultation for a genome-edited soybean variety modified to have increased levels of oleic acid (FDA, 2020a).

The safety of food ingredients derived from GE microorganisms is similarly considered under FDA's existing "food additive" authorities. In practice, developers of these ingredients have employed FDA's GRAS Notification process to establish the safety of their ingredients. Among other things, the GRAS Notification process requires the submission of information showing that there is general recognition among qualified experts that the substance is safe under the conditions of its intended use (*See* 21 U.S.C. § 321(s)). "Safe," in turn, means that "there is a reasonable certainty in the minds of competent scientists that the substance is not harmful under the conditions of its intended use" (21 CFR 170.30(a)). In evaluating the safety of ingredients from GE microorganisms, FDA generally considers the identity of the microorganism, how the substance is produced from that microorganism, as well as the identity, safety, and regulatory status of the substance (FDA, 2010a, 2016). As such, in GRAS Notifications for such ingredients, developers typically characterize the microorganism used to produce the ingredient as well as the manufacturing process, and evaluate the identity and composition of the ingredient as well as potential dietary intake.

Recent developments: Synthetic biology

Historically, the vast majority of ingredients from GE microorganisms that have been commercialized in the United States have been enzymes (or enzyme preparations) (FDA, 2010b, 2018a). Many of these enzymes are considered "incidental additives," as they are used as processing aids or otherwise are present in finished foods at insignificant levels and do not have technical or functional effect in that food.

The landscape, however, is evolving, as advancements in synthetic biology (sometimes referred to as "precision fermentation") have enabled developers to produce proteins and other food ingredients derived from GE microorganisms, for a wide range of technical or functional uses, including for flavor, color, texture, and nutrient purposes. To date, these ingredients have been, or are in the process of being, evaluated under FDA's GRAS Notification process. For example, in July 2018, FDA completed its evaluation of a GRAS Notification for soy leghemoglobin preparation produced from a strain of *Pichia pastoris* for use to optimize flavor in ground beef analogue products (FDA, 2018c). To assess safety, the developer used a "weight of the evidence" approach that evaluated several lines of evidence, including: (1) the safety of *Pichia pastoris* for use as the production microorganism; (2) the history of consumption of hemoglobin proteins in food; (3) the results of bioinformatic analyses comparing soy leghemoglobin and *Pichia pastoris* proteins to known toxins and allergens; (4) the digestibility of soy leghemoglobin preparation proteins in simulated gastric fluid; and (5) publicly available scientific literature. There are other examples of GRAS Notifications for an ingredient from a GE microorganism that have employed an analogous safety approach, tailored to the particular ingredient under evaluation (See e.g., FDA, 2020b).

These GRAS Notifications and FDA's corresponding evaluations help demonstrate that the GRAS Notification paradigm can be applied to ingredients developed using newer forms of biotechnology. These and other Notifications should serve to further inform the regulatory process to better meet the goals set forth in the Coordinated Framework and 2017 National Strategy.

Labeling of "bioengineered" foods

Before 2019, foods marketed in the United States were not required to bear labeling disclosing whether they had been developed using biotechnology. At the time, the applicable statutory provisions required that food labeling be truthful and not misleading, and specified that labeling is misleading if, among other things, it fails to reveal facts that are either: (1) material in light of representations made in the labeling or (2) material with respect to consequences that may result from the use of the food (FDA, 1992). FDA explained that it was not aware of any information showing that GE foods differ from other foods in any meaningful or uniform way, or that, as a class, GE foods present any different or greater safety concern that foods developed by traditional plant breeding (FDA, 1992). Therefore, FDA concluded that the method of development of a GE plant is generally not "material" information and would not usually be required to be disclosed in the labeling for the food (FDA, 1992).

This position remained largely unchanged until July 2016, when Congress passed the National Bioengineered Food Disclosure Standard ("Disclosure Standard"). This action came about, in part, because of growing concern that individual states had passed or were in the process of considering laws that would require labeling disclosing whether foods had been developed using genetic engineering, and food manufacturers would thus be subject to a patchwork of different—and potentially inconsistent—laws. Notably, the Disclosure Standard directs USDA, rather than FDA, to establish mandatory disclosure requirements for "bioengineered food," as defined in the Disclosure Standard, itself. The fact that the Disclosure Standard uses the term "bioengineered," and establishes a particular definition for the term, rather than using "genetically engineered" or "genetically modified" reflects, in part, a negotiation between industry and other stakeholders and Congress on the scope and applicability of the law as well as the use appropriate terminology.

According to the Disclosure Standard, a food is "bioengineered" if it "contains genetic material that has been modified through *in vitro* rDNA techniques," and "for which the modification could not otherwise be obtained through conventional breeding or found in nature" (National Bioengineered Food Disclosure Standard, 2016). Given this definition, it appears that many foods produced using genetic engineering will not require disclosure because rDNA is not "contained in" or present in the food after processing. Interpretation of the law is more complex, however, where rDNA remains present in the food product. For example, in its implementing regulations, USDA's Agricultural Marketing Service ("AMS") added exemptions to the disclosure requirement where the "bioengineered" food can be considered an "incidental additive" under FDA regulations, or is an inadvertent presence at low levels (AMS, 2018).

The Disclosure Standard is still in the process of being interpreted and implemented. AMS's regulations implementing the Disclosure Standard went into effect in February 2019, and the mandatory compliance date for all the regulations is January 1, 2022 (AMS, 2018).

Animal cell culture technology

Another development of significant interest is the use of animal cell culture technology to produce foods. This emerging technology generally involves the controlled growth of cells from livestock, poultry, or fish, their subsequent differentiation into various cell and tissue types, and their collection and processing into food (FDA, 2020c). In March 2019, FDA and USDA's Food Safety and Inspection Service ("FSIS") issued a Formal Agreement outlining a joint regulatory framework for oversight of human food products derived from the cells of livestock and poultry using cell culture technology (FDA, 2019a).

As described in the Formal Agreement, FDA will conduct premarket consultation processes to evaluate production materials/processes and manufacturing controls pursuant to the FDCA (FDA, 2019a). For foods from livestock and poultry cells, FDA will oversee, through routine inspections, the collection, growth, and differentiation of cells until "harvest," at which point regulatory oversight will be transferred to FSIS (FDA, 2019a). FSIS, in turn, will conduct inspections in establishments where cells cultured from livestock and poultry are harvested, processed, packaged, or labeled, pursuant to its authorities under the Federal Meat Inspection Act ("FMIA") and Poultry Product Inspection Act ("PPIA") (FDA, 2019a). In addition, FSIS will require that the labeling of such foods be preapproved and then verified through inspection (FDA, 2019a). Both FDA and USDA will share information to carry out their respective responsibilities.

For foods from the cells of fish (other than Siluriformes fish), FDA will have primary oversight with respect to premarket safety consultations, manufacturing and production, inspections, and labeling (FDA, 2020d).

Both FDA and FSIS have agreed to develop a more detailed joint framework to facilitate coordination of shared regulatory oversight, as well as a joint process to identify any changes needed to effective the framework. In addition, both agencies will develop joint principles for product labeling.

While FDA is currently evaluating a number of premarket consultations for foods developed using animal cell culture, to date, it has not yet completed such a consultation; nor has it issued guidance describing the data and information to be provided in a consultation.

With respect to labeling, in October 2020, FDA requested stakeholder input on the labeling of foods comprised of or containing cultured seafood cells (FDA, 2020d). In general, FDA invited data and other evidence on: (1) appropriate product names or disclosures; (2) consumer understanding of such terms; and (3) how to assess material differences between foods produced using cultured seafood cells, and conventionally produced foods. In September 2021, FSIS issued an Advanced Notice of Proposed Rulemaking on the labeling of meat and poultry products comprised of or containing cultured animal cells in which the agency requested further information on the appropriate labeling for such products. Similarly, FSIS invited input on: (1) appropriate product names or disclosures; (2) consumer understanding of such terms; and (3) how to assess material differences between foods produced using cultured meat or poultry cells, and conventionally produced foods; in addition, FSIS asked for comment on: (4) whether it should issue new standards of identity for foods produced using cultured meat or poultry cells and related benefits and costs to industry; (5) whether FSIS should amend its preexisting regulations to specifically include or exclude such foods; and (6) what claims are likely to be used in the labeling of such products and whether FSIS should develop new regulations or guidance to ensure that such claims are neither false nor misleading (FSIS, 2021).

While rulemaking and other policy developments may take years to complete, the initial premarket consultations for foods developed using animal cell culture along with case-specific labeling determinations made by FDA and USDA will be closely watched and will further shape the regulatory landscape for these products.

Canada's regulatory framework for foods

Regulatory approach

In Canada, the laws governing the research, development, and marketing authorizations for products of biotechnology are federal laws and, depending on the product and its intended use, multiple federal laws may apply. In turn, the government agencies regulating these products are federal agencies.

- Health Canada regulates novel foods under the authority of the *Food and Drugs Act* (FDA, 1985) and Division 28 of the Food and Drugs Regulations (FDR, 1978), known as the "Novel Food Regulations."
- The Canadian Food Inspection Agency (CFIA) regulates products of biotechnology under several statutes, including the *Feeds Act* (Feeds Act, 1985), *Seeds Act* (Seeds Act, 1985), *Plant Protection Act* (PPA, 1990), *Health of Animals Act* (HAA, 1990), and *Fertilizers Act* (Fertilizers Act, 1985).
- Environment and Climate Change Canada (ECCC) regulates new substances, including novel organisms, under the *Canadian Environmental Protection Act* (CEPA, 1999) where an environmental assessment has not been done under "CEPA-equivalent" legislation.
- Other government agencies, such as the Department of Fisheries and Oceans (DFO), may be involved in product reviews where their expertise is required (e.g. DFO, 2013).
- Provincial or local laws may also apply but are excluded for the purposes of this discussion.

Canada does not have specific legislation for products of biotechnology, but instead regulates these products under existing legislation. As with the United States, there is no shared set of definitions between agencies pertaining to products of biotechnology. Canada uses a "product-based" approach in which regulatory oversight is triggered by novelty of traits in food, plants, microorganisms, and animals, rather than the process used to create the product. However, in practice, regulatory and/or data requirements are informed by the process used (i.e., different factors are considered during the safety assessment depending on the type of technology used).

As rapid developments in food and agricultural production techniques expose gaps in legislation and regulatory guidance documents, Canadian regulators are playing catch-up. Regulatory modernization in these areas is consistent with Canada's Federal Sustainable Development Strategy and *Federal Sustainable Development Act* (FSDA, 2008) and the commitment of government agencies such as Health Canada to advance the UN Sustainable Development Goals ("SDGs"). Health Canada has specifically identified the redesign of food regulations to bring innovative products to market as a key action for advancing sustainable food (Health Canada, 2020b).

Recent developments in the regulation of foods derived from genetically modified plants, microorganisms, and animal cell culture technology are discussed below. In particular, the regulation of emerging areas of food production, including gene-edited plants, synthetic biology, and cultured meat, are discussed. Consistent with the United States, the vast majority of foods developed using biotechnology that have been approved in Canada are derived from genetically modified plants and microorganisms. To date, only one genetically modified animal (Atlantic salmon, genetically modified with foreign DNA to promote rapid growth) has been approved for human consumption in Canada (CFIA, 2016; DFO, 2013; Health Canada, 2016a,b). The regulation of genetically modified animals is outside the scope of this chapter and will not be addressed.

Foods derived from genetically modified plants and microorganisms

Foods derived from genetically modified plants

Genetically modified plants which meet certain criteria are subject to premarket notification and assessment under three federal statutes and their associated regulations: (i) the FDA and Novel Food Regulations, if they meet the definition of a "novel food" (FDR, B.28.001); (ii) the *Feeds Act* and *Regulations*, unless it can be shown that the plants will be segregated from the animal feed system; and (iii) the *Seeds Act* and *Regulations*, prior to release into the environment. To increase harmonization, Health Canada and the CFIA have developed a formalized process to coordinate the determination of novelty for new plant varieties or foods and feeds derived from these plants, and joint presubmission consultations are available (Health Canada, 2022a).

Recent developments: Gene editing

While legislation and related guidance documents cover products of biotechnology generally, they were silent regarding more recent technologies, such as gene editing. However, in March 2021, Health Canada and the CFIA undertook consultations to address this gap and in May 2022 Health Canada updated its novel food guidance document to provide greater clarity regarding the regulation of foods derived from gene edited plants (Health Canada, 2022a). Health Canada has also published a notice of intent to develop amendments to the Novel Food Regulations which are consistent with the updated guidance and to develop similar guidance for animals and microorganisms (Health Canada, 2022b). The final amendments to the novel food guidance document point to an overwhelmingly favorable regulatory environment for gene-edited plants and foods, where the Health Canada consultations had concluded: "Through a review of the current scientific knowledge regarding the use of gene editing technologies… Health Canada concludes that the use of gene editing technologies does not present any unique safety concerns compared to other methods of plant breeding." Health Canada also notes that gene-editing technologies are a tool to "assist with global food security and sustainability objectives" (Health Canada, 2021a).

By way of background, the Novel Food Regulations define a "novel food" as:

- **(a)** a substance, including a microorganism, that does not have a history of safe use as food;
- **(b)** a food that has been manufactured, prepared, preserved, or packaged by a process that
  - **(i)** has not been previously applied to that food, and
  - **(ii)** causes the food to undergo a major change; and
- **(c)** a food that is derived from a plant, animal, or microorganism that has been genetically modified such that
  - **(i)** the plant, animal, or microorganism exhibits characteristics that were not previously observed in that plant, animal, or microorganism,
  - **(ii)** the plant, animal, or microorganism no longer exhibits characteristics that were previously observed in that plant, animal, or microorganism, or
  - **(iii)** one or more characteristics of the plant, animal, or microorganism no longer fall within the anticipated range for that plant, animal, or microorganism (FDR, B.28.001).

The term "genetically modify" is broadly defined as a change in the heritable traits of a plant, animal, or microorganism by means of intentional manipulation (FDR, B.28.001).

Health Canada updated its novel food guidance document in May 2022 to clarify when genetically modified plants will meet the definition of "novel food" (Health Canada, 2022a). Notably, plants will not meet the definition of "novel food" unless the genetic modification is due to the insertion of foreign DNA or there are other specified changes in the end-product (e.g., introduction of toxins and allergens, key nutritional changes, or changes in food use). The guidance notes that gene-edited plants will be considered to contain foreign DNA in the unusual circumstances when the gene editing machinery (e.g., CRISPR Cas proteins and associated guide RNAs) is not bred out of the plant. Another favorable development for breeders is the introduction of an expedited review process for products with identical modifications to other approved products, even if produced through a different method such as gene editing (Health Canada, 2021b; 2022a). The guidance seeks to interpret the Novel Food

Regulations in a manner that reflects the importance of innovation, takes into account the purpose of the regulations (i.e., to require premarket assessment of novel foods that may pose a potential risk to the consumer), and more closely aligns with the practices of "Canada's international regulatory partners" (Health Canada, 2021a).

Similarly, the CFIA is proposing amendments to the guidance document for Part V of the *Seeds Regulations* to clarify which plants require authorization before release into the environment (CFIA, 2021e). The proposed amendments follow the CFIA's outcome-based assessment system: when a product of gene editing resembles its nongenetically edited counterpart (i.e., it does not contain foreign DNA, is not a new crop type, and will not negatively impact the environment), it will be exempt from the regulatory requirements under Part V of the *Seeds Regulations*, regardless of the method used in its development (CFIA, 2021a,b). Consistent with the proposed novel food guidance, the proposed *Seeds Act* guidance indicates that gene-edited plants will be considered to contain foreign DNA in the unusual circumstances when the gene editing machinery is not bred out of the plant. At the time of writing, the proposed amendments are under review and subject to change; however, it is expected that they will be finalized in 2022.

Food derived from genetically modified microorganisms

Like plants, foods derived from genetically modified microorganisms may be subject to premarket notification and assessment under (i) the FDA and Novel Food Regulations if they meet the definition of a "novel food" and (ii) the *Feeds Act* and *Regulations* if there is a risk that the food product may enter the animal feed system. Products derived from genetically modified microorganisms that are used in food production, such as enzymes, may be regulated as "food additives" under the FDR (Health Canada, 2014a,b).

Recent developments: Synthetic biology

As discussed in the United States section above, advancements in synthetic biology (sometimes referred to as "precision fermentation") have enabled developers to produce complex organic molecules such as proteins. Use of this food production technique is of increasing interest in the alternative protein and dairy industries. Products recently approved in Canada are described below.

Milk oligosaccharides for infant formula A novel food containing oligosaccharides produced from genetically modified *Escherichia coli* was approved by Health Canada in 2018 for use in infant formula. In its decision summary, Health Canada notes that the product is considered a novel food because it is produced by a genetically modified bacterium, which was modified by the insertion of several gene constructs and chemical mutagenesis (Health Canada, 2018).

Soy *leghemoglobin* In 2020, Health Canada approved a soy leghemoglobin (LegH) preparation as a novel food for use in ground beef analogues (e.g., the Impossible Burger). This product has also been approved in the United States but is still under review in Europe. The LegH protein is produced by a yeast strain (*Pichia pastoris*) genetically modified to express the protein found in soybean root nodules. Health Canada's safety assessment considered the safety of the *P. pastoris* production organism, the manufacturing process for the LegH preparation, the nutritional composition of the LegH preparation, and the potential for the preparation to present a toxic or allergenic concern, among other things. In its decision summary, Health Canada notes that the LegH preparation is considered a "novel food" because it does not have a history of safe use as food (it is found in the root nodules of the soybean plant, which are not typically eaten by humans) (Health Canada, 2020a). Assessment by the CFIA was not required because sale of the product was limited to grocery store outlets and restaurants and segregated from the animal feed system.

From the decision summary, it can be inferred that some food products produced by genetically altered microorganisms will not meet the definition of a novel food. This would be expected where the end-product has a history of safe use as food, the characteristics of the end-product fall within the anticipated ranges, and the processes of manufacture are familiar to the regulator.

Labeling of foods derived from "genetically engineered" plants and microorganisms

There is no requirement to specify whether a food is genetically engineered; however, voluntary labeling is permitted. To facilitate the use of voluntary labeling in Canada, the Canadian Standards Board has developed a national standard for Voluntary Labeling and Advertising of Foods that Are and Are Not Products of Genetic Engineering (PSPC, 2019). The standard governs use of the terms: "genetically engineered (name of product)," "derived from genetically engineered (name of product)," and "product of genetic engineering." Under this standard, the term "genetic engineering" is defined as "techniques by which the genetic material of an organism is changed in a way that does not occur naturally by multiplication and/or natural recombination, e.g., recombinant DNA (rDNA) techniques that use vector systems or other mechanisms involving direct manipulation of the genome." Notably, the standard does not apply to gene editing and other technologies that introduce characteristics that could be achieved through natural recombination and are otherwise indistinguishable from products of natural recombination.

Like all other foods, genetically engineered foods must comply with Canada's general labeling requirements under the FDA, the *Safe Foods for Canadians Act* (SFCA, 2012), and any other sector or product-specific legislation. Labeling must be truthful, verifiable, and not misleading (FDA, s 5(1) and SFCA, s 6(1)). Mandatory labeling is required where foods have significant nutritional and compositional changes, or where potential health and safety risks exist that could be mitigated through labeling (see FDA, s 5(1), and SFCA s 6(1) and the corresponding regulations). For example, the label must state the nature of the nutritional or compositional change or indicate the presence of an allergen but, as described above, it is not mandatory to state that the product is genetically engineered.

In Canada, genetically engineered foods do not currently have traceability requirements over and above the traceability requirements applicable to all foods. The *Safe Foods for Canadian Regulations* (SFCR, 2018) create requirements such that food can be traced one step forward and one step back to ensure unsafe food is quickly removed from the market. Other sector-specific traceability requirements may apply depending on the type of food (e.g., livestock).

Animal cell culture technology

As discussed in the United States section above, animal cell culture technology is an emerging food production technique involving the growth of animal cells in a controlled environment to produce food (e.g., "cultured meat"). Currently, there is no specific legislation or guidance regarding cultured meat in Canada, nor has any cultured meat been approved in Canada. As such, the regulatory pathway and requirements have not been tested in Canada.

As an emerging technology, it is expected that cultured meat will be regulated as a novel food under the FDA and therefore subject to premarket notification and assessment by Health Canada. However, as the industry matures, it is foreseeable that at least some cultured meats will no longer trigger the Novel Food Regulations (i.e., where the end-product has a history of safe use as food, the characteristics of the end-product fall within the anticipated range, and the processes of manufacture are familiar to the regulator). Novel food products may also be subject to premarket notification and assessment by the CFIA as a novel feed under the *Feeds Act* if there is a risk that the food product may enter the animal feed system.

Although beyond the scope of this chapter, cultured cells used in manufacture may be subject to environmental assessment under CEPA or "CEPA-equivalent" legislation, as well as applicable import requirements. Like all other foods, food products produced using animal cell culture technology would be required to meet applicable Canadian food safety requirements, including the provisions of the *SFCA* and *FDA* (see Cellular Agriculture Canada 2020 for detailed review).

Labeling of cultured meat

There is currently no specific guidance regarding labeling of cultured meat in Canada, and cultured meat does not fit neatly under the current definitions within the FDR. The definition of "meat" and "meat by-product" refers to edible parts of an animal that was healthy at slaughter, which appears to exclude cultured meat. The term "simulated meat product," conventionally used for vegetarian products, is defined as "any food that does not contain any meat product…but has the appearance of a meat product." The guidance indicates "consumers must not be misled about the true nature of these products" and the label must include the phrase "contains no meat" (CFIA, 2021c). Further amendment of the regulations and/or guidance may be required to address this gap. The CFIA is expected to publish updated guidance on simulated meat products but has not indicated whether it will address cultured meat (CFIA, 2021d). For labeling of cultured meat products based on genetically altered cells, please also see the section above regarding labeling of genetically altered foods. Finally, like other foods, food products produced using animal cell culture technology would be subject to applicable general labeling requirements in Canada.

Regulatory framework for foods in the European Union

Just as in the United States and in Canada, the European Union ("Union") sees a lot of innovative food business operators using techniques of genome editing or synthetic biology. Although the regulatory framework applicable to the products resulting thereof is in place, the application of this framework in practice sometimes is a struggle. This is, on the one hand, because the regulatory authorities apply, for the time being, a process rather than a product-oriented approach as in the United States and Canada. On the other hand, the general public in the Union seems to be less ready than in the United States and in Canada to embrace products resulting from genome editing or synthetic biology for food purposes. Genetically modified ("GM") foods, for instance, are still considered controversial, as follows, for instance, from the action taken by the French Group Confédération Paysanne, resulting in ECJ C-528/16 (as discussed in "Material requirements under novel food and GMO legislation" section below). As a result, there is a clear split in the regulatory framework between GM foods and non-GM foods. To understand the dynamics around the safety evaluation and enforcement of food products resulting from genome editing or synthetic biology in the Union, this part of the chapter will first introduce (A) the bodies involved in such evaluation and enforcement, followed by a summary of (B) the current regulatory framework on EU food legislation, as well as a window to the future. It will then dive into (C) material requirements applicable to food products resulting from genome editing or synthetic biology followed by product examples.

Bodies involved in EU food legislation and safety evaluation and enforcement

The EU bodies involved in the creation and application of EU food legislation are the European Commission ("Commission"), the European Parliament, the

European Council, the European Food Safety Authority ("EFSA"), the EU Member States via the Committee on Plant, Animals, Food and Feed ("PAFF Committee"), and the European Court of Justice ("ECJ"), respectively.

The Commission is alone responsible for drawing up proposals for new EU legislation, based on shared competence with the Member States, which proposals need to be endorsed by the European Parliament and the Council. Most food products do not require prior market authorization for access to the Union market. Exceptions apply to Novel Foods and GM foods among others. Such market authorization is granted by the Commission based on a safety dossier. For that purpose, the Commission may request a safety opinion from EFSA. Furthermore, the Member States have a say in this process prior to the Commission granting any authorization via the PAFF Committee, consisting of Member States representatives with material knowledge and experience in the field.

Although material laws in the field of food have been largely harmonized at a Union level, enforcement of Union laws is a national affair. This means enforcement of EU food law takes place at Member State level by competent authorities. The competent authorities of the Member States not only check if required authorizations have been obtained, but also if the conditions of use are met, so that food safety can indeed be ensured. Furthermore, they check if the communication on food products is adequate, in terms of both mandatory food information and voluntary information, such as claims. National disparities in enforcement practices regularly end up with the ECJ.

Current regulatory framework on EU food legislation and window to the future

Harmonized EU food law is a relatively young phenomenon: in fact, the BSE-crisis in the 1980s and dioxin-crisis in the 1990s resulted into the General Food Law Regulation (General Food Law Regulation, 2002, hereafter "GFL"). This instrument lays down the general principles and requirements of food law. It forms the statutory basis for EFSA, and it lays down procedures in matter of food safety. The GFL applies to all stages of the production, processing, and distribution of food and also of feed produced for or fed to food-producing animals ("*from farm to fork*"). Products produced under legislation predating the GFL, such as Novel Foods and GM foods, should also satisfy the general food safety criteria formulated in the GFL.

In December 2019, the Commission published the European Green Deal (European Commission, n.d.-a), of which the Farm to Fork Strategy (European Commission, n.d.-b), published in May 2020, forms a vital part. The most important objective of this strategy is to make the EU food system more sustainable, also against the background of the Commission's agenda to achieve the UN Sustainable Development Goals ("SDGs"). This stands in contrast to the United States, which has not incorporated SDGs into regulatory oversight.

One would expect that synthetic biology would hold a prominent place in this policy document. However, the term is not even mentioned once. Avenues identified to enhance the sustainability of the EU food system are reduction of dependency on pesticides and of excess fertilization among others. Whereas EFSA has issued positive scientific opinions regarding crops embodying certain genetic traits removing the need of fertilizers or pesticides (EFSA, n.d.-a), GM foods are not mentioned in the European Green Deal either. Instead, the Commission announces that it will promote greater use of safe alternative ways of protecting harvests from pests and diseases, such as crop rotation and mechanical weeding. Also, the placing on the market of pesticides containing biological active substances will be facilitated.

In its quest for more sustainable livestock production, whereby it is mentioned that 68% of the total agricultural land in the EU is used for animal production, cultured meat is not mentioned either. Likewise, the European Green Deal recognizes that the shift to more sustainable fish and seafood production must be accelerated but does not consider cultured fish as a source of alternative protein.

It is only in the context of research and innovation ("*enabling the transition*") that increasing the availability and source of alternative proteins, such as plant, microbial, insect-based protein, and meat substitutes, are mentioned as a key area of research. Furthermore, it is recognized that for increasing sustainable food production, new innovative techniques, including biotechnology and bio-based products, may play a role, "*provided that they are safe for consumers and the environment, while bringing benefits for society as a whole.*" In this context, the potential of new genomic techniques to improve sustainability along the food supply chain is mentioned.

In April 2021, the Commission published a study on the status of new genomics techniques ("NGTs") under Union law (European Commission, 2021). NGTs are defined as techniques that are capable of altering the genetic material of an organism and that have evolved since 2001 (when the EU GMO Directive became applicable). This study was requested by the Council, further to the ECJ decision on the regulatory status of certain plant breeding techniques (ECJ C-528/16), in which it was held that organisms obtained by new (i.e., post 2001) mutagenesis techniques constituted GMOs under the GMO Directive (GMO Directive, 2001). The Council recognized that the above ECJ decision had brought clarity regarding the regulatory status of mutagenesis in the Union, but at the same time, it raised practical concerns for national competent authorities and for the Union as a whole, in terms of research and beyond. For example, how to ensure equal treatment of EU products vis-à-vis imports, when products obtained with new mutagenesis techniques are not distinguishable from those resulting from natural mutations?

The scope of the Commission study covers the use of NGTs in plants, animals, and microorganisms, in a broad variety of potential applications, including agri-food, medicinal, and industrial sectors. The Commission study establishes there is considerable interest in NGTs in the Union, but most of the development is taking place outside of the EU. The ECJ ruling referred to above has not helped in this respect.

At the same time, the Commission recognizes several of the plant products obtained from NGTs may contribute to the objective of the Farm to Fork strategy and the UN SDGs for a more resilient and sustainable agri-food system. Examples include plants more resistant to diseases and environmental conditions of climate change. Reticence against NGTs is formulated by the EU organic and GMO-free premium market, fearing threats from co-existence with NGTs. They expect that any consideration of NGTs outside the current GMO regulatory framework of the Union would negatively impact their value chain and risk damage to consumer trust in their sector.

However, the study confirms that the current regulatory system in the EU represents implementation and enforcement challenges, especially regarding the detection of NGT products that do not contain any foreign genetic material. Even if existing detection methods may be able to detect small alterations in the genome, this does not necessarily confirm the presence of a regulated product, as the same alterations could have been obtained by conventional breeding.

Moreover, the study recognizes that in light of different regulatory oversight of NGTs in other countries, there could be trade limitations and disruptions that would put the Union at a competitive disadvantage. In sum, the Commission seems to recognize the current EU GMO regulatory framework is no longer fit for purpose and it needs to be made more resilient, future proof, and uniformly applied. The Commission thereby recognizes that a purely risk-based approach may not be enough to promote sustainability and obtain the objectives from the Green Deal.

As a follow-up action, the Commission has published an impact assessment end of September 2021, including a public consultation, to examine potential policy actions. This should eventually lead to a fit-for-purpose regulatory framework. For the time being, the impact assessment will focus on the applications of NGTs in plants (so not on applications in animals or microorganisms even if they were part of the study). The rationale is that according to EFSA, the application of NGTs in plants does not pose additional risks compared to conventionally bred plants (European Commission, n.d.-c). Following this impact assessment a legislative proposal for plants produced by certain new genomic techniques is expected by Q2 2023.

Material requirements under novel food and GMO legislation

Foods resulting from the application of genome editing or synthetic biology will most often be subject to either the Novel Food Regulation (Novel Foods Regulation, 2015) or the GMO legislation, consisting of the GMO Directive, the GMO Regulation on genetically modified food and feed (GMO food and feed Regulation, 2003), and the GMO Traceability and Labeling Regulation

(GMO T&L Regulation, 2003) among others. These two regimes are similar in the sense that both Novel Foods and GM foods require prior market approval based on a comprehensive and lengthy (count between 2 and 3 years) risk assessment, in which the EC, EFSA, and the PAFF Committee are involved. They differ in various ways: (i) *as to the substance*: Novel Foods procedure focuses on food safety exclusively; GMO procedure focuses on animal health and environment as well; (ii) *in the way of requesting authorization*: via e-submission at a central level with EC for Novel Foods, at national Member State level for GM foods; (iii) *as to the aspects involved*: legal and scientific aspects for Novel Foods, also ethical/political aspects for GMOs, and (iv) *as to the term of authorization*: unlimited in time for Novel Foods, 10-year term for GM foods (extension possible).

Product definitions

Novel Foods are defined as foods that were not used for human consumption to a significant degree within the Union before May 15, 1997, and that fall under at least one out of the categories listed therein. The most relevant Novel Food categories for bioengineered food products are foods consisting of, isolated from or produced from (i) microorganisms/fungi/algae; (ii) animals or their parts (except for animals obtained by traditional breeding processes); (iii) cell or tissue culture derived from animals, plants, microorganisms, fungi, or algae; or (iv) foods resulting from a new production process, where "new" is meant to predate 1997.

GM foods include (i) foods *containing or consisting of* a GMO and (ii) foods *produced from* GMOs, but in principle do not include (iii) foods *products with* GMOs. Foods *produced from* GMOs are food products that are derived, in whole or in part, from GMOs, but that do not contain or consist of GMOs (e.g., soy oil). Foods produced with GMOs are foods that have applied GMOs in their production process, the traces of which can no longer be found in the end-product (e.g., food enzyme catalase produced with GM *Aspergillus niger* strain, EFSA, 2021).

GMO is defined as "*an organism (...), in which the genetic material has been altered in a way that does not occur naturally by mating and/or natural recombination.*" The wording "altered in a way" is somewhat ambiguous, as it is not immediately clear if this applies to the process applied or the end-result obtained. In the ECJ decision on the regulatory status of certain plant breeding techniques (ECJ C-528/16), clearly a process-based approach was applied. Shortly after this decision however, the Group of Scientific Advisors advocated that the features of the final product itself must be examined regardless of the underlying technique used to generate that product (European Commission, 2018). This view seems to be confirmed in the study on New Genomic Techniques, published by the European Commission in April 2021.

Authorization regimes

Market authorization for a Novel Food will be granted once an applicant has demonstrated that the product is safe, based on a Novel Food dossier. Such dossier must be submitted to the European Commission, who most often requests a safety evaluation by EFSA. The parameters for demonstrating food safety in a Novel Food dossier comprise, among others, a description of the production process/full compositional data/proposed uses and levels of anticipated intake/ADME/toxicological information/allergenicity. In its application for a Novel Food authorization, the applicant must also make a proposal for how the product will be labeled. Under "Product examples" below, the implications thereof for several food products shall be discussed by way of example.

Market authorization for GM foods comes into play in case of deliberate release of GMOs into the environment. "*Deliberate release*" is defined as any intentional introduction into the environment of a GMO for which no specific containment measures are used to limit their contact with and to provide a high level of safety for the general population and the environment. This is clearly the case when marketing GM foods.

The notification for GM foods must be accompanied by an environmental risk assessment, and it should specify, among other things, the conditions for placing on the market, a monitoring plan, and a proposal for labeling and packaging, as well as a summary of the safety dossier. The competent authority to whom the notification is made shall subsequently forward this to the European Commission, who will forward it to the competent authorities of the other Member States. The Member States participate in the evaluation of the notification by sending comments or reasoned objections to the Commission.

For GM food products, a risk assessment by EFSA forms a mandatory part of the authorization process. Authorization may be granted only if the risk assessment demonstrates that, under its intended conditions of use, the product has no adverse effects for human and animal health and for the environment, does not mislead the user or the consumer and is not nutritionally disadvantageous compared to the food or feed it is intended to replace.

The consent will be granted by the national competent authority to whom the notification for deliberate release of a GMO was made, clearly stating the conditions of use, as well as the labeling and (postmarketing) monitoring requirements. Once consent for the deliberate release of a GMO food product has been granted, Member States may not prohibit, impede, or restrict the placing on the market of GMOs. The GMO Directive, however, contains a safeguard clause, according to which Member States may provisionally restrict or prohibit the use and/or sale of that GMO as or in a product on its territory, if new information reveals it constitutes a risk to human health or the environment.

Under the GMO Traceability and Labeling Regulation, operators placing GMOs and GM food and feed on the market must inform the operators receiving the products, in writing, that the products contain or consist of GMOs. They must provide an indication of each ingredient/material produced from GMOs or, for products without an ingredients list, an indication that the product is produced from GMOs. They must keep that information for 5 years and be able to identify the operator(s) by whom and to whom the products have been made available. The rationale is that traceability requirements allow for close monitoring of potential effects of the product on environment and health, and where necessary for the withdrawal of products if an unexpected risk to human health or to the environment is detected. Furthermore, the GMO Traceability and Labeling Regulation requires that all products consisting of or containing GMOs be labeled as such. This requirement does not apply to food and feed products which contain, consist of, or are produced from GMOs in a proportion no higher than 0.9% of the food ingredients considered individually or food consisting of a single ingredient, provided that this presence is adventitious or technically unavoidable.

Product examples

Cultured meat

Depending on the starting cell types used or the techniques applied, either the Novel Food or the GMO legislation is applicable to this type of food. For instance, if the end-product contains any GMOs, it is produced from GMOs or during the production process GM techniques such as directed mutagenesis have been applied, GMO legislation will be applicable. In Europe, most cultured meat companies try to avoid genetic modification steps in their production process for reasons set out earlier. Therefore, we expect most of the cultured meat products in Europe to obtain market access based on the Novel Foods Regulation.

Under the Novel Foods Regulation, the applicable category will be "*foods consisting of, isolated from or produced from cell or tissue culture from animals*" and potentially "*food resulting from a new production process.*" Any cultured meat company wanting to place cultured meat on the EU market will have to submit a dossier demonstrating the product is safe for food use. The notion of "*placing on the market*" is quite broad; it comprises any offering for sale or any other form of transfer, whether free of charge or not, but also the sale, distribution, and other forms of transfer themselves. As such, many Member States take the view that public tasting sessions of cultured meat prior to obtaining market approval are prohibited (Verzijden and Buijs, 2020). In the Netherlands, a majority of the Dutch Parliament voted in March 2022 in favor of a motion urging the Health and/or Agricultural Ministry to work with the cultured meat companies to permit tasting events in a controlled environment. Also in other EU countries (e.g., France and Germany), tastings of cultured meat products took place without enforcement by the local food safety authorities.

The Novel Food dossier for a cultured meat product should provide information on, among others, the identity of the cells used, and the production process applied (including culture and harvesting conditions, substances used in the manufacturing process, as well as postharvesting conditions).

Any application for a Novel Food authorization of cultured meat must also contain a proposal how the product will be labeled. A burning question in this context is whether the product can actually be called "meat." According to the EU Hygiene Regulation (Hygiene Regulation, 2004), "meat" is defined as "edible parts" of a specific animals. Since cells are not really "edible parts," this question has not been decided yet. We anticipate that the combination of "meat" with a qualifier could be a potential solution. If this is acceptable too for the conventional meat lobby remains to be seen.

Once an authorization for a cultured meat product is granted, this will be published in the Union List, allowing any third party to market the authorized product, provided that the conditions of use are met. To protect competitive interests and investments made, requests for confidential treatment and/or data protection can be filed. If granted, confidential treatment will last for an unlimited period, where data protection is limited to a 5-year term after authorization. Data that are vital for the evaluation for any safety risks are, however, exempt from confidential treatment.

Alternative dairy product using precision fermentation

This example relates to cheese made from animal-free milk proteins. To create such products, selected microorganisms are encoded with DNA sequences to produce milk protein.

Although the production process of this alternative dairy product contains a GM step, this does not necessarily mean that market authorization will need to be obtained based on GMO legislation. The question that needs to be answered here is whether this product is "produced with" or "produced from" GMOs. For that purpose, it is of the essence if the GM step in the production process is sufficiently isolated from successive steps. In the affirmative, the end-product would normally not contain any enabler DNA. Provided that the process-based approach of safety evaluations so far will no longer be leading when evaluating the safety of this product, the Novel Food route would be most likely to apply. The applicable Novel Foods category would be "*foods produced from microorganisms*" or "*food resulting from a new production process.*" If evidence can be found that the microorganism applied at stake was also used in fermentation (which qualifies as a traditional food preparation method in EU law), Novel Food clearance will indeed not be necessary.

Flavoring made from GM yeast strain

This example covers the application of soy leghemoglobin ("heme" in short) as a flavoring in meat analogues produced from the GM *Pichia pastoris* as used by Impossible Foods in its Impossible Burger. So far, this food product has not yet been marketed in Europe, as the GMO application that was filed on September

30, 2019, with the competent authorities of the Netherlands, is still pending at EU level. In parallel, Impossible Foods has applied for authorization under the EU Flavorings Regulation (Flavorings Regulation, 2008).

In its GMO application (EFSA, n.d.-b), Impossible Foods describes its flavoring as a product derived from a GMO in which GMOs capable of multiplication or of transferring genes are not present, but in which newly introduced genes are still present. Based on its status as QPS, *Pichia pastoris* is not expected to have any adverse impact on human or animal health, or on the environment. Considering that all of the genetic modifications have been well characterized through genetic and phenotypic analyses, confirming no unintended changes take place in the production chain, the changes that have been made to the *Pichia pastoris* production strain are not anticipated to result in potential toxic, allergenic, or other harmful effects on human health.

Under "conditions for placement on the market," Impossible Foods specifies only the finished product containing soy leghemoglobin will be imported. Thus, the GM *Pichia pastoris* strain will not enter the EU. The maximum rate of inclusion of soy leghemoglobin in meat analogue products is 0.8 g per 100 g of meat analog, although typical use rates are generally around 0.45 g per 100 g of meat analogue.

As to the potential environmental impact, the GMO application states that the soy leghemoglobin produced from the GM *Pichia pastoris* will be used only internally by Impossible Foods for incorporation into their own meat analogue products. The production process of soy leghemoglobin is conducted in accordance with GMP and under control systems. As such, the potential for inadvertent introduction of recombinant DNA into the environment is negligible.

EFSA acknowledged receipt of Impossible's application on October 21, 2019, but since then no further update is available on public record.

Conclusion

Rapid developments in food science and agricultural production are challenging U.S., Canadian, and European Union regulators to keep pace with innovations that are important in helping to address specific SDGs, including 2 ("End hunger, achieve food security and improved nutrition and promotion of sustainable agriculture"), 12 ("Ensure sustainable consumption and production patterns"), and 13 ("Take urgent action to combat climate change and its impacts"). In each jurisdiction, regulators are actively working to address gaps and uncertainties in the current regulatory frameworks. Thus far, the United States and Canada continue to take a product-based approach, largely focused on the end-product of innovation, while the EU explicitly regulates based on the process of production. While the United States recently updated USDA regulations relating to permit and food labeling requirements, FDA's approach to genome editing and synthetic biology in foods remains a work in progress. For its part, the Canadian government has signaled that additional regulatory and guidance updates are forthcoming as part of a broader, multiyear effort to modernize legislation to reflect emerging developments in biotechnology. As a result, Canada's approach to the regulation of biotechnology remains in flux. Certain parts of EU legislation are recognized no longer to be fit for purpose to regulate products resulting from genome editing. Changes are anticipated but it is not expected these will be rapidly implemented. Other parts are in full shape and the application thereof to bioengineered food will set an important precedent for similar food products. At present, it is thus very much incumbent on product researchers and developers to understand and stay apprised of regulatory developments and to be aware that requirements may differ among jurisdictions.

References

United States

[AMS] Agricultural Marketing Services. 2018 Dec 21. National Bioengineered Food Disclosure Standard. [accessed 2021 Sept 6]; 83 Fed. Reg. 65814. https://www.govinfo.gov/content/pkg/FR-2018-12-21/pdf/2018-27283.pdf.

[APHIS] Animal & Plant Health Inspection Services. 2016 Apr. 16. Letter from Michael J. Firko, Ph.D. to Yinong Yang, College of Agriculture Sciences, Pennsylvania State University. [accessed 2021 Sept 26]. Available from: https://www.aphis.usda.gov/aphis/ourfocus/biotechnology/am-i-regulated/Regulated_Article_Letters_of_Inquiry.

[APHIS] Animal & Plant Health Inspection Services. 2018 May 14. Letter from Michael J. Firko, Ph.D. to Harry J. Klee, University of Florida, Horticulture Sciences. [accessed 2021 Sept 26]. Available from: https://www.aphis.usda.gov/aphis/ourfocus/biotechnology/am-i-regulated/Regulated_Article_Letters_of_Inquiry.

[APHIS] Animal & Plant Health Inspection Services. 2020a May 18. Movement of Certain Genetically Engineered Organisms. [accessed 2021 Sept 6]; 85 Fed. Reg. 29790. https://www.govinfo.gov/content/pkg/FR-2020-05-18/pdf/2020-10638.pdf.

[APHIS] Animal & Plant Health Inspection Services, 2020 May 28. Letter from Bernadette Juarez to Chloe Pavely, Calyxt Inc. APHIS, Riverdale, MD. [Accessed 26 September 2021]. Available from: https://www.aphis.usda.gov/aphis/ourfocus/biotechnology/am-i-regulated/Regulated_Article_Letters_of_Inquiry.

[APHIS] Animal & Plant Health Inspection Services, 2020 Jul 21. Letter from Bernadette Juarez to Gary Bannon, Benson Hill, Inc. APHIS, Riverdale, MD. [Accessed 26 September 2021]. Available from: https://www.aphis.usda.gov/aphis/ourfocus/biotechnology/am-i-regulated/Regulated_Article_Letters_of_Inquiry.

[APHIS] Animal & Plant Health Inspection Services, 2021 Jul 19. Movement of Organisms Modified or Produced Through Genetic Engineering. 86 Fed. Reg. 37988. https://www.govinfo.gov/content/pkg/FR-2021-07-19/pdf/2021-15236.pdf. (Accessed 6 September 2021).

[FDA] Food & Drug Administration, 1992 May 29. Statement of Policy; Foods Derived from New Plant Varieties. 57 Fed. Reg. 22984. https://archives.federalregister.gov/issue_slice/1992/5/29/22970-23005.pdf#page=15. (Accessed 6 September 2021).

[FDA] Food & Drug Administration, 2010 Jul. Guidance for Industry, Enzyme Preparations: Recommendations for Submission of Chemical and Technological Data for Food Additive Petitions and GRAS Notices. FDA, Rockville, MD. https://www.fda.gov/media/79379/download. (Accessed 17 August 2021).

[FDA] Food & Drug Administration, 2010 Nov. 4. Memorandum, Experience with GRAS Notices Under the 1997 Proposed Rule. Docket No. FDA-1997-N-0020: Rulemaking Docket https://www.regulations.gov/docket/FDA-1997-N-0020. (Accessed 1 September 2021).

[FDA] Food & Drug Administration, 2016 Aug 17. Final Rule, Substances Generally Recognized as Safe. 81 Fed. Reg. 54960 https://www.govinfo.gov/content/pkg/FR-2016-08-17/pdf/2016-19164.pdf. (Accessed 6 September 2021).

[FDA] Food & Drug Administration, 2017 Jan 19. Request for Comments, Genome Editing in New Plant Varieties Used for Foods. 82 Fed. Reg. 6564. https://www.govinfo.gov/content/pkg/FR-2017-01-19/pdf/2017-00840.pdf.

[FDA] Food & Drug Administration, 2018 Jan 4. Microorganisms & Microbial-Derived Ingredients Used in Food (Partial List). FDA. https://www.fda.gov/food/generally-recognized-safe-gras/microorganisms-microbial-derived-ingredients-used-food-partial-list. (Accessed 17 August 2021).

[FDA] Food & Drug Administration, 2018 Mar 9. Final Biotechnology Consultations. FDA. https://www.fda.gov/food/consultation-programs-food-new-plant-varieties/final-biotechnology-consultations. (Accessed 17 August 2021).

[FDA] Food & Drug Administration, 2018 Jul 23. Letter from Dennis M. Keefe to Gary L. Yingling Re: GRAS Notice No. GRN 737. FDA, College Park, MD. [Accessed 1 September 2021]. Available from: https://www.fda.gov/media/116243/download.

[FDA] Food & Drug Administration, 2019 Mar 7. Formal Agreement between FDA and USDA Regarding Oversight of Human Food Produced Using Animal Cell Technology Derived from Cell Lines of USDA-amenable Species. FDA. https://www.fda.gov/food/domestic-interagency-agreements-food/formal-agreement-between-fda-and-usda-regarding-oversight-human-food-produced-using-animal-cell. (Accessed 1 September 2021).

[FDA] Food & Drug Administration, 2019 Mar 21. FDA's Voluntary Plant Biotechnology Consultation Program Eases Pathway to Marketplace. FDA. https://www.fda.gov/news-events/fda-voices/fdas-voluntary-plant-biotechnology-consultation-program-eases-pathway-marketplace. (Accessed 17 August 2021).

[FDA] Food & Drug Administration, 2020 Mar 30. Consultation Programs on Food from New Plant Varieties. FDA. https://www.fda.gov/food/food-new-plant-varieties/consultation-programs-food-new-plant-varieties. (Accessed 17 August 2021).

[FDA] Food & Drug Administration, 2020 May 25. Letter from Susan J. Carlson to Melvin Drozen Re: GRAS Notice No. GRN 863. FDA, College Park, MD. [Accessed 1 September 2021]. Available from: https://www.fda.gov/media/136751/download.

[FDA] Food & Drug Administration, 2020 Oct 6. Food Made with Cultured Animal Cells. FDA. https://www.fda.gov/food/food-ingredients-packaging/food-made-cultured-animal-cells. (Accessed 1 September 2021).

[FDA] Food & Drug Administration, 2020 Oct 7. Request for Information; Labeling of Foods Comprised of or Containing Cultured Seafood Cells. 85 Fed. Reg. 63277. https://www.govinfo.gov/content/pkg/FR-2020-10-07/pdf/2020-22140.pdf. (Accessed 6 September 2021).

[FDA] Food & Drug Administration, 2020 Oct 27. Food from New Plant Varieties. FDA. https://www.fda.gov/food/food-ingredients-packaging/food-new-plant-varieties. (Accessed 17 August 2021).

[FDA] Food & Drug Administration, 2020 Dec 21. Animals with Intentional Genomic Alterations. FDA. https://www.fda.gov/animal-veterinary/biotechnology-products-cvm-animals-and-animal-food/animals-intentional-genomic-alterations. (Accessed 26 September 2021).

[FDCA] Food, Drug, & Cosmetics Act. n.d. U.S.C. Sect. 321.

Food Additive Safety, 2020. 2 C.F.R. Sect. 170.30(a).

[FSIS] Food Safety and Inspection Service, 2021 Sept 3. Advance Notice of Proposed Rulemaking; Labeling of Meat or Poultry Comprised of or Containing Cultured Animal Cells. 86 Fed. Reg. 49491. https://www.govinfo.gov/content/pkg/FR-2021-09-03/pdf/2021-19057.pdf. (Accessed 26 September 2021).

Movement of Organisms Modified or Produced through Genetic Engineering. 7 C.F.R. part 340 n.d.

National Bioengineered Food Disclosure Standard, 2016. Pub. Law. No. 114-216, 130 Stat. 834.

[OSTP] Office of Science & Technology Policy, 1986 Jun 26. Coordinated Framework for Regulation of Biotechnology. 51 Fed. Reg. 23303. https://archives.federalregister.gov/issue_slice/1986/6/26/23299-23366.pdf#page=4. (Accessed 16 August 2021).

[OSTP] Office of Science & Technology Policy, 1992 Feb 27. Announcement of Policy, Exercise of Federal Oversight Within Scope of Statutory Authority: Planned Introductions of Biotechnology Products Into the Environment. 57 Fed. Reg. 6753. https://archives.federalregister.gov/issue_slice/1992/2/27/6748-6760.pdf#page=6. (Accessed 6 September 2021).

[OSTP] Office of Science & Technology Policy, 2016 Sept. National Strategy for Modernizing the Regulatory System for Biotechnology Products. https://www.fda.gov/media/102667/download. (Accessed 26 September 2021).

[USDA, FDA, EPA] Food & Drug Administration, 2017. Modernizing the Regulatory System for Plant and Animal Biotechnology Products. FDA. [Accessed 26 September 2021]. Available from https://www.fda.gov/safety/fdas-regulation-plant-and-animal-biotechnology-products/modernizing-regulatory-system-plant-and-animal-biotechnology-products.

Canada

Cellular Agriculture Canada, 2020 Sept. First Steps Towards a Regulatory Framework for Cultured Food Products in Canada. Cellular Agriculture Canada. https://static1.squarespace.com/static/5d2bab7b430eb50001e6b381/t/5f68ce6d347119736c819cf6/1600704113725/First+Steps+Towards+a+Regulatory+Framework+for+Cultured+Food+Products+in+Canada.pdf. (Accessed 12 August 2021).

[CEPA] Canadian Environmental Protection Act, 1999. SC, c 33.

[CFIA] Canadian Food Inspection Agency, 2016 May 19. Decision Document DD2016–117 Determination of the Safety of AquaBounty Technologies Inc.'s Salmon (*Salmo salar* L.) Event. CFIA. https://inspection.canada.ca/plant-varieties/plants-with-novel-traits/approved-under-review/decision-documents/dd2016-117/eng/1463076782568/1463076783145. (Accessed 12 August 2021).

[CFIA] Canadian Food Inspection Agency, 2021 May 19. Draft Guidance for Determining Whether a Plant is subject to Part V of the Seeds Regulations. CFIA. https://inspection.canada.ca/about-cfia/transparency/consultations-and-engagement/share-your-thoughts/draft-guidance/eng/1619540046303/1619540212691. (Accessed 12 August 2021).

[CFIA] Canadian Food Inspection Agency, 2021 May 19. Summary of the Guidance for Determining Whether a Plant is Subject to Part V of the Seeds Regulations. CFIA. https://inspection.canada.ca/about-cfia/transparency/consultations-and-engagement/share-your-thoughts/summary-of-the-guidance/eng/1619736173981/1619739540987. (Accessed 12 August 2021).

[CFIA] Canadian Food Inspection Agency, 2021 Jun 4. Labelling Requirements for Meat and Poultry Products: Simulated Meat and Simulated Poultry Products. CFIA. https://inspection.canada.ca/about-cfia/transparency/consultations-and-engagement/proposed-regulatory-framework-for-livestock-feeds/what-we-heard-report/eng/1625073587806/1625073588369#a4. (Accessed 12 August 2021).

[CFIA] Canadian Food Inspection Agency, 2021 Jul 9. What We Heard Report—Consultation on Canada's Proposed Guidelines for Simulated Meat and Poultry. CFIA. https://inspection.canada.ca/about-cfia/transparency/consultations-and-engagement/proposed-regulatory-framework-for-livestock-feeds/what-we-heard-report/eng/1625073587806/1625073588369#a4. (Accessed 12 August 2021).

[CFIA] Canadian Food Inspection Agency, 2021 Jul 19. Share Your Thoughts: Guidance for Determining Whether a Plant is Subject to Part V of the Seeds Regulations. https://inspection.canada.ca/about-cfia/transparency/consultations-and-engagement/share-your-thoughts/eng/1619740964754/1619741042405. (Accessed 12 August 2021).

[DFO] Fisheries and Oceans Canada, 2013. Summary of the Environmental and Indirect Human Health Risk Assessment of AquAdvantage Salmon. National Capital Region: DFO Canadian Science Advisory Secretariat Response 2013/023. https://waves-vagues.dfo-mpo.gc.ca/Library/361091.pdf. (Accessed 12 August 2021).

[FDA] Food and Drugs Act, 1985. RSC, c F-27.

[FDR] Food and Drugs Regulations, 1978. CRC, c 870, Division 28 (also known as "Novel Food Regulations").

Feeds Act, 1985. RSC, c F-9.

Feeds Regulations, 1983. SOR/83-593.

Fertilizers Act, 1985. RSC, c F-10.

[FSDA] Federal Sustainable Development Act, 2008. R.S.C. c. 33.

[HAA] Health of Animals Act, 1990. SC, c 21.

Health Canada, 2014 Aug. Policy for Differentiating Food Additives and Processing Aids. Health Canada. https://www.canada.ca/en/health-canada/services/food-nutrition/reports-publications/policy-differentiating-food-additives-processing-aids-2008.html. (Accessed 12 August 2021).

Health Canada, 2014 Aug 11. Enzymes used in Food Processing. Health Canada. https://www.canada.ca/en/health-canada/services/food-nutrition/food-safety/food-additives/enzymes-used-processing.html. (Accessed 12 August 2021).

Health Canada, 2016 May 19. Health Canada and Canadian Food Inspection Agency Approve AquAdvantage Salmon. Health Canada, Ottawa, ON. https://www.canada.ca/en/health-canada/news/2016/05/health-canada-and-canadian-food-inspection-agency-approve-aquadvantage-salmon.html?_ga=2.119741064.1617568643.1505365662-2077739461.1502341793. (Accessed 12 August 2021).

Health Canada, 2016 May 19. Novel Food Information—AquAdvantage Salmon. Health Canada, Ottawa, ON. https://www.canada.ca/en/health-canada/services/food-nutrition/genetically-modified-foods-other-novel-foods/approved-products/novel-food-information-aquadvantage-salmon.html. (Accessed 12 August 2021).

Health Canada, 2018 Dec 5. Novel Food Information—2′-Fucosyllactose (2'-FL) from *Escherichia coli* BL21 (DE3) Strain #1540. Health Canada, Ottawa, ON. https://www.canada.ca/en/health-canada/services/food-nutrition/genetically-modified-foods-other-novel-foods/approved-products/2-fucosyllactose-escherichia-coli-bl21/technical-summary.html. (Accessed 12 August 2021).

Health Canada, 2020 Jan 6. Soy Leghemoglobin (LegH) Preparation as an Ingredient in a Simulated Meat Product (i.e., The Impossible™ Burger) and Other Ground Beef Analogues. Health Canada, Ottawa, ON. https://www.canada.ca/en/health-canada/services/food-nutrition/genetically-modified-foods-other-novel-foods/approved-products/soy-leghemoglobin/document.html. (Accessed 12 August 2021).

Health Canada, 2020 Feb 10. Frequently Asked Questions: AquAdvantage Salmon. Health Canada. https://www.canada.ca/en/health-canada/services/food-nutrition/genetically-modified-foods-other-novel-foods/approved-products/frequently-asked-questions-aquadvantage-salmon.html. (Accessed 12 August 2021).

Health Canada, 2021 Mar. Consultation Document: Proposed Changes to Health Canada Guidance on the Interpretation of Division 28 of Part B of the Food and Drug Regulations (The Novel Food Regulations): When Is a Food that Was Derived from a Plant Developed Through Breeding a "Novel Food"?.

Health Canada, 2021 Mar. Consultation Document: Proposed Health Canada Guidance on the Pre-Market Assessment of Foods Derived from Retransformants Under of Division 28 of Part B of the Food and Drug Regulations (The Novel Food Regulations).

Health Canada. 2022a May. Guidelines for the Safety Assessment of Novel Foods. [place unknown]: Health Canada; s. 3.2. [accessed 2022 Jun 11]. https://www.canada.ca/en/health-canada/services/food-nutrition/legislation-guidelines/guidance-documents/guidelines-safety-assessment-novel-foods-derived-plants-microorganisms/guidelines-safety-assessment-novel-foods-2006.html.

Health Canada. 2022b. Notice of intent to propose amendments to Division 28 of the Food and Drug Regulations (Novel Foods). Ottawa (ON): Health Canada; [accessed 2022 Jun 11]. https://www.canada.ca/en/health-canada/services/food-nutrition/public-involvement-partnerships/notice-intent-propose-amendments-division-28-food-drug-regulations-novel-foods.html.

[PPA] Plant Protection Act, 1990. SC, c 22.

[PSPC] Public Services and Procurement Canada, 2019 Nov 6. Voluntary Labelling and Advertising of Foods that Are and Are Not Products of Genetic Engineering. CAN/CGSB-32.315-2004, PSPC, p. iii. https://www.tpsgc-pwgsc.gc.ca/ongc-cgsb/programme-program/normes-standards/internet/032-0315/index-eng.html. (Accessed 12 August 2021).

Seeds Act, 1985. RSC, c S-8.

Seeds Regulations, n.d. CRC, c 1400.

[SFCA] Safe Foods for Canadians Act, 2012. SC, c 24.

[SFCR] Safe Foods for Canadians Regulations, 2018. SOR/2018-108.

European Union

EC, 17.4.2001. Directive 2001/18/EC of the European Parliament and of the Council of 12 March 2001 on the deliberate release into the environment of genetically modified organisms and repealing Council Directive 90/220/EEC. OJ L 106., p. 1. https://eur-lex.europa.eu/legal-content/EN/TXT/PDF/?uri=CELEX:02001L0018-20080321&qid=1655189926934&from=EN.

EFSA.n.d.-a Scientific Opinion Assessment of Genetically Modified Oilseed Rape 73496 for Food and Feed Uses, Under Regulation (EC) No 1829/2003 (application EFSA-GMO-NL-2012-109). [17 June 2021]. https://efsa.onlinelibrary.wiley.com/doi/10.2903/j.efsa.2021.6610.

EFSA. n.d.-b Scientific Evidences Inventory for EFSA-Q-2019-00651; [20 Aug 2021]. https://open.efsa.europa.eu/study-inventory/EFSA-Q-2019-00651.

European Commission, 13 Nov 2008. Statement by the Group of Chief Scientific Advisors—A Scientific Perspective on the Regulatory Status of Products Derived from Gene Editing and the Implications for the GMO Directive. 8 pp https://ec.europa.eu/info/sites/default/files/2018_11_gcsa_statement_gene_editing_1.pdf.

European Commission, 29 April 2021. Study on the Status of New Genomic Techniques Under Union Law and in Light of the Court of Justice Ruling in Case C-528/16. Brussels. 116 pp https://ec.europa.eu/food/system/files/2021-04/gmo_mod-bio_ngt_eu-study.pdf.

European Commission n.d.-a A European Green Deal. [20 August 2021]. https://ec.europa.eu/info/strategy/priorities-2019-2024/european-green-deal_en.

European Commission n.d.-b. Farm to Fork strategy; [21 August 2021]. https://ec.europa.eu/food/horizontal-topics/farm-fork-strategy_en.

European Commission. n.d.-c New Techniques in Biotechnology: The Roadmap (Inception Impact Assessment) for the Initiative on Plants Obtained by Targeted Mutagenesis and Cisgenesis has Now Been Published and is Open for Feedback; [26 September 2021]. https://ec.europa.eu/food/plants/genetically-modified-organisms/new-techniques-biotechnology_en.

European Commission. n.d.-d New Genomic Techniques: the EU Paving the Way Ahead; [20 August 2021]. https://ec.europa.eu/newsroom/sante/items/710964/en.

European Food Safety Authority, 2021. Scientific opinion: scientific evaluation of the food enzyme catalase from the genetically modified *Aspergillus niger* strain DP-Azw-58. Adopted: 8 July 2021, EFSA J. 19 (8), 6787. https://efsa.onlinelibrary.wiley.com/doi/epdf/10.2903/j.efsa.2021.6787.

Judgement of the European Court of Justice of 25 July 2018 in Case C-528/16, 2018. https://eur-lex.europa.eu/legal-content/EN/TXT/PDF/?uri=CELEX:62016CJ0528&from=NL.

Regulation (EC) No 178/2002 of the European Parliament and of the Council of 28 January 2002 laying down the general principles and requirements of food law, establishing the European Food Safety Authority and laying down procedures in matters of food safety, OJ L 31, 1.2.2002, p. 1–24 https://eur-lex.europa.eu/legal-content/EN/TXT/PDF/?uri=CELEX:02002R0178-20140630&qid=1654200372741&from=EN.

Regulation (EC) No 1829/2003 of the European Parliament and of the Council of 22 September 2003 on genetically modified food and feed. OJ L 268, 18.10.2003, p. 1. https://eur-lex.europa.eu/legal-content/EN/TXT/PDF/?uri=CELEX:02003R1829-20080410&qid=1654200432890&from=EN.

Regulation (EC) No 1830/2003 of the European Parliament and of the Council of 22 September 2003 concerning the traceability and labelling of genetically modified organisms and the traceability of food and feed products produced from genetically modified organisms and amending Directive 2001/18/EC. OJ L 268, 18.10.2003, p. 24. https://eur-lex.europa.eu/legal-content/EN/TXT/PDF/?uri=CELEX:02003R1830-20081211&qid=1654200622243&from=EN.

Regulation (EC) No 853/2004 of the European Parliament and of the Council of 29 April 2004 laying down specific hygiene rules for food of animal origin. OJ L 139, 30.4.2004, p. 55. https://eur-lex.europa.eu/legal-content/EN/TXT/PDF/?uri=CELEX:02004R0853-20160401&qid=1654200733768&from=EN.

Regulation (EC) No 1334/2008 of the European Parliament and of the Council of 16 December 2008 on flavourings and certain food ingredients with flavouring properties for use in and on foods and amending Council Regulation (EEC) No 1601/91, Regulations (EC) No 2232/96 and (EC) No 110/2008 and Directive 2000/13/EC. OJ L 354, 31.12.2008, p. 34. https://eur-lex.europa.eu/legal-content/EN/TXT/PDF/?uri=CELEX:02008R1334-20181126&qid=1654200897504&from=EN.

Regulation (EU) 2015/2283 of the European Parliament and of the Council of 25 November 2015 on novel foods, amending Regulation (EU) No 1169/2011 of the European Parliament and of the Council and repealing Regulation (EC) No 258/97 of the European Parliament and of the Council and Commission Regulation (EC) No 1852/2001. OJ L 327, 11.12.2015, p. 1 https://eur-lex.europa.eu/legal-content/EN/TXT/PDF/?uri=CELEX:02015R2283-20210327&qid=1654200987168&from=EN.

Verzijden, K.E., Buijs, A.J., 2020. Meat 3.0—how cultured meat is making its way to the market. Eur. Food Feed. Law Rev. 2 (2020), 96–107.

Chapter 12

Crop biotech: Creating the crops to face the future

María Andrea Uscátegui-Clavijo and Sherly Montaguth-González
Agro-Bio Andean Region, Bogotá, Colombia

Plant breeding

All crops grown today, which are where most of our food and fiber come from, have been genetically modified by one method or another and, therefore, are not the same as those found by our ancestors. Nature itself has been in charge of selecting the most resistant crops that have been best adapted to the conditions of the planet. Likewise, man has selected them and, with the continuous advance of knowledge, science and technology have developed various methods to obtain more varieties and improve characteristics in terms of yields, nutritional content, growth under different conditions, and/or safer products.

The so-called conventional and widely used methods include crosses and hybridizations between plants, mutagenesis induced by chemical agents or radiation, among others. Modern biotechnology is one more method of the continuum process to develop techniques to improve crops in a more precise way. Through genetic engineering, it is possible to: (a) determine genes' functions, (b) transfer to another organism or modify a gene, and (c) express a desired characteristic in a plant that provides a benefit.

In this chapter we will focus on two of the main techniques of modern biotechnology: transgenesis and gene editing.

Transgenesis

A method by which one or more genes are transferred between unrelated species in order to suppress characteristics or obtain new ones. The product is known as a transgenic, genetically modified (GMO), or biotechnological organism. Although the technology can be applied to any organism, currently (2022) only 14 crops are commercially available with the following characteristics:

Genomics and the Global Bioeconomy. https://doi.org/10.1016/B978-0-323-91601-1.00014-6

Crop	Commercial trait	First approval for cultivation country and year
Soybean	- Resistant to insects	Brazil, 2010
	- Tolerant to herbicides	United States, 1996
	- Healthier fats	Canada, 2009
	- Drought tolerant	Argentina, 2015
Corn	- Insect resistant	United States, 1996
	- Herbicide tolerant	United States, 1997
	- Drought tolerant	Canada, 2010
Cotton	- Resistant to insects	United States, 1995
	- Tolerant to herbicides	United States, 1995
Canola	- Herbicide tolerant	Canada, 1997
Sugar beet	- Tolerant to herbicides	United States, 1998
Alfalfa	- Tolerant to herbicides	United States, 2005
Squash	- Viral disease resistant	United States, 1994
Papaya	- Viral disease resistant	United States, 1996
Eggplant	- Resistant to insects	Bangladesh, 2013
Potato	- Late browning	United States, 2014
Apple	- Late browning	United States, 2014
Beans	- Viral disease resistant	Brazil, 2011
Pineapple	- Lower levels of the enzymes that convert lycopene to beta-carotene	Costa Rica, 2011
Rice	- Enriched with beta-carotene	Philippines, 2021

Gene editing

A group of breeding technologies that uses molecular biology to introduce a site-directed mutation or deletion into the genome of a plant. Among gene editing methods, CRISPR (regularly interspaced clustered short palindromic repeats) has been the most popular to date, as it is a tool that can recognize specific DNA sequences and which carries an enzyme (CAS) that is in charge of cutting the DNA in the recognized sequence. This makes it possible to edit (delete, insert, or modify) targeted genes to improve crops in a process that is more efficient, precise, and predictable than other methods. The results can be changes that also happen through conventional methods and natural process, but now are reached in shorter periods of time.

Genome editing has opened doors and brought many opportunities to plant breeders to do research and development of crops without such restrictive and expensive regulations as that of transgenic crops. Actually, the biggest challenge facing agriculture is to produce more and better food in a sustainable and safe way for a constantly growing population. Facing the impact of climate change on agriculture, as well as making good use of and protecting natural resources, will require the development of better crops, for which modern biotechnology is undoubtedly an essential tool.

The role of modern agricultural biotechnology on the sustainable development goals

Although agriculture is not in itself one of the Sustainable Development Goals (SDGs), it does play a key role in many of them and modern biotechnology is a tool that, at present allows and in the future, will help face the challenges of agriculture to continue producing more and better food in a sustainable way.

SDG 1: No poverty

Modern crop biotechnology helps farmers improve their income and quality of life by:

- Minimizing the loss of crops
- Reducing input costs for better control of pests and weeds
- Increasing the yields of crops
- Offering crops that are more resilient to climate change or that can be developed under special conditions

From 1996 to 2018, more than 18 million small farmers and their families planted biotech crops, increasing their production and their incomes. In 2018, farmers in developing countries received \$4.42 as extra income for each extra dollar invested in GM crop seeds, whereas farmers in developed countries received \$3.24 as extra income for each extra dollar invested in GM crop seeds. The net farm-level economic benefit was just under \$19 billion in 2018, equal to an average increase in income of \$103/ha (Brookes and Barfoot, 2020b).

Colombian farmers, for each US \$1 spent on GM seeds relative to conventional seed, have gained an additional US \$3.09 in extra income from growing GM cotton and an extra US \$5.25 in extra income from growing GM maize. These income gains have mostly arisen from higher yields (+30.2% from using stacked—herbicide-tolerant and insect-resistant—cotton and +17.4% from using stacked maize) (Brookes, 2020).

In Bangladesh, insect-resistant eggplant, also called Bt plants (due to *Bacillus thuringiensis* genes transferred to them to express proteins against some specific insects), has, on average, increased by 21.7%/ha the revenue of farmers by protecting the plants, reducing the costs to control pests. and increasing yields (Shelton et al., 2020).

SDG 2: Zero hunger

Modern biotechnology contributes to producing more and better food to feed a growing population by:

Between 1996 and 2018, GM crops have had an additional production of 824 million tons of food, feed, and fiber worldwide (Brookes and Barfoot, 2020b), among them: 277.63 million tons of soybean, 497.74 of maize, 14.07 of canola, and 1.59 of sugar beet.

It is important to recognize that Bt technology has been a great ally to protect crops against insects and illness, and avoid food loses, which means higher yields to feed a growing population. Bt technology has also been shown to improve the quality of maize by protecting against pest damage of the kernel and by reducing the mycotoxins level (Yu et al., 2020). Mycotoxins in maize are secondary metabolites, mainly known as fumonisins and aflatoxins, produced by *Fusarium verticillioides*, *Fusarium proliferatum*, *Aspergillus flavus*, and *Aspergillus parasiticus*, and can be toxic, carcinogenic, or both to humans and animals.

Fruit browning is a major problem that contributes to waste. The genetically modified "Arctic apple" developed by Okanagan (Maxnem, 2017) Specialty Fruits delays the browning process after being cut and prolongs shelf life, which allows the fruit to be consumed for a longer period of time and reduces waste. The United States was the first country to approve these modified apples in 2015, and currently, they are also commercialized in Canada.

Likewise, modern biotechnology can provide better and healthier products for consumers such vitamin and minerals enriched rice to prevent malnutrition,

SDG 3: Good health and wellbeing

Advances in agriculture biotechnology can help to supply more nutritious and healthier food by:

- Improving nutritional value
- Reducing allergens

The rice known as "Golden Rice"[a] is biosynthesized with beta-carotene, a vitamin A precursor that can help people with deficiency of this vitamin. Blindness and a weakened immune system are the most common conditions caused by the deficiency of vitamin A and affect an estimated of 190 million children in the world.

In 2021, the Philippines became the first country in the world to approve commercial cultivation of Golden Rice, so its farmers are now allowed to plant it to provide to thousands of people in the country that suffer vitamin A deficiency and help to improve their health. Food agencies in other countries, such as the United States (2018), Canada (2018), New Zealand (2017), and Australia (2017), have already approved its use for human consumption.

High oleic soybean (Plenish[b]) produces healthier oils that can be more stable and do not need hydrogenation, meaning that consumers could avoid trans fats. And what about a gluten-free wheat or an allergen-free peanut? Even though they are not yet in the market, these will arrive soon to help those with celiac disease and allergic reactions.

SDG 5: Gender equality

In a male dominated environment like agriculture, women are an important engine for this field and different studies (Zambrano et al., 2009) have shown that having access to technological and innovative tools such as genetically modified seeds has:

- Empower and allowed them to play a more relevant and productive role to advance gender equality.

SDG 6: Clean water and sanitation

Genetically modified crops can help to preserve the nonrenewable resource that is water and try to keep it clean.

- Optimizing the use of water in agriculture
- Reducing soil erosion to pollute waterways

Insect resistant genetically modified crops has shown reduction in the applications of insecticides (Brookes and Barfoot, 2020b) therefore optimizing the use of water needed for it.

---

a. https://www.goldenrice.org/.

b. https://www.pioneer.com/us/products/soybeans/plenish.html.

No tillage or conservation tillage with herbicide-tolerant biotech crops has shown to reduce soil erosion, conserving moisture, nutrients and its structure to clog it run into and no pollute waters.

SDG 8: Decent work and economic growth

Access to biotech crops has brought millions of farmers around the world:

- Economic growth to farmers, the agricultural sector, and countries
- Better yields with fewer inputs

Brookes and Barfoot concluded in their paper (Brookes and Barfoot, 2020b) that the economic benefits of GM crops adoption at the farm level amounted to $225.1 billion for the period 1996–2018 in 28 countries. These gains have been divided 52% to farmers in developing countries and 48% to farmers in developed countries. 72% of the gains have derived from yield and production gains with the remaining 28% coming from cost savings.

SDG 9: Industry innovation and infrastructure

The biotech sector works on the research and development of innovative seeds for more resilient crops, which contribute to facing global challenges such as food security, climate change, and biodiversity use and conservation.

SDG 11: Sustainable cities and communities

Modern biotechnology in plants has shown that it can increase productivity and crop yields, as well as improving the income of farmers in communities and rural areas while reducing the ecological footprint of agriculture on the environment.

SDG 12: Responsible consumption and production

Producing in a sustainable way is one of the great contributions and most important lines of research and development in genetically modified crops. Crop biotechnology has reduced the use of nonrenewable resources such as water and soil.

SDG 13: Climate action

More sustainable practices with GM crops, such as reduced tillage, less use of fossil fuels, and being more efficient per area, have reduced greenhouse gas emissions. If GM crops had not been grown in 2018, for example, an additional 23 billion kilograms of carbon dioxide would have been emitted into the atmosphere, which is the equivalent of adding 15.3 million cars to the roads (Brookes and Barfoot, 2020b). Furthermore, scientists and crop breeders work with modern biotechnology tools to obtain crops that can adapt to climate change by being tolerant to abiotic stresses such as drought or floods, or under special conditions of nutrients in the soil such as alkalinity/acids or salt concentrations.

SDG 15: Life on land

GM crops have proven to be more environmentally friendly because:

- They are more efficient per area (ton/ha), therefore reducing the pressure to use new land in agriculture or deforestation
- Bt technology has reduced the use of insecticides and broad-spectrum active ingredients while protecting biodiversity
- As mentioned above, practices such as reduced tillage allow the preservation of nutrients and soil moisture, preventing its erosion

SDG 17: Partnership for the goals

Modern biotechnology requires alliances, good will, and work between academic, private, and public institutions, NGOs, and governments that allows the development of innovative products but also makes them more accessible and available to reach for those who need them. This is the way to continue sustainable development that benefits farmers, consumers, and the environment.

Biotechnology in agriculture: Some developments and impacts

As stated before, humans have performed genetic modification of virtually every food since we learned how to grow crops. However, things have changed in terms of technology and, with new techniques, innovations in crop biotechnology continue to emerge.

Since scientists discovered in the 1980s the mechanism through which a soil bacteria was capable of using a plasmid to transfer its genes into a plant's nuclear DNA, and then managed to use it as a transformation tool; this innovation in plant breeding using genomic tools became a new driving force of research and innovation Back in 1982, when the first transgenic plant was developed—which was a tobacco plant resistant to an antibiotic—both the industry and academia could foresee how this could be used to transfer important traits to a plant of commercial interest that would eventually help farmers win the fight against pests and some other stresses.

From then on, the list of developments and innovations has kept growing and has diversified to the point where we now find products capable of huge impact on large scale agriculture as well on a local scale, demonstrating that plant biotechnology is primarily a wonderful set of tools that can help us achieve sustainable agriculture, as well as improving small communities' livelihoods and health.

New seeds that revolutionized agriculture

By 1985, new traits such as insect resistance were being tested. Genes encoding the active form of Cry proteins (entomocidal δ-endotoxin) from *B. thuringiensis* (Bt) were introduced to help plants survive the attack of certain pest insects. Such proteins bind to specific receptors in the epithelial cells inside the insect's midgut, forming pores in the cells' membranes, provoking an inflow of ions and water causing the cells to swell so much the membranes eventually break and the insect dies. Such a result can be named death by colloid osmotic lysis and is very specific against target insects; the detailed investigation on how Cry proteins work is fundamental to avoid effects on nontarget insects.

The crops that carry this technology—known as Bt crops—have been commercialized since the mid-1990s, with maize and cotton being the first to be available for farmers. Since then, the impact of Bt crops in global agriculture is undeniable, showing benefits in improved pest control and the overall quality of the products. One of the top Bt crops worldwide is insect-resistant (IR) maize. By 2018, the farm-level cumulative impact of using GM IR maize was $43.3 billion. This farm income gain has mostly derived from less pest damage; in some countries, farmers have also reported less use of insecticides, which ultimately reduces the money spent on insecticides. The figure for IR cotton is just as relevant, reporting an income gain of $66.58 billion (Brookes and Barfoot, 2020a).

Benefits for the environment are also a reality. In 25 years since the adoption of biotech crops, farmers have been able to apply more sustainable practices resulting in a reduction of greenhouse gas emissions. Only in 2018, the reduction in fuel coming from less frequent applications of herbicides and insecticides plus the implementation of less tillage to no-tillage systems meant an estimated 23 billion kilograms of carbon dioxide were not emitted into the atmosphere (Brookes and Barfoot, 2020a). Such practices help cut soil erosion, improve its quality, and help create a better soil environment for crops.

The productivity gained by adopting biotech crops has also helped farmers to reduce the need to expand agricultural borders to have a more profitable business, thus saving an estimated amount of 231 million hectares of arable land since 1996. By needing less land, farmers are also helping reduce the impact on the biodiversity that is lost when wildlands are wiped out to plant commercial crops.

Innovation that saved an industry

It is a mistake to believe that plant biotechnology can only be executed by the biggest seed corporations. Although extreme regulations make the approval of a GM crop quite an expensive and long process, experience has shown many times that the government involvement can make it happen when need overcomes fear.

Such is the case of papaya in Hawaii, a once booming industry that came close to extinction because of a virus. The papaya ringspot virus (PRSV) was first spotted on the island in the 1940s and within a decade it began affecting crop yields. In papaya, the disease provokes the leaves to distort, turn yellow, and exhibit severe mosaic. Spots and streaks soaked in oil or water appear, and the fruit presents spots in the form of a ring that gives the disease its common name.

Production then started moving from Oahu Island to one district and then another as farmers desperately tried to avoid the disease, but the efforts were unfruitful. In each region, the virus would destroy plantations. By the late 1990s, PRSV had almost destroyed what once was one of the top three crops in Hawaii. Planting papaya was not a good business anymore, farmers abandoned infected lands, and overall papaya production dropped dramatically to the point that the industry almost disappeared completely. Then, a local biotechnology scientist named Dennis Gonsalves decided to work on a solution to this problem.

Inspired by previous work on virus-resistant tobacco, Gonsalves and his team inserted the virus's coat protein (CP) gene into papaya's genome through the biolistic method. The trials produced a line of "Sunset" variety plants that showed no symptoms of the disease, except two that showed symptoms on the side shoots; apart from that, the line was capable of delivering transgenic virus-resistant plants with fruits that grew normally, thus improving the quality of the product (Gonsalves, 2006).

This important result served as germplasm for the development of "SunUp" and "Rainbow" transgenic varieties that, once released in 1998, were rapidly adopted and virtually saved a dying industry deeply affected by the ringspot virus. As of 2018, 77% of papaya commercially grown in Hawaii was genetically modified. The impact of this technology has been positive for farmers with higher yields compared to conventional varieties. China also adopted the GM papaya, reporting 9600 ha in 2018.

Achieving the impossible for better health

Years of misinformation about biotech crops have certainly impacted negatively on public opinion. Despite a vast quantity of literature demonstrating the benefits of these types of crops, still, hundreds of people around the globe—from ordinary people to authorities and decision-makers—believe genetically modified crops and derived products have catastrophic impacts on the environment and human health.

Far from that belief, the new breeding techniques are showing that technology can also be helpful on public health issues in certain communities. That is the case with Golden Rice (GR), one of the most known and mentioned transgenic inventions, which took almost two decades to get the green light because of the pressure from activists that relentlessly campaigned against it for years.

Rice is a staple food in many countries of the world. Nine out of 10 top consumers are located in Asia. According to The Food and Agriculture Organization, 90% of the global rice is produced and consumed in the Asia-Pacific region, but nearly a fourth of the Asian population is still poor and has considerable unmet demand for rice. Meanwhile, the Asian population will continue growing.

One of the countries where poor communities depend on rice consumption is the Philippines. Citing the Philippines Statistics Authorities (PSA), a Filipino consumes a total of 118.81 kg annually; this is equivalent to 325.5 g of milled rice daily. However, rice is not particularly known for its nutritional value. Aside from being rich in carbohydrates that act as a source of energy and providing basic nutrients, milled rice does not offer health benefits for consumers. Filipinos who depend mostly on rice for nourishment have unmet nutritional requirements, such as vitamin A, a molecule important for growth and development as well as vision. Vitamin A, combined with a protein called opsin, helps the human eye to absorb light and is very much needed for both color and low-light vision. Given this, a severe lack of vitamin A in the first years of life can cause blindness and a fragile immune system, which can ultimately cause death.

The World Health Organization estimates that around 250,000–500,000 children become blind each year because of vitamin A deficiency (VAD); half of them die within 12 months after losing their vision. In the Philippines, VAD has been an issue for decades; in 2018, the prevalence of vitamin A deficiency among children aged 6 months to 5 years was 16.9%, meaning VAD can be considered a moderate public health concern (DOST-FNRI, 2020). Considering the number of children that suffer from VAD in Asian communities living in poverty, the Golden Rice Project was born as a new approach to help tackle the problem with genetically modified rice capable of producing beta-carotene (vitamin A precursor) in the grains. Maybe one day, children that depend on rice could receive the amount of vitamin A they need to avoid deficiency.

Any rice variety is capable of producing beta-carotene in the grains. However, despite having the metabolic route for it, the pathway is active only in the leaves, which is where the plant needs it for photosynthetic tissues. The goal for Ingo Potrykus and Peter Beyer, the European scientists that led this project, was to help rice complete the route so the plant could be able to produce grains with beta-carotene content.

What was impossible to achieve through conventional breeding became a reality with biotechnology: in 1999, the team obtained the first version of Golden Rice (GR1) by inserting two transgenes that helped rice to complete the pathway. The first was the PSY gene from the narcissus flower, which encodes phytoene synthase, an enzyme involved in the biosynthesis of carotenoids. The other gene was CRT I from the bacterium *Erwinia uredovora*, which encodes the enzyme phytoene desaturase that synthetizes carotenoids in bacteria, fungi, and archaea. The insertion of both genes helped rice complete the metabolic pathway that prevented beta-carotene production; however, the final amount of beta-carotene was still very low (1.6 μg/g of carotenoids) and was not enough to provide the daily provitamin A requirements of the target population.

The new GM rice sparked international recognition for both scientists. Along with the media frenzy came the activists, who found a new GM product to protest against. In the meantime, a new version of Golden Rice (GR2) was in the making. By changing the PSY gene from narcissus for a PSY gene from corn, the GR2 produced grains with 23 times more carotenoids than the previous version. Further crossbreeding was performed to transfer the trait into local varieties.

The GR was expected to be available for farmers in 2002, but years passed without it reaching the most vulnerable communities that would benefit from it. The pressure coming from activist groups was strong and persistent. Finally, 20 years after the first version was created, the Philippines became the first country to give the green light for commercialization.

The technology was created for humanitarian reasons in developing countries, it is intended to cost no more than the local varieties of rice, and replanting is allowed. Any country can also use the technology in breeding projects. Regulatory agencies in the United States, Canada, Australia, New Zealand, and the Philippines have declared GR safe for human consumption. By 2022, Bangladesh could be the second country to approve commercialization. By itself, Golden Rice will not provide a permanent solution to poverty and malnourishment in these countries, but it certainly represents a newly available option that can have a positive impact on the lives of children whose parents cannot afford a diverse and healthy diet.

Small communities, huge benefits

Through word of mouth, farmers have got to know their benefits and have turned GM crops into the fastest adopted technology in modern agriculture. For Arnulfo Cupitra, a maize farmer of indigenous descent in the Andean region of Colombia, genetically modified crops have helped him collect successful harvests: "I used to plant conventional maize and was getting 2 to 3 tons of corn. Meanwhile, my neighbors were getting 5 to 6 tons, but it took me almost six years to make the change from conventional to transgenic seed because I didn't want to give up on tradition. Seeing others helped me make the decision. Weeds and pests used to be an issue but not since the technology came."

But the development of biotech crops has not only impacted medium and large farmers by making agriculture a more profitable and sustainable business; it has also impacted small farmers in remote areas of the world. Such is the case of eggplants (brinjal) in Bangladesh, a staple food just as important as potato and rice in this country. With the commercialization of the Bt eggplant in 2014 and the subsequent insertion of the trait into nine local varieties, Bangladeshi farmers now have the technology to fight the attack of the eggplant fruit and shoot borer, the most destructive insect pest for eggplants in Asia. This Lepidoptera larva bores inside shoots and also feeds on young and maturing fruit, damaging the harvest and making the eggplant completely unsuitable for the market. With the rapid adoption of this development, Bt brinjal farmers saved 61% of pesticides compared to conventional brinjal farmers (Shelton et al., 2018); they also have not experienced crop loss due to the attack of the pest and their income has increased six times.

New developments such as GM cassava with resistance to brown streak disease in Kenya, the GM drought-tolerant wheat in Argentina, and GM beans with resistance to golden mosaic virus in Brazil are just a few examples of recently approved crops that were designed to solve local problems that can become global in the future.

A new generation of biotech crops

Developments created to benefit farmers have been the common approach when it comes to biotech crops, but a new generation of innovations marketed for consumers is showing how genetic engineering is evolving with different techniques that do not necessarily imply the insertion of foreign DNA. Consumers that are concerned about healthy food, sustainable habits, saving some time when cooking, eating fresh produce, or that would like to post unique-looking preparations on social media, can now consciously buy GMOs to satisfy their quest for a better, more interesting lifestyle.

One of these GMOs is the nonbrowning sliced apples developed for people that prefer more convenient food that remains fresh over time. While some apple cultivars present slow enzymatic browning, the genetically modified variety called Arctic apple exhibits a complete absence of oxidation reaction that gives the consumers a fruit that does not brown when bitten, bruised, or sliced, which provides a longer shelf life for the product and better preservation of nutrients. Using the RNA interference (RNAi) technique, which is a biological process that recognizes and destroys specific RNA, developers silenced the expression of polyphenol oxidase (PPO), the enzyme that reacts with the phenolic content and oxygen, browning the fruit after being cut.

Given that food waste is one of the UN Sustainable Development Goals, demand for these kinds of innovations in crop science is likely to increase in the coming years.

Using the same technology, another company created the "PinkGlow" pineapple, a sweeter variety with pink flesh that became a trend on social media upon release. To achieve this result, it was necessary to diminish the production of beta-carotene and increase the production of lycopene by silencing the β-cyclase (b-Lyc) and lycopene ε-cyclase (e-Lyc) genes in the pineapple, causing the degradation of the enzymes needed to convert lycopene into beta-carotene. Once the lycopene starts accumulating, the flesh acquires a pink color that makes this variety distinctive and unique. Although lycopene tends to be mentioned as an antioxidant useful for aging and even cancer prevention, there is no scientific evidence to sustain such claims. Consequently, the pink pineapple offers no extraordinary health benefits but can give the consumer a fun product that can be used to create memorable food and drinks that look good in pictures.

Picture of margarita cocktails prepared with PinkGlow pineapple by Sarah Knoebel (@sarah_licious_eats on Instagram), a food content creator from Austin, Texas (United States). 2022.

With the arrival of the genome-editing tool called CRISPR-Cas9, plant transformation gained a new push, since the technique is considered cheaper and faster, and its products lack the overregulation that transgenic crops have to endure to get to the farmer's fields. GE is often portrayed to the public as a scissor that allows scientists to get a desired trait by cutting specific sequences within the genome, similar to the cut-paste-delete tools found in text editors, with CRISPR being the most famous of all. Plants obtained by GE are subject to the same regulation as conventional products in some countries. However, this is just one of many techniques for genome editing and it is possible to obtain transgenic products with it. Nevertheless, given the less strict regulation, it is more convenient to develop an edited crop that will not take decades and significant amounts of money spent in field trials for it to get commercialization approval from regulatory agencies.

With CRISPR-Cas9 as a new tool for plant breeding, the biotechnology industry is blossoming with new projects that aim to help farmers through climate change and to meet consumers' needs. The very best example is the world's first gene-edited crop approved for commercialization: the Sicilian Rouge High GABA tomato, which contains four to five times higher levels of gamma-aminobutyric acid (GABA), an inhibitory neurotransmitter believed to improve stress, although there is no scientific consensus on this topic yet. The GE tomato was developed by a Japanese company and was approved for commercialization in Japan in 2021.

Genome-edited tomatoes with increased levels of GABA. Sanatech Seed. 2020.

Many other biotech crops are being developed. From more climate-resilient to nutrient-enriched products, this new generation appears to offer a once-in-a-lifetime opportunity to change the public perception of genetically modified organisms for good.

Biotech crops in Latin America: Between highs and lows

Globally, despite almost all countries being signatories to the Cartagena Protocol—an international agreement that regulates living organisms that have been modified through modern biotechnology—each one is entitled to its own regulation for genetically modified crops. Unlike regions such as North America, where biotech crops are fully adopted, or Europe, where biotech crops are not allowed almost anywhere, the status in Latin America is far from harmonized and is still debated. Ranging from top producers, such as Brazil, and full bans on GM seeds, as is the case of Ecuador, the region presents a panorama as polarized and heated as its politics.

Brazil and Argentina figure as the top two and top three countries in the world in terms of the biggest biotech crop areas. With 52.8 and 23.9 million hectares, respectively, both are reaping the benefits of GM soybeans and GM maize industries that have reached over 90% adoption rates. In Paraguay, Uruguay, Bolivia, Mexico, and Colombia, the number of hectares of GM crops keeps growing at a steady pace. Some countries have declared war on GM crops while others are making efforts to establish biosafety regulations. For a few, such as Venezuela, the subject is not being discussed at all.

However, despite the numbers and the benefits at the farm level and even to the environment, the debate over the safety of these crops is far from over and still feeds from old debunked myths about GMOs. It is safe to say that in Latin America this debate is entirely political, it is mostly focused on GM plants, and is driven by a large number of activists groups that persistently demand a ban on

GM crops throughout the entire region; a few of them manage to influence congressmen and politicians along the way. For instance, Ecuador declared itself a territory free of GMOs in 2008. Through a new constitution led by former president Rafael Correa, the country placed a ban on transgenic crops and seeds for good. Years later, Correa himself acknowledged this as a mistake, stating that he lacked the integrity to oppose this "infantile environmentalism." He also praised the huge potential of biotechnology in creating more resilient varieties, but it was too late—changing the constitution is a hard task to achieve in a polarized country.

In Peru, a moratorium against genetically modified organisms was established in 2011. Before it was over, the Peruvian Congress promulgated an extension of another 15 years, arguing lack of infrastructure and the precautionary principle. The new ban sparked heated debates between advocates and anti-GMO activists online but not within Congress, where a total of 15 bills seeking the extension were settled, making clear that most members were in favor of extending the moratorium on GM seeds and were not interested in including science and technical experts in the discussion.

In countries such as Colombia, Bolivia, and Mexico, the debate is very much alive, as the efforts to ban transgenic seeds are still on the political agenda. For example, in Colombia, a country with a well-defined regulatory framework, with over 100,000 ha of GM maize and cotton, small to medium farmers benefit from the technology and local scientists work with new breeding techniques, a different bill from a very small group of congressmen emerges every legislative session, sparking the discussion for a certain amount of time, but with no success yet.

In Bolivia, where herbicide-tolerant soybean is commercially legal and plantings total 1.4 million hectares (ISAAA, 2019), the adoption of genetically modified crops is subject to the political turmoil the country has faced in recent years, swinging between intentions to expand the offer of GM seeds to going back to stricter regulations, depending on who is in charge. Meanwhile, in Mexico, a country with 0.2 million hectares of GM cotton (ISAAA, 2019), the government led by Manuel López Obrador enacted a decree in 2020 to phase out the use of glyphosate, ban the cultivation of GM maize, and end the importation of GM corn grains by 2024, arguing health risks that have no scientific grounds. Corn is of great importance for Mexicans' diets and deep-rooted culture, which allows for traditionalist narratives to dominate the debate.

However, bans have not stopped countries from importing genetically modified commodities, and in a few cases, have not stopped farmers from planting GM seeds either. According to a report released in 2020 by the Ministry of Environment, Peruvian farmers have been planting and breeding imported genetically modified grain in their fields. The report stated that regulators detected transgenes in 88.3% of the fields inspected, all samples coming from local varieties (MINAM, 2019). Seed companies have never sold GM seeds in Peru and these plantations are completely illegal under Peruvian law.

Despite being politically instrumentalized and used to polarize public opinion, there are high hopes for biotech crops in the region as new local developments arise. Argentinian scientists created the world's first transgenic wheat tolerant to drought and ammonium glufosinate herbicide that was approved by local regulatory authorities. Brazil developed and recently approved a GM bean with a resistance trait for the bean golden mosaic virus (BGMV), a disease that can destroy more than half of a farmer's plantation. In Colombia, an alliance between a public university and a private union made it possible to create its first off-patent GM maize.

In a biodiverse region with so much disparity and poverty, outdated myths about genetically modified crops can lead to restrictive regulations that subject farmers to old practices and force scientists to flee to biotechnology-friendly countries to be able to work. However, genome editing tools represent a new hope in such a heated landscape. As new CRISPR innovations appear, more decision-makers in regulatory strict countries are turning their heads at the possibility of allowing edited crops to enter. It is expected that, at some point, gene editing can be used to solve local problems that surely can have a relevant impact on the quality of life of people and the economy of each country.

References

Brookes, G., 2020. Genetically modified (GM) crop use in Colombia: farm level economic and environmental contributions. GM Crops Food 11 (3), 140–153. https://doi.org/10.1080/21645698.2020.1715156.

Brookes, G., Barfoot, P., 2020a. GM Crop Technology Use 1996-2018: Farm Income and Production Impacts. PG Economics Limited, United Kingdom. 213 pp.

Brookes, G., Barfoot, P., 2020b. GM crop technology use 1996-2018: farm income and production impacts. GM Crops Food 11 (4), 242–261. https://doi.org/10.1080/21645698.2020.1779574.

Department of Science and Technology-Food and Nutrition Research Institute (DOST-FNRI), 2020. Philippine Nutrition Facts and Figures: 2018 Expanded National Nutrition Survey (ENNS). FNRI Bldg., DOST Compound, Gen. Santos Avenue, Bicutan, Taguig City, Metro Manila, Philippines. 360 pp.

Gonsalves, D., 2006. Transgenic papaya: development, release, impact and challenges. In: Advances in Virus Research. Elsevier, pp. 317–354, https://doi.org/10.1016/S0065-3527(06)67009-7.

ISAAA, 2019. Global Status of Commercialized Biotech/GM Crops in 2019: Biotech Crops Drive Socio-Economic Development and Sustainable Environment in the New Frontier. ISAAA Brief No. 55. ISAAA, Ithaca, NY.

Maxnem, A., 2017. Genetically Modified Browning Resistant Appels Reaches U.S. Stores. Scientifica American. https://www.scientificamerican.com/article/genetically-modified-browning-resistant-apple-reaches-u-s-stores/.

Ministerio del Ambiente (MINAM), 2019. Acción de vigilancia en el cultivo de maíz en las provincias de Piura y Sechura, región Piura, Informe N° 04-2019-MINAM, Lima, Perú. 21 pp https://bioseguridad.minam.gob.pe/acciones_vigilancia/vigilancia-n-04-2019-minam/. (Accessed 1 October 2021).

Shelton, A.M., Hossain, M.J., Paranjape, V., Azad, A.K., Rahman, M.L., Khan, A.S.M.M.R., Prodhan, M.Z.H., Rashid, M.A., Majumder, R., Hossain, M.A., Hussain, S.S., Huesing, J.E.,

McCandless, L., 2018. Bt eggplant project in Bangladesh: history, present status, and future direction. Front. Bioeng. Biotechnol. 6, 106. https://doi.org/10.3389/fbioe.2018.00106.
Shelton, A.M., Sarwer, S.H., Hossain, M.J., Brookes, G., Paranjape, V., 2020. Impact of Bt Brinjal cultivation in the market value chain in five districts of Bangladesh. Front. Bioeng. Biotechnol. 8, 498. https://doi.org/10.3389/fbioe.2020.00498.
Yu, J., Hennessy, D.A., Wu, F., 2020. The impact of Bt corn on aflatoxin-related insurance claims in the United States. Sci. Rep. 10, 10046. https://doi.org/10.1038/s41598-020-66955-1.
Zambrano, P., Fonseca, L., Cardona, I., Magalhaes, E., 2009. Insect Resistant Cotton in Colombia: Impact on Farmers. Paper presented at: 13th ICABR Conference, Ravello, Italy,.

Chapter 13

Bioeconomy policy: Beyond genomics R&D

Jim Philp
Organisation for Economic Cooperation and Development (OECD), Paris, France

Introduction

The bioeconomy is a political construct that has seen over 50 countries develop national strategies or similar policy documents. The OECD (2009) described the bioeconomy as "*the set of economic activities in which biotechnology contributes centrally to primary production and industry, especially where the advanced life sciences are applied to the conversion of biomass into materials, chemicals and fuels.*" The conversions depend on a range of developments in biotechnology, perhaps with origins in metabolic engineering. The most recent elaboration of these technologies is synthetic or engineering biology, with its distinctive infrastructure, the biofoundry (Kitney et al., 2019), the success of which depends largely on what has been termed "digital biology." These developments would not have been possible without the various genomics technologies, which in turn sprang from next-generation sequencing.

The bioeconomy can be seen as an integral part of an ongoing transition from high dependence on fossil resources toward carbon neutrality, fueled by political action on climate change mitigation. At present, around 110 countries have signed up to carbon neutrality by 2050 (United Nations, 2020). Most of the relevant policy has been directed toward energy. This is directly relevant to manufacturing as 80% of production costs relate to oil and gas as feedstock and energy (Arns, 2018). It will be apparent, however, that much more industrial production will also require biomass and carbon recycling as feedstocks instead of crude oil and gas. A very important component will be carbon capture and storage (CCS), including direct air capture (DAC), with value-adding carbon capture and utilization (CCU) assuming greater importance as the century wears on (Ghiat and Al-Ansari, 2021; International Energy Agency, 2020).

Genomics and the Global Bioeconomy. https://doi.org/10.1016/B978-0-323-91601-1.00008-0

A complex policy environment

To place the bioeconomy within this transition of monumental importance to the future requires an approach to public policy that has not been seen before. Previous transitions, especially from wood to coal during the First Industrial Revolution, and then coal to oil during the 20th century, took several decades to achieve (Bennett and Pearson, 2009). In this current transition, climate change is forcing change in a compressed timeframe. There is little room for error, requiring a systemic approach to policymaking.

Bioeconomy is driven by innovation, and innovation policy generally follows a sequence from "ideas to market" (European Commission, 2020; World Bank Group, 2020). A policy framework for the bioeconomy needs the integration of its specific policy instruments with the wider policy arena that encompasses, for example, carbon pricing and taxation. Thus, there is a meeting of transition, climate, bioeconomy, and innovation policy areas that will need careful management and coordination. It is well known that policy coherence (OECD, 1996) is needed between families of policy to prevent lock-ins that put some policies, and perhaps sectors, at crossed purposes (Meijer and Hekkert, 2007). Given the large number of sectors included in the bioeconomy (Kuosmanen et al., 2020), this is especially relevant and prescient. Finally, all bioeconomy strategies are immersed in the need for sustainability, which will become a mode of governance for the bioeconomy with its own set of policy challenges (Vogelpohl, 2021).

Supply-side, demand-side, and cross-cutting measures

There are several critical policy areas, many of them under innovation policy that can be readily identified. In Table 13.1, these are grouped under three categories, which can roughly be translated to supply-side, demand-side, and a mixture of both supply- and demand-side policies (i.e., cross-cutting measures). This is consistent with the view that both supply- and demand-side policies are needed for effective innovation (OECD, 2011a).

Demand is a major potential source of innovation, yet the critical role of demand as a key driver of innovation is not universally recognized in government policy (Edler and Georghiou, 2007). Historically, governments have tended to rely on macroeconomic policies (e.g., monetary and fiscal policy) and framework conditions (e.g., competition, tax or entrepreneurship policies) to support market demand and avoid distortion. In recent years, however, OECD countries and emerging economies such as Brazil and China have used more targeted demand-side innovation policies. These include measures such as public procurement, regulation, standards, consumer policies, user-led innovation initiatives, and lead market initiatives to address market and system failures in areas in which social needs are pressing (OECD, 2011a).

TABLE 13.1 Policy inputs for a bioeconomy framework.

Feedstock/ technology push	Market pull	Cross-cutting
Local access to feedstocks	Targets and quotas	Standards and norms
International access to feedstocks	Mandates and bans	Certification
R&D subsidy	Public procurement	Skills and education
Pilot and demonstrator support	Labels and raising awareness	Regional clusters
Flagship financial support	Direct financial support for bio-based products	Public acceptance
Tax incentives for industrial R&D	Tax incentives for bio-based products	Definitions and terminology
Improved investment conditions	Incentives related to GHG emissions (e.g., ETS)	
Technology clusters	Taxes on fossil carbon	
Governance and regulation	Removing fossil fuel subsidies	

Modified from OECD, 2018. Meeting Policy Challenges for a Sustainable Bioeconomy. OECD Publishing, Paris.

Moreover, experience in OECD member countries has shown that the use of such demand-side policies remains limited to areas in which societal needs are not met by market mechanisms alone (e.g., environment) or in which private and public markets intersect (e.g., energy supply). Bioeconomy policy goals are both environmentally and energy-driven. This focus on the demand side also reflects a general perception that traditional supply-side policies—despite refinements in their design over recent decades—have not been able to bring innovation performance and productivity to desired levels.

Local access to feedstocks: Supply and value chains in the distributed manufacturing model

There are several policy advantages to making use of local feedstocks that are currently attractive to policymakers. Nevertheless, there are major challenges ahead. One of the challenges is the complexity of biorefinery value chains.

In Europe alone, there are some 16 million forest owners and 14 million farm owners (Hetemäki, 2014). In Finland for example, the average forest size is 30 ha, and there are hundreds of thousands of forest owners. The biorefinery at Bazancourt-Pomacle in France involves ten thousand farmers. The distributed manufacturing "glocal" model means establishing many interconnected local production plants that are integrated with other nearby industries to ensure that residues and wastes are fully utilized in different processes (Luoma et al., 2011).

Government programs are promoting research and development across supply and value chains, but supply markets receive little attention (Knight et al., 2015). This creates one of the conditions that deters investors. This lack of attention to supply markets possibly reflects reluctance by governments to be seen to be intervening in markets and potentially contravening anticompetitive practices (Institute of Risk Management and Competition and Markets Authority, 2014).

The stakeholders concerned are so different that they would never come into regular contact with each other in the fossil economy. However, to make biorefining economically sustainable, this silo mentality needs to be overcome. There are roles to be played by policymakers to prevent this communication process from being random, ad hoc, and inefficient. Analysis points to the potential importance of buyer cooperatives and other forms of supply market intermediaries (Knight et al., 2015).

This is consistent with the activities of publically funded regional clusters in industrial biotechnology becoming involved with supply and value chains (Kircher, 2012). Regional clusters are well positioned to evaluate local options, to build capacity in the regions, and then to look beyond the regions (Philp and Winickoff, 2017). Building capacity at the local level depends on quality local business networks, e.g., agricultural and forestry machine rings and business relationships of trust. Building beyond the clusters can exploit the expertise gained at regional level to expand and join up with other regions.

Encouraging software design to improve decision making would be a relatively low-cost public sector intervention. For example, a database developed by Black et al. (2016) for the assessment of biomass supply chains for biorefinery development covers origin, logistics, technical, and policy aspects. This is assuming greater importance as the need to establish bespoke biomass supply chains is becoming a reality. Industrial developers will face many business-critical decisions on the sourcing of biomass and the location of biorefineries. Software of this type could simplify decision making. It could be developed in an open-source manner with regional clusters to make it sufficiently bespoke.

International access to feedstocks: Biomass potential and sustainability

Large quantities of biomass are already being shipped around the globe (BP-EBI, 2014). Clearly, the use of biomass globally is increasing and is going to increase further in the future (e.g., Klein et al., 2014; Schmitz et al., 2014). Biomass potential and its cost could become crucial factors that affect overall climate change mitigation costs (Rose et al., 2013). Despite the fact that over 180 billion tons of lignocellulosic biomass is produced annually (Dahmen et al., 2018), it is also recognized at the international level that production of biomass can have major negative impacts that need to be addressed in policy and business (Knudsen et al., 2015). Thus, metrics and indicators of sustainability are essential (Kardung et al., 2021); otherwise, companies can use sustainability as a marketing tool without objective scrutiny.

Need for metrics of biomass sustainability and potential

The meaning of sustainability is intuitive, but an internationally agreed method of its measurement is elusive. It is often described in terms of three pillars—economic, environmental, and social. There is no internationally agreed framework for measuring biomass sustainability: no agreed criteria and no agreed tools for measurement (Bracco et al., 2018). If there are no such tools and criteria, then it is not possible to measure how much biomass can be grown, harvested, and used (the biomass potential) in the bioeconomy. When criteria have been suggested, it has been extremely difficult to get international consensus (Fig. 13.1).

Generally, van Dam and Junginger (2011) saw that the criteria that were given the highest importance are focused on climate and energy issues,

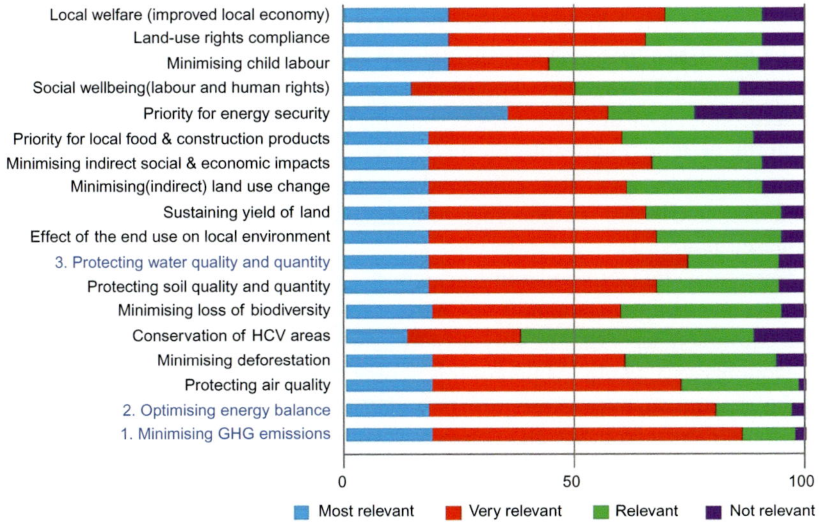

FIG. 13.1 Consensus building on sustainability metrics is a challenge. From a questionnaire, 473 responses were received from 25 EU member countries and 9 non-European countries; 285 could be used for further processing. *(Modified from van Dam, J., Junginger, M., 2011. Striving to further harmonization of sustainability criteria for bioenergy in Europe: recommendations from a stakeholder questionnaire. Energy Policy 39, 4051–4066.)*

followed by water conservation. There was no clear preference among respondents for including biodiversity, environment, or socioeconomic criteria in a certification system. Comparing different bio-based products environmental impact studies, Broeren et al. (2017) showed that while around 70% of studies include a resource indicator for energy, only 15%–25% of studies report water and land indicators. Sustainability is probably most often associated with reduction of emissions. Environmental sustainability is most readily measured by life cycle analysis (LCA). However, LCA is not optimal for measuring biomass or bio-based product sustainability as it does not account for the economic or social pillars. Social criteria are the most difficult to measure and tend to be given lower priority as a result.

Sustainability applies to all stages of bioeconomy production processes, e.g., feedstock pretreatment processes (Islam et al., 2020). It is regarding biomass that the issues are of greatest policy focus (Bosch et al., 2015). Biomass potential refers to how much biomass can be grown. The sustainability of biomass potential is a linchpin of bioeconomy policy: if biomass cannot be grown and harvested sustainability, then negative environmental impacts are inevitable, e.g., overexploitation, soil erosion, deforestation. Seidenberger et al. (2008) have attempted to compile global biomass potential ranges from 18 different studies, and noted huge variability in the results. Overall in the literature, Schueler et al. (2016) observed a range of technically available potentials between 50 and 500 EJ/year by mid-century.

There has been a concerted effort of well over a decade in the United States to discover the national biomass potential, resulting in the first "*Billion Ton Report*," completed in 2005 (USDOE, 2005), with updates in 2011 and 2016 (USDOE, 2011, 2016). The basics remain the same throughout these reports: that the United States, depending on assumptions made, may be able to produce one billion tons of dry biomass per annum, thus substituting 30% of gasoline requirement with renewable biofuels. The authors estimate that the United States currently uses 365 million dry tons of agricultural crops, forestry resources, and waste to generate biofuels, renewable chemicals, and other bio-based materials.

Currently, the UN Food and Agriculture Organization (FAO) is at the forefront on assessing sustainability of bio-based products. It recommends a stepwise approach to sustainability assessment of bio-based products and the bioeconomy as a whole. As part of this approach, countries or bio-based product producers and manufacturers are provided with a long list of scientifically robust indicators, from which to choose a limited number of indicators that suits their needs and circumstances (FAO, 2019). The limitation of indicators may be essential for affordability for industry.

R&D subsidy

The central biotechnology for bio-based materials production is metabolic engineering. It can take 50–300 person years and many millions of dollars to bring a metabolically engineered product to the market (Hong and Nielsen, 2012). However, there is emerging evidence that success rates would be higher if governments paid more attention to R&D subsidy for convergence, e.g., of industrial biotechnology with green chemistry (e.g., Dusselier et al., 2015), and a higher level of systems integration through IT/computation convergence with synthetic biology and metabolic engineering (e.g., Carbonell et al., 2020; Whitehead et al., 2020).

The bottlenecks in bioproduction

The commercial successes in metabolic engineering are dwarfed by the research successes, creating a large bottleneck in the bio-based chemical production pipeline (Opgenorth et al., 2019). Therefore, for governments, this could look like a poor return on investment. Korea is established as a nation with advanced capability in metabolic engineering. In response to the lack of commercial success, researchers have recently suggested ten general strategies of systems metabolic engineering to successfully develop industrial microbial strains (Lee and Kim, 2015). Systems metabolic engineering differs from conventional metabolic engineering by incorporating traditional metabolic engineering approaches along with tools of other fields, such as systems biology, synthetic biology, and molecular evolution. What is evident is that there are many groups competent in one or more of these specialisms, but very few companies that integrate them all into a production process. There is a clear need for better collaboration between academia and industrial biotechnology companies (Pronk et al., 2015), and far more rapid transfer of knowledge between the public and private sectors.

There is a need for biotechnology to have its own high-level programming language(s) and software to transform its design-build-test-learn (DBTL) engineering cycle (Sadowski et al., 2016). Phenotype evaluation is a major rate-limiting step in metabolic engineering (Wang et al., 2014). Because the price of DNA synthesis has plummeted recently, the design and build stages are bottlenecked by the test part of the cycle. Improving this throughput by mechanical or electronic automation is going to be limited as the orders of magnitude of improvement needed are so high. The advances necessarily have to come from biology itself, but will be vastly facilitated by the integration of computational tools.

Therefore a new data analysis pipeline that simplifies the interrogation of phenotype-sequence relationships is urgently required. In the age of machine learning, ultimately the data should inform the next iteration of design without human intervention (Rogers and Church, 2016). For example, AutoBioCAD promises to design genetic circuits for *Escherichia coli* with virtually no human user input (Rodrigo and Jaramillo, 2013). Thus, algorithms are needed that incorporate machine learning to correlate data from different datasets for the purpose of linking genes, proteins, and pathways without a priori knowledge (Wurtzel and Kutchan, 2016).

Production facility support

Biorefineries at demonstration scale are difficult to bankroll because the level of production is not high enough to influence a market. Full-scale biorefineries are also very difficult to build for various reasons, most relating to uncertainty—of technology, of supply and value chains and of policy. The private sector is unwilling to shoulder the entire financial burden, and this has necessitated public-private partnerships (PPPs) to de-risk private investments.

Building biorefineries needs the largest investments in the innovation chain for a bioeconomy (Fig. 13.2). The most common form of financing for such technologies in the United States is a hybrid of equity, teamed with either federal grants or a federally backed loan guarantees. This streamlines the approval steps and the control. To build biorefineries, both the USDA and USDOE have favored 20-year loan guarantees.

Loan guarantees have not been common in Europe, but have become available. InnovFin consists of a series of integrated and complementary financing tools and advisory services offered by the European Investment Bank (EIB) Group, covering the entire value chain of research and innovation (R&I). InnovFin aims to improve access to risk finance for research and innovation projects, research infrastructures, public-private partnerships, and special-purpose projects promoting first-of-a-kind, industrial demonstration projects (Scarlat et al., 2015).

The biofoundry and biological resource centers (BRCs): Missing links in bio-based production

In the chain from idea to industrial production, it may be that the biofoundry is a missing link that at last improves reproducibility and reliability in

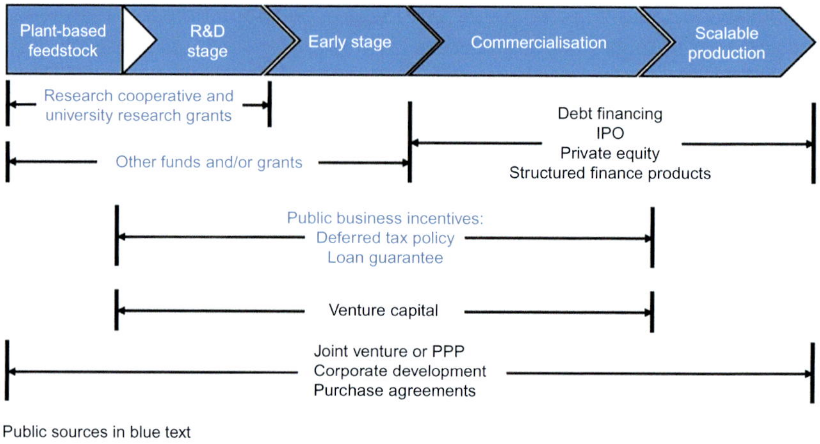

FIG. 13.2 Funding mechanisms in bio-based production. *(Modified from Milken Institute, 2013. Unleashing the Power of the Bio-Economy. Milken Institute, Washington, DC.)*

biotechnology and the life sciences (Kitney et al., 2019). Biofoundries are highly automated facilities characterized by the coordinated use of laboratory robots (Fig. 13.3). Biofoundries are based on information infrastructures that allow the robots and other equipment within the biofoundries to be programmed to follow detailed, complex workflows (Chao et al., 2017). Further digitalization of biology will advance the field in the near future. Overcoming some of the limitations of biology in industrial processes will require workflows to determine how to split a process between biochemical conversion and chemical catalysis (Robinson et al., 2020; Whitehead et al., 2020), allied to the automation of synthetic chemistry (Coley et al., 2019). Eventually, the automated generation of robotic build worklists and the integration of machine learning algorithms will collectively allow fully automated smart biomanufacturing without human intervention (Carbonell et al., 2020).

Public biofoundries are risk-reducing facilities, often open to the private sector for public-private cooperation, of which there is a growing number. The Global Biofoundry Alliance (GBA) is a formalized alliance of public biofoundries (Hillson et al., 2019), one of the objectives of which is to build a synthetic biology industry and accelerate commercialization of engineering/synthetic biology and biomanufacturing process engineering with broad public benefits.

A particularly powerful combination could be the interaction between biofoundries and the modern biological resource centers (BRCs). This is the combination of the known biology (from sequencing in the BRCs) with the new biology (from biofoundries) (Fig. 13.4). Sequencing initiatives such as the

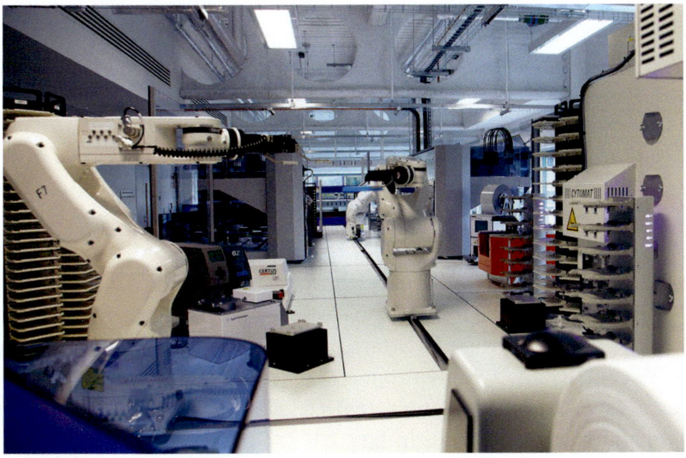

FIG. 13.3 The Edinburgh biofoundry, a public biofoundry driven by robotic workflows. *(Courtesy of the Edinburgh Genome Foundry, University of Edinburgh.)*

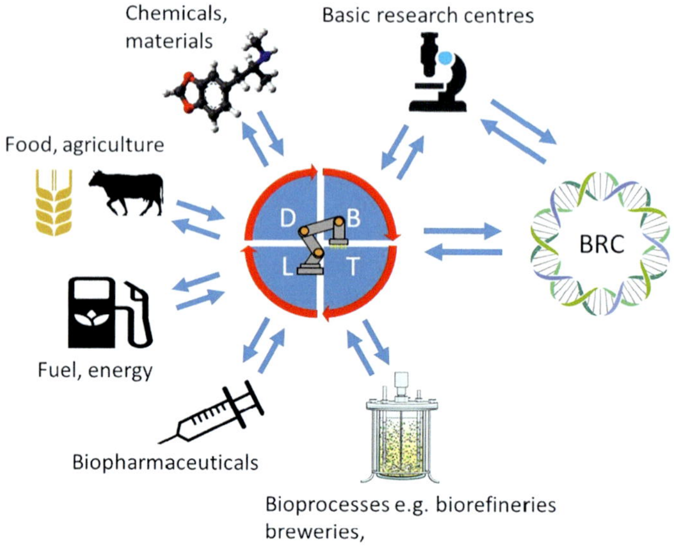

FIG. 13.4 Engineering biology platforms address a wide range of sectors. *BRC*, Biological resource center; *DBTL*, design-build-test-learn cycle.

Earth Biogenome Project, which aims to sequence the genomes of all known eukaryotic organisms (Lewin et al., 2018), will mean a large increase in the availability of genome sequence for use in products over a wide range of sectors.

Biological resource centers (BRC) have been defined: "*They consist of service providers and repositories of the living cells, genomes of organism, and information relating to heredity and the functions of biological systems. BRCs contain collections of culturable organisms (e.g., microorganisms, plant, animal and human cells), replicable parts of these (e.g., genomes, plasmids, viruses, cDNAs), viable but not yet culturable organisms, cells and tissues, as well as databases containing molecular, physiological and structural information relevant to these collections and related bioinformatics*" (OECD, 2001).

Subsequent to the earlier OECD work, the microbiology community followed up with the Global Biological Resource Centre Network demonstration project, which focused on collections of living microorganisms. European member states in the European Strategy Forum for Research Infrastructures (ESFRI) accepted the pan-European research infrastructure Microbial Resources Research Infrastructure (MIRRI) on to their roadmap for 2010 and its preparatory phase began in late 2012 (Smith et al., 2014).

Other instruments

Other innovative instruments may become more important in the future. Green Bonds enable capital-raising and investment for new and existing projects with environmental benefits. The Green Bond Principles[a] instrument is a mechanism to raise large capital sums, with the financing and management of project risks undertaken by the project sponsors, not the investors that might or might not have the capacity to manage said risks. Akomea-Frimpong et al. (2021) reviewed various instruments for green banking.

A government-backed green investment bank differs from a typical fund in that it should not just disburse government money, but as a "bank," it should be able to raise its own finance and fill a gap in the market for government-backed bonds, bring in banking expertise, and offer a range of commercially driven interventions—loans, equity, and risk reduction finance. A green investment bank is a publicly capitalized entity established specifically to facilitate private investment into domestic low-carbon and climate-resilient infrastructure and other green sectors such as water and waste management. These dedicated green investment entities have been established at national level and at state and even city level (OECD, 2016).

Tax incentives for industrial R&D

Tax incentives reduce the marginal cost of R&D and innovation spending, and they are usually more neutral than direct support. Tax incentives have become the number one policy tool that governments use to encourage companies to invest in research and development (OECD, 2020b). Governments should consider the position of small domestic companies in bioproduction in relation to MNEs, which are becoming attracted to bioproduction. Nevertheless, R&D tax incentives seem to boost R&D for smaller firms, but not necessarily because of firm size per se; rather, it is because these firms perform less R&D on average.

Clusters

The regional bioeconomy cluster is frequently used in Europe to build capacity (Stegmann et al., 2020). The main rationale for public policies to promote technology clusters through infrastructure and knowledge-based investments, networking activities, and training is an increase in knowledge spillovers among actors in clusters. This is purported to generate a collective pool of knowledge that results in higher productivity, more innovation, and an increase in the competitiveness of firms.

---

a. http://www.icmagroup.org/Regulatory-Policy-and-Market-Practice/green-bonds/green-bond-principles/.

A successful cluster in France with tangible results and benefits is *Industries and Agro-Resources* (IAR). With over 200 members, this cluster in the Champagne-Ardenne and Picardy regions of France unites stakeholders from research, education, industry, and agriculture in France around a shared goal: to optimize the added value from the exploitation of biomass. It has regional roots where biorefining has been particularly successful; it also has a global mission by integrating external know-how through international strategic alliances.

Government support for SMEs

SMEs are the backbone of economies. In Europe, they provide 85% of all new jobs.[b] All high-technology SMEs face challenges in their specific sectors. In stark contrast to IT, biotechnology SMEs can face many years of high-risk research without revenues (Pisano, 2010), requiring expensive specialist facilities and complex market entry (Jernström et al., 2017). Additionally, in bioproduction, they may well be in competition with some of the world's largest oil and petrochemistry firms that have established markets, stable supply and value chains, proven technology, and fully amortized production facilities.

Technology and regional clusters are a leading support mechanism for SMEs (Wilde and Hermans, 2021), providing a range of services, such as access to venture capital and other finance routes; business advice on the strategic use of standards, labels, certificates, assistance with specific LCA and sustainability tools, access to demonstration and testing facilities. National government programs can provide a wide range of support mechanisms, especially exemptions from tax and national insurance payments.

Interestingly, an article in The Economist (2018) speculated that instead of SME biotechnology companies owning and running their own expensive laboratories, biofoundries in the future could do this for them, allowing scientists to concentrate more on experimental design than repetitive wet laboratory work. The current model would be the public biofoundry, such as the Agile Biofoundry of the USDOE (OECD, 2019a), but increasingly in the future, the role would be taken over wholly or partly by private companies, such as Ginkgo Bioworks and Zymergen (Kim, 2019).

Access to national expertise in bioprocess design

One of the greatest challenges facing the SMEs in bio-based production is scale-up (Gatto and Re, 2021). In most countries, risk capital is limited, and the cost of moving from laboratory pilot to demonstration scale is beyond the financial scope of most of these SMEs. As the bioeconomy grows and these applications of bio-based production expand, greater efficiency of public investments could

b. http://www.ultraeuropeannetwork.eu/news/increasing-competitiveness-growth-eu-program-smes/.

be made through the public finance of regional bio-based production facilities at demonstrator and/or pilot scale. Models already exist, e.g., the Centre for Process Innovation (CPI)[c] in the United Kingdom. CPI helps SMEs understand the commercial feasibility of products or processes in a way that reduces risk to the companies and their investors. Such facilities could maximize their benefits by offering a range of ancillary business services, such as training, quality management, and certification (Schieb and Philp, 2014).

Demand-side (market pull) instruments

The OECD analysis on demand-side policies (OECD, 2011a) identified several general principles for governments to consider when applying demand-side policies. One that it is particularly relevant when applied to industrial biotechnology and the bioeconomy more generally relates to timing and alignment to make sure that policies realize their maximum efficiency.

Mandates and targets

Mandates and targets exemplify the different approaches to the introduction of biofuels in Europe and the United States. Incorporation targets (i.e., targets of percentages of biofuels blended into gasoline and diesel) have been approved voluntarily by several EU member states as national initiatives, not an obligation from the EU. The US biofuels policy has specified absolute production quantities through a mandate rather than a less-binding incorporation target (Ziolkowska et al., 2010). Mandates and targets for biofuels production have become standard for introduction of biofuels. By contrast, there is virtually no such mechanism available for bio-based chemicals and materials, although suggestions have been made about how to make them work (e.g., Philp, 2015).

Arguably, the best-known mandate in bioproduction was created in the US *Energy Independence and Security Act* (EISA) through the Renewable Fuels Standard (RFS2) (Federal Register, 2010). It set high production volume mandates for biofuels. Together with blending mandates, a comprehensive policy support regime for biofuels came into being in the 21st century in the United States.

Lapan and Moschino (2012) found biofuels production mandates to be more revenue-neutral than tax and excise reductions. They derived that an ethanol volume mandate is equivalent to a combination of an ethanol production subsidy and a fossil fuel (petrol) tax that is revenue-neutral. They conclude that the (optimal) ethanol mandate yields higher welfare than the (optimal) ethanol subsidy.

However, if mandates do not distinguish among biofuels according to their feedstock or production methods, despite wide differences in environmental

c. https://www.uk-cpi.com/work-with-us/smes/technology-focus.

costs and benefits, governments could end up supporting a fuel that is more expensive than its corresponding petroleum product and with poorer environmental protection credentials (Global Subsidies Initiative, 2007). A key to preventing such a mistake in bio-based materials production support is, in the short term, harmonizing life cycle analysis within the industry, and in the longer term developing robust and internationally coherent sustainability assessment. There is a conspicuous need for more empirical studies on the effectiveness of biofuels mandates, including their unintended consequences on other areas, such as land use (Kesan et al., 2017).

Public procurement

Public procurement affects a substantial share of world trade flows. It accounts for 13% of GDP on average in OECD member countries (OECD, 2012). It is also being increasingly used to drive innovation and economic growth (OECD, 2019b). While there seem possibilities in public procurement for facilitating market entry for bioeconomy innovative products, there are intrinsic obstacles on both sides of the market, not least being a lack of a substantive theoretical research grounding (Tremblay and Boyle, 2018).

On the supply side, only a small proportion of the possible plethora of products from the bioeconomy addresses the business-to-consumer (B2C) market in which public procurers normally operate (e.g., fuel and consumer products). The larger share of bio-based products is chemicals and intermediates, which are only interesting to private industry in a business-to-business (B2B) market.

On the demand side, public procurement is a highly fragmented landscape: in the EU, there are over 250,000 public authorities, and public procurement accounts for 14% of EU GDP, or two trillion Euro per annum (Núñez Ferrer, 2020). Public procurers naturally tend to be very price-sensitive, which is a barrier for any innovative product.

The USDA BioPreferred Program specifically aims to increase the purchase and use of bio-based products (Fig. 13.5). It has a catalogue containing around 16,000 bio-based products. There are broadly two main activities of the program: mandatory federal purchasing and a voluntary labeling initiative (USDA, 2018).

Promote standards and certification

Stringent standards and certification give confidence to consumers and industry as they provide credibility to claims of performance and sustainability (Dammer and Carus, 2015), such as "bio-based," "renewable raw material," "biodegradable," "recyclable," or "reduced greenhouse gas impact." They help verify claims such as biodegradability and bio-based content that will promote market uptake (OECD, 2011b). Claims ought to be verifiable by consumers, waste

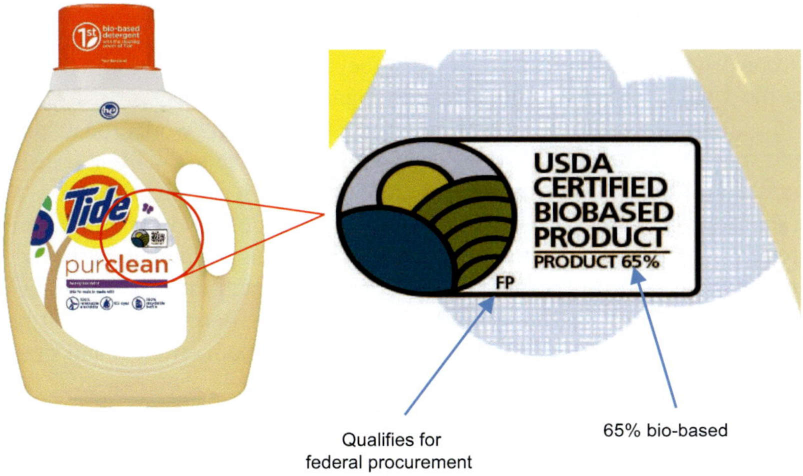

FIG. 13.5 The BioPreferred Program has a voluntary label that shows the consumer that the product is bio-based to some extent.

management authorities, and legislators. Third-party verification is a means to prevent unwarranted claims and green washing.

According to the British Standards Institution (BSI), a surprisingly large proportion of annual productivity growth can be attributed to standards (BSI, 2015). Companies can use matching or beating a standard as an R&D and marketing tool, which then spurs competitors to innovate further, driving technical advances and delivering efficiencies. They are developed in close cooperation between industry, research, and policymakers, which is essential to create the right environment for new products and technologies to grow to full-scale deployment.

Standards and certification schemes are also joining-up measures between policy frameworks and practical implementation. Standards provide the necessary scientific basis for implementing legislation by demonstrating compliance with legal requirements. They can also be used to verify that policy goals and targets are being met.

Standards in various forms are needed in synthetic biology. Technical standards of high priority are those that automate methods, the description and assembly of components, and documentation of the performance of engineered microbial strains. de Lorenzo and Schmidt (2018) argued that the adoption of standards will accelerate the transition to a future advanced bioeconomy, driven by bio-based manufacturing. Synthetic biology standards need user-friendly distribution platforms, ideally maintained by governmental support, for their implementation (Tas et al., 2020).

Fossil carbon taxes and emissions incentives

Carbon taxation and the removal of fossil fuel subsidies are necessities for meeting the United Nations sustainable development goals (SDGs) (El-Chichakli et al., 2016). The purpose of carbon pricing policy frameworks today should be to send clear and credible price signals that foster the low-carbon transition over the medium to long term (OECD, 2015a). Explicit carbon prices can either be set through a carbon tax, expressed as a fixed price per ton of emissions, or through cap-and-trade systems, where an emissions reduction target is set through the issuance of a fixed number of permits, and the price is set in the market through supply and demand. Recently, vom Berg et al. (2021) have argued that a fossil carbon tax, a raw material tax, is more efficient and simpler to implement than an end-of-pipe CO_2 emissions tax.

As of 2019, there were 57 carbon prices either in practice or in development. This represents some 11 gigatons of CO_2 equivalent, or 20% of global emissions per annum, and the figure is steadily rising (International Bank for Reconstruction and Development/The World Bank, 2019). However, the level of tax is nowhere near high enough: the average price of emissions worldwide is only $2 per tons, when a realistic level would be around $70 per tons (IMF, 2019).

There are essentially two methods for using revenues from these taxes to help grow the bioeconomy. In the first, revenues are added to the general budget of a government, and that government can choose to use these revenues for climate-friendly purposes. Alternatively, the revenues can be legally earmarked or hypothecated for specific projects or purposes, rather than being added to the general budget. Both approaches have advantages: adding to the general budget minimizes the cost for new administration, while earmarking is more direct, transparent, and perhaps easier to gain public acceptance (Marvik and Philp, 2020). This could be a cost-effective way to support long-term, higher-risk research, and more targeted short- to mid-term R&D, precompetitive or near-market.

Fossil fuel subsidies reform

Subsidies for emerging industries are supposed to end at some stage when the industry is considered to be self-supporting. Yet fossil fuel subsidies represent perhaps the largest subsidy system of all time—and the industry is over a century old. Most of these subsidies are inefficient and wasteful, but political backlash against their removal makes reform difficult (OECD, 2020a). The environmental and social costs of fossil fuel subsidies (Whitley and van der Burg, 2015) are unlikely to be obvious to the public and may even be masked for finance ministers (Edenhofer, 2015). It is estimated that on average, only 7% of the benefits from fossil fuel subsidies reach the poorest 20% of the population. Governments could use the money saved to fund decarbonization projects and technologies (Martin, 2016), as exemplified by the bioeconomy.

Fossil energy received a staggering USD 5.3 trillion, or 6.5% of global GDP, in posttax subsidies in 2015 (Coady et al., 2017). The IMF estimated that eliminating posttax subsidies in 2015 could raise government revenue by USD 2.9 trillion (3.6% of global GDP), cut global CO_2 emissions by more than 20%, and cut premature air pollution deaths by more than half.

Cross-cutting measures

Definitions and terminology

Definitions are necessary in any economic activity to gather data that are comparable across regions, countries, and globally. This will be evident from the discussions on biomass sustainability. Robust data are needed to build metrics for the performance of a bioeconomy. "Bioeconomy" itself means different things in different countries. A definition of "bio-based product" is needed as a standard for public procurement and business development (OECD, 2021). There are several types of biorefinery in operation and under development: each needs a definition to differentiate one from the other.

The debate over "waste or resource" (e.g., House of Lords, 2014) is an important one for the bioeconomy. A mixture of terms and a lack of standardized definitions make it very difficult to truly assess the volumes of different waste materials that can be used in biorefining. For example, gathering data on "agricultural residues" suffers from this. This compares poorly with the easily identifiable volumes that would be available from crop feedstocks, such as volumes of sugar cane or sugar beet: these figures are collected internationally and are readily comparable.

A key objective of biorefining, especially for second-generation biofuels and bio-based materials, is the valorization of waste (Fava et al., 2015). "Bio-waste" is acquiring greater important in biorefining, and tonnages should be known when formulating biorefinery roadmaps. However, any definition that excludes agricultural or forestry residues drastically changes available estimates of tonnages. The definition of "waste disposal" could be changed to allow collection, transportation, sorting in view of its conversion in biorefineries. If a material is to be converted in a biorefinery, then it should no longer be regarded as a waste but as a secondary raw material (Schreck and Wagner, 2017).

Ultimately, the integration of actors across sectors and hence the creation of new value chains is limited by disparity and lack of control of terminology and standards. In short what is called for is *commonly agreed vocabulary throughout value chains, from feedstock suppliers to biorefining to downstream actors in the application sectors.*[d]

d. http://cordis.europa.eu/programme/rcn/700101_en.html.

In the wider context, the key roles of definitions are also evident in carbon neutrality policy. As adumbrated by Rogelj et al. (2021), different definitions and pathways to "net zero" can have drastically differing outcomes. This conundrum is evident through bioeconomy and sustainability. Thus at the micro-level of "bio-based product" all the way to the strategy that will determine how we tackle the climate emergency, definitions are absolutely necessary, even if they appear lumpen and boring.

Design skills and education to train the workforce of the future

For some, synthetic biology is a field of engineering, not of biology (Andrianantoandro et al., 2006). Synthetic biologists must be grounded in one or several core disciplines: genetics, systems biology, microbiology, or chemistry. The need for a specially trained workforce for the synthetic biology sector is considered by many in industry to be a key pinch point for the industrial development of the area. The industrial view, regarding personnel, is that for every engineer, mathematician, and computer scientist, there are at least 30 PhDs in biology and biochemistry available (Kitney et al., 2021). Crucial for the development of the synthetic biology industry is the training of many more engineers, mathematicians, and computer scientists with an understanding of biology.

It is entirely feasible that apprenticeships and day-release education will play a prominent part in developing this workforce. Higher education is rising to the challenges with a range of solutions from technician training, undergraduate degrees, Masters and interdisciplinary PhD programs, massive open online courses (MOOCs), and business management courses (Delebecque and Philp, 2018; Hallinan et al., 2019).

Governance and regulation

OECD analysis suggests that innovation heavily depends on issues of governance and implementation (OECD, 2015b). Governance matters in innovation policy due to various levels of authority and policy competencies involved. Budgetary resources are distributed across various levels of government when horizontal policy is created. Regionalization and decentralization have made regional and local governments more powerful and has increased their capacity to operate their own development policies. This is particularly important in industrial biotechnology and bioeconomy. But bioeconomy also spans national and global connections, creating the need for multilevel governance, which is not easily achieved (Vogelpohl et al., 2021).

Poor regulatory policy can do more harm than good

Regulation refers to the implementation of rules by public authorities and governmental bodies to influence the behaviors of private actors in the economy.

The primary purpose of regulation in innovation policy should be to stimulate innovation, although the opposite is undeniably possible. Complex and time-consuming regulation is far more damaging to small bio-based companies than it is for large companies. Governments could act to reduce this impact.

A study for the government of the Netherlands (Sira Consulting, 2011) identified around 80 regulatory barriers to the bioeconomy. These were assigned different categories:

- *Fundamental constraints.* These call for a political and policy approach (e.g., import duties, level playing field, certification, and financial feasibility);
- *Conflicting constraints.* These barriers cannot be removed, but governments can help the companies to meet the regulations (e.g., REACH regulations);
- *Structural constraints.* These require adjustment to regulations, but do not demand policy or political action;
- *Operational constraints.* Here, the regulation itself is not the problem but its implementation by, for example, local authorities. Especially for SMEs, these lead to substantial barriers to investment in the bioeconomy.

Communication and raising awareness

Information campaigns for consumers can strengthen the demand for bio-based materials when they convey to the consumers that bio-based products possess many ecological advantages. In this case, official labels for bio-based products (see Fig. 13.5) would strengthen the public awareness of bio-based plastics and their products and would strengthen the trust placed in such products, therefore supporting the market introduction and establishment. Communication is vital for public acceptance (McKinsey Global Institute, 2020): up to 70% of the potential of the "Bio Revolution" may depend on consumer, societal, and regulatory acceptance. Policy action is urgent as bio-based production and synthetic biology have already been associated with negative connotations (Mankad et al., 2021).

Systemic and transition considerations in policy development

One of the primary confrontations for the policymaker is that bioeconomy value chains are new and untried and are characterized by an interdependency between multiple stakeholders creating a systemic business risk. Systemic risk discourages investments and needs to be tackled in policy development. Without a holistic approach and understanding of the complex interactions of value chains involved in the social carbon cycle, policies may fail to deliver on their sustainability objectives.

Systemic business risk in value chains

A value chain can be defined as "*a set of interlinked activities that deliver products/services by adding value to bulk material (feedstock)*" (Lokesh et al., 2018). In the emerging bioeconomy, a value chain would typically comprise a cascading series of manufacturing processes, spanning biomass production, pretreatment, and conversion, through to the manufacture and marketing of bio-based products. Lewandowski et al. (2019) describe the bioeconomy value chains concept in more detail.

Typically, many of these individual processing and manufacturing steps are all new and untried and various public and private actors need to work together to create new industrial ecosystems. The complexity of the bioeconomy web can easily be underestimated, resulting in, for example, unforeseen shortages of critical material(s) (National Academies of Sciences, Engineering, and Medicine, 2020).

Systems thinking in sustainability policies

Conversion of waste streams is seen as the basis for policies aiming to foster a cyclic or cascading resource utilization (Keegan et al., 2013; Klitkou et al., 2019). However, given the fact that the carbon cycle is fundamentally an energy cycle, it is important to avoid implementing policies for local recycling of carbon that do not make sense from an overall energy perspective (Hernandez and Cullen, 2019) and which may in fact suboptimize energy consumption within a larger system. While most industry processes are validated by calculating the mass and energy balances of the process (Larsson, 1992), there is in fact a need to put bioeconomy policies to a similar test at the systemic level.

Replacing fossil resources with fresh biomass will put huge pressure on agriculture and forestry. Shortage of arable land, water, and fertilizers have already led to conflicts between different sustainability goals (D'Amato et al., 2017) related to, for instance food, energy and biodiversity and major concerns from associated land-use change and deforestation (Searchinger et al., 2018). One would expect policies to prioritize the use of renewable carbon in those value chains where no alternatives are available, e.g., food, chemistry, and materials, while in fact, public policy attention has mainly been directed toward bioenergy. This indicates a need to better balance the policies (Philp, 2015), in consideration of potentially conflicting sustainability goals.

Dietz et al. (2018) identified the political management of conflicting goals as one of the major challenges for a sustainable bioeconomy governance framework. While it is generally agreed that human primary needs, such as food security, have to be prioritized in the bioeconomy, food production per se is typically not the major cause of malnutrition and famine, but rather inefficiencies in food management, distribution, and wastage (Berners-Lee et al., 2018). In a sustainable bioeconomy, measures to improve logistics in food supply chains, thereby reducing waste and extending shelf life, are very important (FAO, 2017). Seen from this angle, plastic packaging is an integral aspect of food security (Claudio, 2012); hence, crops grown for either food or for bioplastics both have justifiable claims when they compete for arable land. This is a further illustration of the complexity to be tackled in the bioeconomy transition.

Temporal aspects in policy development

Table 13.1 may have limited utility as it does not imply a temporal strategy and a progression path for policymakers, i.e., it lacks any conception of a sequence of policy implementation. Marvik and Philp (2020) refined the approach by describing the mix of specific and general measures in a widely accepted innovation policy sequence from "ideas to market" (European Commission, 2019) (Table 13.2), an approach familiar in other sectors, e.g., energy (International Energy Agency, 2009) and nanotechnology (Lim et al., 2015). The specific four-step matrix shown in Fig. 13.5 was used to develop the Norwegian national bioeconomy strategy (The Ministry of Trade, Industry and Fisheries (Norway), 2018). It may give policymakers a broader idea of how to construct a strategy that will connect supply- and demand-side drivers to achieve a stronger and more robust effect on the economic system. Specifically, this matrix may guide different ministries and agencies to know when and where their roles are required, or how and when they need to work together.

For many important bioeconomy value chains, market pull is the weakest driver, especially for those in which bio-based production is intended to replace long-standing petrochemical production. It is argued that the success of demand-side policies depends on several temporal factors such as coordination with supply-side measures and the duration of government intervention: too long can lead to market distortion and too short may not enable the desired effect.

Trade-offs and unintended consequences

History is replete with detrimental unintended consequences of well-meaning policies, often leading from too great a focus on intended consequences (Ehrlinger and Eibach, 2011; Herrero et al., 2020). As an example, environmental regulations to preserve wilderness and wildlife can result in increased greenhouse gas (GHG) emissions (Severnini, 2019). The use of food crops as feedstocks for bio-based products can have several unwanted consequences, such as ambiguities around the food versus fuel argument (e.g., Wesseler and Drabik, 2016) and effects on indirect land use change (ILUC) (Searchinger et al., 2008) and water use (Brizga et al., 2020). Thus, a further complication for holistic, systemic policy development is that efforts should be made to identify and model the most likely trade-offs involved—or, as

TABLE 13.2 A bioeconomy innovation policy matrix.

Feedstock	Technology	Industrialisation	Market
Objectives			
Stimulate availability of bioresources	Strengthen skills and technology base	Trigger investments in new manufacturing	Increased sustainability and value creation
Value chain specific policies			
Resource regulations and permits	Targeted R&D grant program	Public technology scale-up and pilot facilities	Product standards and norms
Transportation and logistics infrastructure	Specific education and training program	Financial support for flagship projects	Price subsidies and product tax policies
Feedstock specific trade regulations	Technology cluster and network support	Targeted government investments program	Product mandates and bans policies
Generic policies			
Biomass sustainability assessment studies	Broad scope R&D grant program	Start-up and SME support	Sustainability labels and communication
Governance and regulation efficiency	Tax incentives for applied R&D	Industry-oriented education program	Public awareness and acceptance campaigns
Waste management policies	Stimulate international partnerships	Techno-economic feasibility studies	Tax on CO_2 emissions and fossil fuel subsidy reform
International trade agreements	Exchange program and apprenticeship	Private investment stimulating policies	Public procurement of bio-based products

Modified from Marvik, O.J., Philp, J., 2020. The systemic challenge of the bioeconomy: a policy framework for transitioning towards a sustainable carbon cycle economy. EMBO Rep. 21 e51478.

expressed by Kotchen (2018), "*offsetting goods and bads*." A promising approach is to couple supply chain models to feedstock conversion models (De Buck et al., 2020; Ulonska et al., 2018).

Concluding remarks

Awkwardly, this chapter concludes where much of the content could have begun—the Human Genome Project (HGP). From a public policy perspective, the HGP was a towering success. Even as far back as 2011, it was claimed that every $1 of federal HGP investment contributed to generating $141 in the economy (Battelle, 2011). It launched the genomics age and set off the race toward next-generation sequencing. It may be surprising that in a chapter on genomics policy that this huge story has not been dealt with in greater detail. The reason can be found in the title of this chapter—the aim was to look beyond genomics R&D.

Public R&D subsidy is absolutely needed to bring forth the ideas that change the world. However, in terms of technology readiness levels (TRLs) (NASA, 2012), R&D subsidy is very far upstream from a technology on the market: on a scale from 1 to 9, TRL 1 represents the transition from scientific research to applied research. R&D subsidy can often be one of the least expensive parts of the chain to technology launch, involving a small number of highly specialized stakeholders, typically publicly funded scientists and engineers. The steps beyond R&D are much more complicated and expensive.

Public policy has many roles to play, and many represent risk with taxpayers' money. This chapter attempts to show the roles of public policy in the bioeconomy transition. Ideally, it would guide policymakers through a process that would help minimize risk to the taxpayer. The emphasis on value chains is important as the technology and its infrastructure is one link in the chain. As with any chain, the value chain is only as strong as its weakest link, and despite large potential for societal benefits, a single failure in the value chain might have the overall effect that the system will not work technically, logistically, or financially. In other words, if policy simply acts on individual parts of a complex industrial system, then there is a substantial risk of wasted resources and efforts. This underscores the need for coordination of different policy families along value chains, as well as across disciplines and sectors (Weber and Rohracher, 2012). Bioeconomy value chains are almost completely unproven, highlighting the risk to the public purse.

Nevertheless, if we are serious about a transition to carbon neutrality, then there is an inevitability of a large expansion of the bioeconomy contemporaneous with the phasing out of fossil resources. Carbon neutrality is not about "zero carbon" but net-zero (International Energy Agency, 2021), and this creates a major role for biomass as an industrial feedstock of the future. Standing over all policy aspects is sustainability. For politicians, the fundamental justification for public intervention in the bioeconomy is increased sustainability. It is how they respond to this justification that is the basis of this chapter.

Disclaimer statement

The views expressed are those of the author and not necessarily those of the OECD or of the governments of its member countries.

References

Akomea-Frimpong, I., Adeabah, D., Ofosu, D., Tenakwah, E.J., 2021. A review of studies on green finance of banks, research gaps and future directions. J. Sustain. Finance Invest. https://doi.org/10.1080/20430795.2020.1870202.

Andrianantoandro, E., Basu, S., Karig, D.K., Weiss, R., 2006. Synthetic biology: new engineering rules for an emerging discipline. Mol. Syst. Biol. 2. https://doi.org/10.1038/msb4100073.2006.0028.

Arns, D., 2018. Chemical Raw Materials in Europe—Trends & Challenges. https://www.energy-community.org/dam/jcr:cf9322f6-9725-45dc-9db7-f15388ecccb2/OF092018_Petrochemicals.pdf.

Battelle, 2011. The Economic Impacts of Human Genome Project. The Battelle Memorial Institute, Columbus, OH. https://www.battelle.org/docs/default-source/misc/battelle-2011-misc-economic-impact-human-genome-project.pdf?sfvrsn=6.

Bennett, S.J., Pearson, P.J.G., 2009. From petrochemical complexes to biorefineries? The past and prospective co-evolution of liquid fuels and chemicals production in the UK. Chem. Eng. Res. Des. 87, 1120–1139.

Berners-Lee, M., Kennelly, C., Watson, R., Hewitt, C.N., 2018. Current global food production is sufficient to meet human nutritional needs in 2050 provided there is radical societal adaptation. Elementa 6, 52.

Black, M.J., Sadhukhan, J., Day, K., Drage, G., Murphy, R.J., 2016. Developing database criteria for the assessment of biomass supply chains for biorefinery development. Chem. Eng. Res. Des. 107, 253–262.

Bosch, R., van de Pol, M., Philp, J., 2015. Define biomass sustainability. Nature 523, 526–527.

BP-EBI, 2014. Biomass in the Energy Industry. An Introduction. Pub. BP plc, London.

Bracco, S., Calicioglu, O., Juan, M.G.S., Flammini, A., 2018. Assessing the contribution of bioeconomy to the total economy: a review of national frameworks. Sustainability 10, 1698.

Brizga, J., Hubacek, K., Feng, K., 2020. The unintended side effects of bioplastics: carbon, land, and water footprints. One Earth 3, 45–53.

Broeren, M.L.M., Zijp, M.C., der Loop, S.L.W.-v., Heugens, E.H.W., Posthuma, L., Worrell, E., Shen, L., 2017. Environmental assessment of bio-based chemicals in early-stage development: a review of methods and indicators. Biofuels Bioprod. Biorefin. 11, 701–718.

BSI, 2015. The Economic Contribution of Standards to the UK Economy. British Standards Institution, London.

Carbonell, P., Le Feuvre, R., Takano, E., Scrutton, N.S., 2020. *In silico* design and automated learning to boost next-generation smart biomanufacturing. Synth. Biol., ysaa020. https://doi.org/10.1093/synbio/ysaa020.

Chao, R., Mishra, S., Si, T., Zhao, H., 2017. Engineering biological systems using automated biofoundries. Metab. Eng. 42, 98–108.

Claudio, L., 2012. Our food: packaging & public health. Environ. Health Perspect. 120, A232–A237.

Coady, D., Parry, I., Sears, L., Shang, B., 2017. How large are global fossil fuel subsidies? World Dev. 91, 11–27.

Coley, C.W., Thomas III, D.A., Lummiss, J.A.M., Jaworski, J.N., Breen, C.P., Hart, V.T., Fishman, J.S., Rogers, L., Gao, H., Hicklin, R.W., Plehiers, P.P., Byington, J., Piotti, J.S., Green, W.H., Hart, A.J., Jamison, T.F., Jensen, K.F., 2019. A robotic platform for flow synthesis of organic compounds informed by AI planning. Science 365, eaax1566. https://doi.org/10.1126/science.aax1566.

Dahmen, N., Lewandowski, I., Zibek, S., Weidtmann, A., 2018. Integrated lignocellulosic value chains in a growing bioeconomy: status quo and perspectives. GCB Bioenergy 11, 107–117.

D'Amato, D., Droste, N., Allen, B., Kettunen, M., Lähtinen, K., Korhonen, J., Leskinen, P., Matthies, B.D., Toppinen, A., 2017. Green, circular, bio economy: a comparative analysis of sustainability avenues. J. Clean. Prod. 168, 716–734.

Dammer, L., Carus, M., 2015. Standards, norms and labels for bio-based products. In: Aeschelmann, F., Carus, M., Baltus, W., Blum, H., Busch, R., Carrez, D., von Pogrell, H. (Eds.), Bio-Based Building Blocks and Polymers in the World. Capacities, Production and Applications: Status Quo and Trends Towards 2020. Nova-Institut GmbH, Chemiepark Knapsack, Köln, Germany. report 2015-05.

De Buck, V., Polanska, M., Van Impe, J., 2020. Modelling biowaste biorefineries: a review. Front. Sustain. Food Syst. 4, 11. https://doi.org/10.3389/fsufs.2020.00011.

de Lorenzo, V., Schmidt, M., 2018. Biological standards for the knowledge-based bioeconomy: what is at stake. New Biotechnol. 40, 170–180.

Delebecque, C.J., Philp, J., 2018. Education and training for industrial biotechnology and engineering biology. Eng. Biol. 3, 6–11.

Dietz, T., Börner, J., Förster, J.J., von Braun, J., 2018. Governance of the bioeconomy: a global comparative study of national bioeconomy strategies. Sustainability 10, 3190. https://doi.org/10.3390/su10093190.

Dusselier, M., Van Wouwe, P., Dewaele, A., Jacobs, P.A., Sels, B.F., 2015. Shape-selective zeolite catalysis for bioplastics production. Science 349, 78–80.

Economist, T., 2018. Robotic Labs for High-Speed Genetic Research Are on the Rise. The Design of Synthetic Lifeforms Could Become a New Industry. The Economist. March 01/2018.

Edenhofer, O., 2015. King coal and the queen of subsidies. Science 349, 1286–1287.

Edler, J., Georghiou, L., 2007. Public procurement and innovation—resurrecting the demand side. Res. Policy 36, 949–963.

Ehrlinger, J., Eibach, R.P., 2011. Focalism and the failure to foresee unintended consequences. Basic Appl. Soc. Psychol. 33, 59–68.

El-Chichakli, B., Von Braun, J., Lang, C., Barben, D., Philp, J., 2016. Five cornerstones of a global bioeconomy. Nature 535, 221–223.

European Commission, 2019. Innovation Policy, Fact Sheets of the European Union. https://www.europarl.europa.eu/factsheets/en/sheet/67/innovation-policy.

European Commission, 2020. Innovation Policy. Fact Sheets on the European Union—2020. www.europarl.europa.eu/factsheets/en.

FAO, 2017. The Future of Food and Agriculture—Trends and Challenges. UN Food and Agriculture Organization, Rome.

FAO, 2019. Indicators to Monitor and Evaluate the Sustainability of Bioeconomy. UN Food and Agriculture Organization, Rome.

Fava, F., Totaro, G., Diels, L., Reis, M., Duarte, J., Carioca, O.B., Poggi-Varaldo, M., Ferreira, B.S., 2015. Biowaste biorefinery in Europe: opportunities and research & development needs. New Biotechnol. 32, 100–108.

Federal Register, 2010. Regulation of fuels and fuel additives: changes to renewable fuel standard program; Final rule. Fed. Regist. 75 (58), 14669–15320. FRL–9112–3. Book 2 of 2 Books.

Gatto, F., Re, I., 2021. Circular bioeconomy business models to overcome the valley of death. A systematic statistical analysis of studies and projects in emerging bio-based technologies and trends linked to the SME instrument support. Sustainability 13 (1899).

Ghiat, I., Al-Ansari, T., 2021. A review of carbon capture and utilisation as a CO2 abatement opportunity within the EWF nexus. J. CO_2 Util. 45, 101432.

Global Subsidies Initiative, 2007. Biofuels—At What Cost? Government Support for Ethanol and Biodiesel in Selected OECD Countries., ISBN: 978-1-894784-03-0.

Hallinan, J.S., Wipat, A., Kitney, R., Woods, S., Taylor, K., Goñi-Moreno, Á., 2019. Future-proofing synthetic biology: educating the next generation. Eng. Biol. 3, 25–31.

Hernandez, A.G., Cullen, J.M., 2019. Exergy: a universal metric for measuring resource efficiency to address industrial decarbonisation. Sustain. Prod. Consum. 20, 151–164.

Herrero, H.P.K., Thornton, D., Mason-D'Croz, J., Palmer, B.L., Bodirsky, P., Pradhan, C.B., Barrett, T.G., Benton, A., Hall, I., Pikaar, J.R., Bogard, G.D., Bonnett, B.A., Bryan, B.M., Campbell, S., Christensen, M., Clark, J., Fanzo, C.M., Godde, A., Jarvis, A.M., Loboguerrero, A., Mathys, C.L., McIntyre, R.L., Naylor, R., Nelson, M., Obersteiner, A., Parodi, A., Popp, K., Ricketts, P., Smith, H., Valin, S.J., Vermeulen, J., Vervoort, M., van Wijk, H.H.E., van Zanten, P.C., West, S.A.W., Rockström, J., 2020. Articulating the effect of food systems innovation on the sustainable development goals. Lancet Planet. Health 5, e50–e62.

Hetemäki, L. (Ed.), 2014. Future of the European Forest-Based Sector. European Forest Institute, ISBN: 978-952-5980-17-2.

Hillson, N., Caddick, M., Cai, Y., Carrasco, J.A., Wook Chang, M., Curach, N.C., Bell, D.J., Le Feuvre, R., Friedman, D.C., Fu, X., Gold, N.D., Herrgård, M.J., Holowko, M.B., Johnson, J. R., Johnson, R.A., Keasling, J.D., Kitney, R.I., Kondo, A., Liu, C., Martin, V.J.J., Menolascina, F., Ogino, C., Patron, N.J., Pavan, M., Poh, C.L., Pretorius, I.S., Rosser, S.J., Scrutton, N.S., Storch, M., Tekotte, H., Travnik, E., Vickers, C.E., Yew, W.S., Yuan, Y., Zhao, H., Freemont, P.S., 2019. Building a global alliance of biofoundries. Nat. Commun. 10, 2040.

Hong, K.K., Nielsen, J., 2012. Metabolic engineering of *Saccharomyces cerevisiae*: a key cell factory platform for future biorefineries. Cell. Mol. Life Sci. 69, 2671–2690.

House of Lords, 2014. Waste or Resource? Stimulating a Bioeconomy. Science and Technology Select Committee, 3rd Report of Session 2013–14. Published by the authority of the House of Lords, The Stationery Office Limited, London.

IMF, 2019. Putting a Price on Pollution. Carbon-Pricing Strategies Could Hold the Key to Meeting the world's Climate Stabilization Goals. International Monetary Fund, Washington, DC.

Institute of Risk Management and Competition and Markets Authority, 2014. Competition Law Risk, A Short Guide. Crown copyright, London.

International Bank for Reconstruction and Development/The World Bank, 2019. Using Carbon Revenues. Technical Note 16, Publishing and Knowledge Division, The World Bank, Washington, DC.

International Energy Agency, 2009. Ensuring Green Growth in a Time of Economic Crisis: The Role of Energy Technology. IEA Publishing, Paris.

International Energy Agency, 2020. Energy Technology Perspectives 2020. Special Report on Carbon Capture, Utilisation and Storage. IEA Publishing, Paris.

International Energy Agency, 2021. Net Zero by 2050. A Roadmap for the Global Energy Sector. IEA Publishing, Paris.

Islam, M.K., Wang, H., Rehman, S., Dong, C., Hsu, H.-Y., Lin, C.S.K., Leu, S.-Y., 2020. Sustainability metrics of pretreatment processes in a waste derived lignocellulosic biomass biorefinery. Bioresour. Technol. 298, 122558.

Jernström, E., Karvonen, V., Kässib, T., Kraslawski, A., Hallikas, J., 2017. The main factors affecting the entry of SMEs into bio-based industry. J. Clean. Prod. 141, 1–10.

Kardung, M., Cingiz, K., Costenoble, O., Delahaye, R., Heijman, W., Lovri'c, M., van Leeuwen, M., M'Barek, R., van Meijl, H., Piotrowski, S., Ronzon, T., Sauer, J., Verhoog, D., Verkerk, P.J., Vrachioli, M., Wesseler, J.H.H., Zhu, B.X., 2021. Development of the circular bioeconomy: drivers and indicators. Sustainability 13, 413.

Keegan, D., Kretschmer, B., Elbersen, B., Panoutsou, C., 2013. Cascading use: a systematic approach to biomass beyond the energy sector. Biofuels Bioprod. Biorefin. 7, 193–206.

Kesan, J.P., Yang, H.-S., Peres, I.F., 2017. An empirical study of the impact of the renewable fuel standard (RFS) on the production of fuel ethanol in the U.S. Utah Law Rev. 2017 (1), 4.

Kim, H., 2019. AI, big data, and robots for the evolution of biotechnology. Genomics Inform. 17 (4), e44.

Kircher, M., 2012. The transition to a bio-economy: national perspectives. Biofuels Bioprod. Biorefin. 6, 240–245.

Kitney, R., Adeogun, M., Fujishima, Y., Goñi-Moreno, Á., Johnson, R., Maxon, M., Steedman, S., Ward, S., Winickoff, D., Philp, J., 2019. Enabling the advanced bioeconomy with engineering biology. Trends Biotechnol. 37, 917–920.

Kitney, R.I., Bell, J., Philp, J., 2021. Build a sustainable vaccines industry with synthetic biology. Trends Biotechnol. https://doi.org/10.1016/j.tibtech.2020.12.006. in press.

Klein, D., Humpenöder, F., Bauer, N., Dietrich, J.P., Popp, A., Bodirsky, B.L., Bonsch, M., Lotze-Campen, H., 2014. The global economic long-term potential of modern biomass in a climate-constrained world. Environ. Res. Lett. 9, 074017. https://doi.org/10.1088/1748-9326/9/7/074017.

Klitkou, A., Fevolden, A.M., Capasso, M. (Eds.), 2019. From Waste to Value Valorisation Pathways for Organic Waste Streams in Circular Bioeconomies. Routledge, London, https://doi.org/10.4324/9780429460289.

Knight, L., Pfeiffer, A., Scott, J., 2015. Supply market uncertainty: exploring consequences and responses within sustainability transitions. J. Purch. Supply Manag. 21, 167–177.

Knudsen, M.T., Hermansen, J.E., Thostrup, L.B., 2015. Mapping Sustainability Criteria for the Bioeconomy. Report of Aarhus University, Department of Agroecology. October 2015.

Kotchen, M.J., 2018. Environment, Energy, and Unintended Consequences. NBER Reporter 3, National Bureau of Economic Research (NBER), Cambridge, MA, pp. 10–13.

Kuosmanen, T., Kuosmanen, N., El Meligi, A., Ronzon, T., Gurria Albusac, P., Iost, S., M'barek, R., 2020. How Big is the Bioeconomy. EUR 30167 EN, Publications Office of the European Union, Luxembourg, https://doi.org/10.2760/144526.

Lapan, H., Moschino, G.C., 2012. Second-best biofuel policies and the welfare effects of quantity mandates and subsidies. J. Environ. Econ. Manag. 63, 224–241.

Larsson, J.E., 1992. Model-based measurement validation using MFM. IFAC Proc. 25, 127–132.

Lee, S.Y., Kim, H.U., 2015. Systems strategies for developing industrial microbial strains. Nat. Biotechnol. 33, 1061–1072.

Lewandowski, I., Bahrs, E., Dahmen, N., Hirth, T., Rausch, T., Weidtmann, A., 2019. Biobased value chains for a growing bioeconomy. GCB Bioenergy 11, 4–8.

Lewin, H.A., Robinson, G.E., Kress, W.J., Baker, W.J., Coddington, J., Crandall, K.A., Durbin, R., Edwards, S.V., Forest, F., Gilbert, M.T.P., Goldstein, M.M., Grigoriev, I.V., Hackett, K.J., Haussler, D., Jarvis, E.D., Johnson, W.E., Patrinos, A., Richards, S., Castilla-Rubio, J.C., van Sluys, M.-A., Soltis, P.S., Xu, X., Yang, H., Zhang, G., 2018. Earth BioGenome project: sequencing life for the future of life. Proc. Natl. Acad. Sci. U. S. A. 115, 4325–4333.

Lim, J.S., Shin, K.M., Yoon, J.S., Bae, S.H., 2015. Study of US/EU national innovation policies based on nanotechnology development, and implications for Korea. J. Inf. Sci. Theory Pract. 3, 50–65.

Lokesh, K., Ladu, L., Summerton, L., 2018. Bridging the gaps for a 'circular' bioeconomy: selection criteria, bio-based value chain and stakeholder mapping. Sustainability 10, 1695.

Luoma, P., Vanhanen, J., Tommila, P., 2011. Distributed Bio-Based Economy: Driving Sustainable Growth. SITRA, Helsinki, Finland, ISBN: 978-951-563-790-1.

Mankad, A., Hobman, E.V., Carter, L., 2021. Effects of knowledge and emotion on support for novel synthetic biology applications. Conserv. Biol. 35, 623–633.

Martin, T., 2016. Hasten end of dated fossil-fuel subsidies. Nature 538, 171.

Marvik, O.J., Philp, J., 2020. The systemic challenge of the bioeconomy: a policy framework for transitioning towards a sustainable carbon cycle economy. EMBO Rep. 21, e51478.

McKinsey Global Institute, 2020. The Bio Revolution: Innovations Transforming Economies, Societies, and Our Lives. May 2020.

Meijer, I., Hekkert, M.P., 2007. Managing uncertainties in the transition towards sustainability: cases of emerging energy technologies in the Netherlands. J. Environ. Policy Plan. 9, 281–298.

Ministry of Trade, Industry and Fisheries (Norway), 2018. Familiar Resources—Undreamt of Possibilities. The Government's Bioeconomy Strategy. https://www.regjeringen.no/contentassets/5b2dc02e8dd047adba138d7aa8b4dcc1/nfd_biookonomi_strategi_engelsk_uu.pdf.

NASA, 2012. Technology Readiness Level. https://www.nasa.gov/directorates/heo/scan/engineering/technology/technology_readiness_level.

National Academies of Sciences, Engineering, and Medicine, 2020. Safeguarding the Bioeconomy. The National Academies Press, Washington, DC.

Núñez Ferrer, J., 2020. The EU's Public Procurement Framework. How is the EU's Public Procurement Framework Contributing to the Achievement of the Objectives of the Paris Agreement and the Circular Economy Strategy? Policy Department for Economic, Scientific and Quality of Life Policies. Report number PE 648.770, April 2020.

OECD, 1996. Building Policy Coherence: Tools and Tensions. OECD Publishing, Paris.

OECD, 2001. Biological Resource Centres: Underpinning the Future of Life Sciences and Biotechnology. OECD Publishing, Paris.

OECD, 2009. The Bioeconomy to 2030: Designing a Policy Agenda. OECD Publishing, Paris.

OECD, 2011a. Demand-Side Innovation Policies. OECD Publishing, Paris.

OECD, 2011b. Future Prospects for Industrial Biotechnology. OECD Publishing, Paris.

OECD, 2012. Progress Made in Implementing the OECD Recommendation on Enhancing Integrity in Public Procurement. OECD Publishing, Paris.

OECD, 2015a. Aligning Policies for a Low-carbon Economy. Produced in Cooperation with the International Energy Agency, International Transport Forum, and Nuclear Energy Agency, OECD Publishing, Paris.

OECD, 2015b. The Innovation Imperative. Contributing to Productivity, Growth and Well-Being. OECD Publishing, Paris, ISBN: 978-92-64-23980-7.

OECD, 2016. Green Investment Banks: Scaling Up Private Investment in Low-carbon, Climate-Resilient Infrastructure. Green Finance and Investment, OECD Publishing, Paris, ISBN: 978-92-64-24511-2.

OECD, 2019a. Innovation ecosystems in the bioeconomy. In: OECD Science, Technology and Industry Policy Papers 76. OECD Publishing, Paris.

OECD, 2019b. Productivity in Public Procurement a Case Study of Finland: Measuring the Efficiency and Effectiveness of Public Procurement. OECD Publishing, Paris.

OECD, 2020a. Designing Fossil Fuel Subsidy Reforms in OECD and G20 Countries: A Robust Sequential Approach Methodology. OECD Publishing, Paris.

OECD, 2020b. The Effects of R&D Tax Incentives and Their Role in the Innovation Policy Mix: Findings From the OECD microBeRD Project, 2016-19. OECD Science, Technology and Industry Policy Papers, No. 92, OECD Publishing, Paris.

OECD, 2021. Recommendation of the Council on Assessing the Sustainability of Bio-Based Products. OECD/LEGAL/0395.

Opgenorth, P., Costello, Z., Okada, T., Goyal, G., Chen, Y., Gin, J., Benites, V., de Raad, M., Northen, T.R., Deng, K., Deutsch, S., Baidoo, E.E.K., Petzold, C.J., Hillson, N.J., Martin, H. G., Beller, H.R., 2019. Lessons from two design – build – test – learn cycles of dodecanol production in *Escherichia coli* aided by machine learning. ACS Synth. Biol. 8, 1337–1351.

Philp, J.C., 2015. Balancing the bioeconomy: supporting biofuels and bio-based materials in public policy. Energy Environ. Sci. 8, 3063–3068.

Philp, J., Winickoff, D.E., 2017. Clusters in industrial biotechnology and bioeconomy: the roles of the public sector. Trends Biotechnol. 35, 682–686.

Pisano, G.P., 2010. The Evolution of Science-Based Business: Innovating How We Innovate. Prepared for Industrial and Corporate Change, Special Issue in Honour of Alfred D. Chandler, Working Paper 10-062.

Pronk, J.T., Lee, S.Y., Lievense, J., Pierce, J., Palsson, B., Uhlen, M., Nielsen, J., 2015. How to set up collaborations between academia and industrial biotech companies. Nat. Biotechnol. 33, 237–240.

Robinson, C.J., Carbonell, P., Jervis, A.J., Yan, C., Hollywood, K.A., Dunstan, M.S., Currin, A., Swainston, N., Spiess, R., Taylor, S., Mulherin, P., Parker, S., Rowe, W., Matthews, N.E., Malone, K.J., Le Feuvre, R., Shapira, P., Barran, P., Turner, N.J., Micklefield, J., Breitling, R., Takano, E., Scrutton, N.S., 2020. Rapid prototyping of microbial production strains for the biomanufacture of potential materials monomers. Metab. Eng. 60, 168–182.

Rodrigo, G., Jaramillo, A., 2013. AutoBioCAD: full biodesign automation of genetic circuits. ACS Synth. Biol. 2, 230–236.

Rogelj, J., Geden, O., Cowie, A., Reisinger, A., 2021. Net-zero emissions targets are vague: three ways to fix. To limit warming, action plans from countries and companies must be fair, rigorous and transparent. Nature 591, 365–368.

Rogers, J.K., Church, G.M., 2016. Multiplexed engineering in biology. Trends Biotechnol. 34, 198–206.

Rose, S.K., Kriegler, E., Bibas, R., Calvin, K., Popp, A., van Vuuren, D.P., Weyant, J.P., 2013. Bioenergy in energy transformation and climate management. Climate Change 123, 477–493.

Sadowski, M.I., Grant, C., Fell, T.S., 2016. Harnessing QbD, programming languages, and automation for reproducible biology. Trends Biotechnol. 34, 214–227.

Scarlat, N., Dallemand, J.-F., Monforti-Ferrario, F., Nita, V., 2015. The role of biomass and bioenergy in a future bioeconomy: policies and facts. Environ. Dev. 15, 3–34.

Schieb, P.-A., Philp, J.C., 2014. Biorefinery policy needs to come of age. Trends Biotechnol. 32, 496–500.

Schmitz, C., Van Meijl, H., Kyle, P., Nelson, G.C., Fujimori, S., Gurgel, A., Havlik, P., Heyhoe, E., D'croz, D.M., Popp, A., Sands, R., Tabeau, A., Van Der Mensbrugghe, D., Von Lampe, M.,

Wise, M., Blanc, E., Hasegawa, T., Kavallari, A., Valin, H., 2014. Land-use change trajectories up to 2050: insights from a global agro-economic model comparison. Agric. Econ. 45, 69–84.

Schreck, M., Wagner, J., 2017. Incentivizing secondary raw material markets for sustainable waste management. Waste Manag. 67, 354–359.

Schueler, V., Fuss, S., Steckel, J.C., Weddige, U., Beringer, T., 2016. Productivity ranges of sustainable biomass potentials from non-agricultural land. Environ. Res. Lett. 11, 074026. 074026. https://doi.org/10.1088/1748-9326/11/7/074026.

Searchinger, T., Heimlich, R., Houghton, R.A., Dong, F., Elobeid, A., Fabiosa, J., Tokgoz, S., Hayes, D., Yu, T.-H., 2008. Use of US croplands for biofuels increases greenhouse gases through emissions from land-use change. Science 319, 1238–1240.

Searchinger, T.D., Wirsenius, S., Beringer, T., Dumas, P., 2018. Assessing the efficiency of changes in land use for mitigating climate change. Nature 564, 249–253.

Seidenberger, T., Thrän, D., Offermann, R., Seyfert, U., Buchhorn, M., Zeddies, J., 2008. Global Biomass Potentials—Investigation and Assessment of Data, Remote Sensing in Biomass Potential Research, and Country Specific Energy Crop Potentials. German Biomass Research Centre.

Severnini, E., 2019. The unintended impact of ecosystem preservation on greenhouse gas emissions: evidence from environmental constraints on hydropower development in the United States. PLoS One 14, e0210483.

Sira Consulting, 2011. Botsende belangen in de biobased economy. Een inventarisatie en een analyse van de belemmeringen in de transitie naar een biobased economy. Sira Consulting, Den Haag.

Smith, D., McCluskey, K., Stackebrandt, E., 2014. Investment into the future of microbial resources: culture collection funding models and BRC business plans for biological resource centres. Springerplus 3, 81.

Stegmann, P., Londo, M., Junginger, M., 2020. The circular bioeconomy: its elements and role in European bioeconomy clusters. Resour. Conserv. Recycl. X 6, 100029.

Tas, H., Amara, A., Cueva, M.E., Bongaerts, N., Calvo-Villamañán, A., Hamadache, S., Vavitsas, K., 2020. Are synthetic biology standards applicable in everyday research practice? Microb. Biotechnol. 13, 1304–1308.

Tremblay, P., Boyle, A., 2018. Literature Review on Public Procurement: Theories, Evidence and Implications for Regional Australia. Charles Darwin University, Northern Institute Report, Darwin. https://www.cdu.edu.au/sites/default/files/the-northern-institute/lit_review_on_public_procurement_for_regions.pdf.

Ulonska, K., König, A., Klatt, M., Mitsos, A., Viell, J., 2018. Optimization of multiproduct biorefinery processes under consideration of biomass supply chain management and market developments. Ind. Eng. Chem. Res. 57, 6980–6991.

United Nations, 2020. Press Release: Net-Zero Emissions Must be Met by 2050 or COVID-19 Impact on Global Economies will Pale Beside Climate Crisis, Secretary-General Tells Finance Summit. Press Release SG/SM/20411, 12 November.

USDA, 2018. An Economic Impact Analysis of the U.S. Biobased Products Industry (2018). https://www.biopreferred.gov/BPResources/files/EconomicReport.pdf.

USDOE, 2011. U.S. Billion-Ton Update: Biomass Supply for a Bioenergy and Bioproducts Industry. R.D. Perlack and B.J. Stokes (Leads). ORNL/TM-2011/224, Oak Ridge National Laboratory, Oak Ridge, TN.

USDOE, 2016. 2016 Billion-Ton Report: Advancing Domestic Resources for a Thriving Bioeconomy, Volume 1: Economic Availability of Feedstocks. Langholtz, M.H., B.J. Stokes and L.M. Eaton (Leads). ORNL/TM-2016/160, Oak Ridge National Laboratory, Oak Ridge, TN.

USDOE (Department of Energy), 2005. Biomass as Feedstock for a Bioenergy and Bioproducts Industry: The Technical Feasibility of a Billion-Ton Annual Supply. DOE/GO-102995-2135 or ORNL/TM-2005/66, Oak Ridge National Laboratory, Oak Ridge, TN.

van Dam, J., Junginger, M., 2011. Striving to further harmonization of sustainability criteria for bioenergy in Europe: recommendations from a stakeholder questionnaire. Energy Policy 39, 4051–4066.

Vogelpohl, T., 2021. Transnational sustainability certification for the bioeconomy? Patterns and discourse coalitions of resistance and alternatives in biomass exporting regions. Energy Sustain. Soc. 11, 3.

Vogelpohl, T., Beer, K., Ewert, B., Perbandt, D., Töller, A.E., Böcher, M., 2021. Patterns of European bioeconomy policy. Insights from a cross-case study of three policy areas. Environ. Polit. https://doi.org/10.1080/09644016.2021.1917827.

vom Berg, C., Carus, M., Dammer, L., Babayan, T., Porc, O., 2021. A Tax on Fossil Carbon is More Effective than a Tax on CO_2 Emissions. A Tool for Elegantly Pricing the True Cause of Global Warming. Nova Institute Report. 21-05-25.

Wang, B.L., Ghaderi, A., Zhou, H., Agresti, J., Weitz, D.A., Fink, G.R., Stephanopoulos, G., 2014. Microfluidic high-throughput culturing of single cells for selection based on extracellular metabolite production or consumption. Nat. Biotechnol. 32, 473–478.

Weber, K.M., Rohracher, H., 2012. Legitimizing research, technology and innovation policies for transformative change: combining insights from innovation systems and multi-level perspective in a comprehensive 'failures' framework. Res. Policy 41, 1037–1047.

Wesseler, J., Drabik, D., 2016. Prices matter: analysis of food and energy competition relative to land resources in the European Union. NJAS Wagening. J. Life Sci. 77, 19–24.

Whitehead, T.A., Banta, S., Bentley, W.E., Betenbaugh, M.J., Chan, C., Clark, D.S., Hoesli, C.A., Jewett, M.C., Junker, B., Koffas, M., Kshirsagar, R., Lewis, A., Li, C.-.T., Maranas, C., Papoutsakis, E.T., Prather, K.L.J., Schaffer, S., Segatori, L., Wheeldon, I., 2020. The importance and future of biochemical engineering. Biotechnol. Bioeng. 117, 2305–2318.

Whitley, S., van der Burg, L., 2015. Fossil Fuel Subsidy Reform: From Rhetoric to Reality. New Climate Economy, London and Washington, DC. http://newclimateeconomy.report/misc/working-papers.

Wilde, K., Hermans, F., 2021. Deconstructing the attractiveness of biocluster imaginaries. J. Environ. Policy Plan. 23, 227–242.

World Bank Group, 2020. A Practitioner's Guide to Innovation Policy. Instruments to Build Firm Capabilities and Accelerate Technological Catch-Up in Developing Countries. World Bank Group, Washington, DC.

Wurtzel, E.T., Kutchan, T.M., 2016. Plant metabolism, the diverse chemistry set of the future. Science 353, 1232–1236.

Ziolkowska, J., Meyers, W.H., Meyer, S., Binfield, J., 2010. Targets and mandates: lessons learned from EU and US biofuels policy mechanisms. AgBioforum 13, 398–412.

Index

Note: Page numbers followed by *f* indicate figures and *t* indicate tables.

S

T

U

V

W

Y

Z

Printed in the United States
by Baker & Taylor Publisher Services